Photorefractive Materials, Effects, and Applications

Critical Reviews Series

Applications of Electronic Imaging (v.1082, J. C. Urbach)
Applications of High-Power Lasers (v.527, R. R. Jacobs)
Characterization of Very High Speed Semiconductor Devices and Integrated Circuits (v.795, R. Jain)
Design of Optical Systems Incorporating Low-Power Lasers (v.741, D. C. O'Shea)
Diagnostic Imaging Applications (v.516, E. S. Beckenbach)
Diffractive and Miniaturized Optics (CR49, S. H. Lee)
Digital Image Processing (v.528, A. G. Tescher)
Digital Optical Computing (CR35, R. Athale)
Electro-Optical Imaging Systems Integration (v.762, R. Wight)
Fiber Optic Communication Technology (v.512, C. W. Kleekamp)
Fiber Optic Sensors (CR44, E. Udd)
Flashlamp-Pumped Laser Technology (v.609, F. Schuda)
Free Electron Lasers (v.728 B. E. Newnam)
Geometrical Optics (v.531, R. E. Fischer, W. H. Price, W. J. Smith)
Gradient-Index Optics and Miniature Optics (v.935, J. D. Rees, D. C. Leiner)
Highly Parallel Signal Processing Architectures (v.614, K. Bromley)
Holographic Nondestructive Testing: Status and Comparison with Conventional Methods (v.604, C. M. Vest)
Holography (v.532, L. Huff)
Image Pattern Recognition: Algorithm Implementation, Techniques, and Technology (v.755, F. J. Corbett)
Industrial Optical Sensing (v. 961, J. Carney, E. Stelzer)
Infrared Detectors (v. 443, W. L. Wolfe)
Infrared Detectors and Arrays (v.930, E. L. Dereniak)
Infrared Optical Design and Fabrication (CR38, R. Hartmann, W. J. Smith)
Infrared Thin Films (CR39, R. P. Shimshock)
Integrated Optics and Optoelectronics (CR45, K.-K. Wong, M. Razeghi)
Interferometric Metrology (v.816, N. A. Massie)
Laser Processing of Semiconductors and Hybrids (v.611, E. J. Swenson)
Laser Wavefront Control (v.1000, J. F. Reintjes)
Lens Design (CR141, W. J. Smith)
Machine Vision Systems Integration (CR36, B. G. Batchelor, F. M. Waltz)
Modeling of the Atmosphere (v.928, L. S. Rothman)
Optical Component Specifications for Laser-Based Systems and Other Modern Optical Systems (v.607, R. E. Fischer, W. J. Smith)
Optical Computing, (v.456, J. A. Neff)
Optical Pattern Recognition (CR40, J. L. Horner, B. Javidi)
Optical Inspection and Testing (CR46, J. D. Trolinger)
Optical Specifications: Components and Systems (v.406, W. J. Smith, R. E. Fischer)
Optical Technologies for Aerospace Sensing (CR47, J. E. Pearson)
Optomechanical and Electro-Optical Design of Industrial Systems (v.959, R. J. Bieringer)
Optomechanical Design (CR43, P. R. Yoder, Jr.)
Photorefractive Materials, Effects, and Applications (CR48, P. Yeh, C. Gu)
Photovoltaics (v.543, S. K. Deb)
Radiation Effects on Optical Materials (v.541, P. W. Levy)
Real-Time Signal Processing for Industrial Applications (v.960, B. Javidi)
Remote Sensing (v.476, P. N. Slater)
Robotics and Robot Sensing Systems (v.442, D. Casasent, E. L. Hall)
Spatial Light Modulators and Applications III (v.1150, U. Efron)
Standards for Electronic Imaging Systems (CR37, M. Nier, M. E. Courtot)
Technology of Stratified Media (v.387, R. F. Potter)
The Aviation Security Problem and Related Technologies (CR42, W. H. Makky)

Critical Reviews
of Optical Science
and Technology

Volume CR48

Photorefractive Materials, Effects, and Applications

Pochi Yeh
University of California/Santa Barbara

Claire Gu
The Pennsylvania State University
Editors

Proceedings
of a conference held
11–12 July 1993
San Diego, California

Sponsored by
SPIE—The International Society for Optical Engineering

SPIE OPTICAL ENGINEERING PRESS

A Publication of SPIE—The International Society for Optical Engineering
Bellingham, Washington USA

Library of Congress Cataloging-in-Publication Data

Photorefractive materials, effects, and applications : proceedings of a conference held 11–12 July 1993, San Diego, California / Pochi Yeh, Claire Gu, editors ; sponsored by SPIE—The International Society for Optical Engineering.
p. cm. — (Critical reviews of optical science and technology; v. CR48)
ISBN 0-8194-1290-2 (hardcover) : ISBN 0-8194-1289-9 (softcover)
1. Optoelectronics—Materials—Congresses. 2. Photorefractive materials—Congresses. 3. Holography—Congresses. 4. Integrated Optics—Materials—Congresses. 5. Image processing—Congresses. I. Yeh, Pochi, 1948– . II. Society of Photo-Optical Instrumentation Engineers. III. Series.
TA1750.P472 1993
621.36—dc20 93-47389
CIP

Published by
SPIE—The International Society for Optical Engineering
P.O. Box 10, Bellingham, Washington 98227-0010 USA
Telephone 206/676-3290 (Pacific Time) • Fax 206/647-1445

Printed in the United States of America.

Cover illustration: Holographic associative memory using time-division multiplexing. From the paper on "Optical neural networks using photorefractive crystals and microlasers" by E. G. Paek (p. 297).

Contents

Preface

The photorefractive effect has been under extensive investigation for many years. Owing to their unique properties such as real-time response, low-intensity operation, nonlinear energy coupling, and volume holographic capacity, photorefractive materials have become important in optical computing, storage, information processing, interconnections, and neural networks. We are pleased to have leaders and well-known researchers in this field to review previous and current works on photorefractives and discuss practical applications.

The first three papers address several fundamental issues of photorefractive physics including properties of grating diffraction and mutually pumped phase conjugation by Dr. R. Saxena from Rockwell International Science Center and Dr. Q. B. He from University of California at Santa Barbara, respectively, and an interesting comparison between photorefractive crystals and liquid crystal materials by Prof. I.-C. Khoo from The Pennsylvania State University. The next three papers present important development and practical availability of various photorefractive materials. These include an overview of photorefractive tungsten bronze materials by Dr. R. R. Neurgaonkar et al. from Rockwell International Science Center, discussion on recent developments of photorefractive fibers by Prof. L. Hesselink et al. of Stanford University, as well as a report on fixing and infrared response of doped photorefractive $BaTiO_3$ by Prof. T. W. McNamara et al. from the Massachusetts Institute of Technology.

Following the review and discussion on photorefractive physics and materials, there are nine papers on the applications of photorefractive media. As photorefractive holographic storage is becoming closer and closer to practical application, a great deal of attention has been focused on research in this area. Within the nine application papers, three of them address various issues in three-dimensional optical storage. These include research performed by Prof. F. T. S. Yu et al. from The Pennsylvania State University, Prof. S. H. Lee et al. from University of California at San Diego, and Dr. C. Alves et al. from Institut d'Optique Théorique et Appliquée.

Optical neural networks are also an attractive topic. Prof. D. Psaltis et al. from the California Institute of Technology, and Dr. E. G. Paek from Bell Communications Research give two excellent presentations on optical neural networks. In addition, traditional image processing techniques combined with the applications of photorefractive materials provide the ability of real-time operation. Examples of image processing are given by Prof. K. Hsu et al. from

National Chiao Tung University, and Dr. L.-J. Cheng (Jet Propulsion Laboratory) et al. Two of the nine application papers describe some of the important works conducted at Rockwell International Science Center, i.e., optical interconnections by Dr. A. E. Chiou and acoustic signal processing by Dr. J. H. Hong.

Although a wide range of previous and current research effort has been included in this critical review, there are still many important and interesting works that are not covered here. We hope that continuing development in materials and configurations will make photorefractive nonlinear optics even more important to practical applications.

Pochi Yeh
Claire Gu

Photorefractive Materials I

Liquid Crystal Nonlinear Optics vs. Photorefractives

I. C. Khoo

Department of Electrical Engineering
The Pennsylvania State University
University Park, PA 16802

Abstract

We present a discussion of the physics and nonlinear optical properties of liquid crystals in processes and phenomena where photorefractive crystals are currently being applied. The broadband birefringence and nonlinearities of liquid crystals spanning the visible to the far infrared, and recent successful observation of novel nonlinear optical phenomena in bulk liquid crystalline film or optical fiber, show that they are potentially very useful materials for technological applications.

2. Introduction

Liquid crystals are composed of fairly large organic molecules with a typical chemical structure as depicted in fig. 1. From a physicist's or engineer's point of view, they can be thought of as cylindrically symmetrical cigar-shaped molecules [1]. Thermotropic liquid crystals exist in three typical distinct phases (c.f., figs. 2a-c) as a function of the temperature. In the nematic phase (fig. 2a) the molecules are directionally correlated but are positionally random. The direction of the macroscopic sample is described by the so-called director $\hat{n}$, which coincides with the optical axis of the molecule. In the smectic phase there is an extra ordering because of the positional correlation; molecules exhibit layered structures (c.f., figure 2b). There are actually eight known distinct smectic phases [2]. Cholesteric liquid crystals may be viewed upon as twisted nematics, and exhibit helical structures in their alignment. In the isotropic phase, generally liquid crystals behave just as other anisotropic organic liquids, although they exhibit so-called pretransitional phenomena in the vicinity of the phase transition temperature T_c.

Most liquid crystal molecules are uniaxial, centrosymmetrical and nonpolar, although there is a class of smectic liquid crystals which is ferroelectric and possesses large permanent dipole moments [2]. The present discussion will be limited to the former class, where because of the centro-symmetry, the physical properties of the crystals are the same in the $+\hat{n}$ and the $-\hat{n}$ directions. As a result of this uniaxiality, the bulk properties of liquid crystals can be expressed in terms of two components, one parallel to (//) and the other perpendicular to (⊥) the director. The main bulk physical properties that play an important role in the steady-state and transient optical responses of liquid crystals are the refractive indices ($n_{/\!/}$ and $n_{\perp}$) c.f., figure 3a-b and the optical dielectric anisotropy $\Delta\varepsilon = (\varepsilon_{/\!/} - {}_{/\!/\perp})$, the diffusion coefficients ($D_{/\!/} - D_{\perp}$) the magnetic susceptibilities ($\chi^m{}_{/\!/} - \chi^m{}_{\perp}$), and various flow coefficients and elastic constants that will be defined presently.

As a result of their unique and complex physical structures, liquid crystals are also optically highly nonlinear materials. Their physical properties (temperature, molecular orientations, density, etc.) are perturbed by an applied optical field in complicated manners, depending on the time scales.[3-5]

In this paper, our emphasis is on the nematic phase, which exhibits birefringent crystalline properties just as photorefractives. Furthermore, we will limit our discussion to those processes that are characterized by large optical nonlinearities. For nonlinear optical effects involving microsecond or shorter (nanosecond) laser pulses, isotropic phase liquid crystals have also been recently shown to be promising candidates. Some discussions of the isotropic phase will be included in the last section dealing with optical limiting.

3. Equations of Motion and Time Scales

Optical nonlinearities of substantial magnitudes for applications in the nanosecond or longer time scales are molecular orientation/reorientations, thermal and density effects. These equations have been derived elsewhere [3b, 5, 6, and are listed below.

(a) Orientational:

Nematic Phase

$$\frac{\mu\partial^2\Delta\theta}{\partial t^2} + \gamma\frac{\partial\Delta\theta}{\partial t} + \frac{1}{2}(\gamma_1 - \gamma_2\sin 2\Delta\theta)\frac{\partial V}{\partial z} -$$

$$K\frac{\partial^2\Delta\theta}{\partial z^2} + \frac{\Delta\varepsilon E_{op}{}^2}{8\pi}\sin 2(\beta+\Delta\theta) \qquad (1a)$$

Induced Index Change:

$$\Delta n \sim \frac{dn}{d\vartheta}\cdot\Delta\vartheta \qquad (1b)$$

Isotropic Phase

Induced Order:

$$\eta\frac{\partial Q_{\alpha\beta}}{\partial t} + AQ_{\alpha\beta} = \Delta\chi(3E_\alpha^* E_\beta - |E|^2\,\delta_{\alpha\beta})/9 \qquad (2)$$

(b) Thermal and Density:

$$\rho_0 C_p\frac{\partial(\Delta T)}{\partial t} - \lambda_T\nabla^2(\Delta T) = \frac{\alpha nc}{4\pi}Re(\vec{E}_{op}\cdot\vec{E}^*_{op}) \qquad (3)$$

$$\frac{\partial^2\Delta\rho}{\partial t^2} + \frac{V_s^2}{\gamma}\nabla^2(\Delta\rho) + \frac{V_s^2 B\rho_0}{\gamma}\nabla^2(\Delta T) +$$

$$\frac{\eta}{\rho_0}\frac{\partial\nabla^2(\Delta\rho)}{\partial t} = \frac{\gamma^e}{8\pi}\nabla^2 E_{OP}^2 \qquad (4)$$

Induced Index Change:

Nematic: $$\Delta n = \frac{dn}{dS}\frac{dS}{dT}\cdot\Delta T + \frac{dn}{d\rho}\Delta\rho \qquad (5a)$$

Isotropic: $$\Delta n = \frac{dn}{d\rho} \cdot \Delta\rho \qquad (5b)$$

where μ is the molecule's moment of inertia, γ's and η are viscosity coefficients, $\mathbf{v}$ is the liquid crystal flow velocity, K is the effective elastic constant, β is the angle between the optical field and the liquid crystal director axis, c.f., figure 4, $Q_{\alpha\beta}$ is the tensorial order parameter, $\Delta\chi$ is the laser induced susceptibility anisotropy, ρ_0 is the density, C_p is the specific heat, λ_T the thermal conductivity, α the linear absorption, V_s the velocity of sound, B the bulk Modulus, and γ^e the electrostrictive coefficient.

Since liquid crystal molecules are anisotropic, most of these physical parameters are also anisotropic. The correct values to be used in quantitative analysts are governed by the interaction geometry. Solutions of these equations are also highly dependent on the interaction geometry, and the temporal and spatial field/intensity profiles of the impinging optical wave(s).

Consider the case where the imparted optical intensity function is a sinusoidal function as in most optical wave mixing phenomena. The optical intensity $E_{op}^2(\vec{r}, t)$ is given by

$$|E_{op}|^2 \sim E^2(1 + \cos \vec{q} \cdot \vec{r}) \qquad (6)$$

where $\vec{q}$ is the grating wave vector. Solution of the above equations yield the relaxation time scales, and magnitude of the nonlinear index change coefficients n_2 (defined by n_2 = index change/optical intensity) [2b, 5, 6].

Specifically, we have the following:

<u>Nematic Phase</u>

Orientational Relaxation Time Constant τ_θ:

$$\tau_\theta = \frac{\gamma}{K} \left(\frac{\pi^2}{d^2} + q^2\right)^{-1} \qquad (7)$$

Temperature Decay Time Constant τ_{th}:

$$\tau_{th} = \rho_o C_p / \lambda_T q^2 \qquad (8)$$

Density (Brillouin) Decay Time τ_B

$$\tau_B = 2\rho_o / \eta q^2 \qquad (9)$$

Isotropic Phase

Orientational Relaxation Time τ [3b-c]:

$$\tau = \eta / a(T - T^*) \qquad (10)$$

The following are the typical values of liquid crystal (c.f., Table 1) parameters [ignoring their anisotropies]: refractive index $\eta \sim 1.5$, $\eta \sim 7 \times 10^{-2}$ $k_g m^{-1} s^{-1}$ (or ~ 0.1 poise in Cgs unit, velocity of sound $v_s \sim 1.5 \times 10^3$ m/s; $\rho_o C_p \sim 2$ Joule/K; $D = \lambda_T/\rho_o C_p \sim \sim 1.2 \times 10^{-3}$ cm^2/sec; elastic constant $K \sim 10^{-6}$ dynes (or 10^{-11} kg. m. s^{-1} in m.k.s. unit). Using these values, and a grating constant $\Lambda_g = 2\pi q^{-1}$ of 20 μm, for example, we have typically [3b]

$\tau_\theta \sim 40$ ms (for $\Lambda_g \ll c$)
$\tau_{th} \sim 100$ μs
$\tau_B \sim 100$ ns

Notice that these time scales are about three orders of magnitude different from each other. Also, they are proportional to the square grating constant, Λ_g^2. Accordingly, the relaxation times can be much faster for smaller grating constant.

Thermal and density relaxation times in the isotropic phase are similar to the nematic. The orientational relaxation time, however, is much shorter. Typical τ is on the order of a few nanoseconds for temperature far above T_c, and lengthen to a few 10^2 ns as the temperature approaches T_c [3].

Table 1 *Visible → 0.5μm; near IR→CO wavelength ≈ 5μm
IR → 10μm

Parameters	Value	Nematic
Absorption	17, visible	E7
Constant (α in cm^{-1})	23, near IR	E7
	40-100, IR	E7
	69, IR	5CB
	44, IR	E46 (EM Chemicals)
Diffusion Constant	$D_{\parallel} = 1.95 \times 10^{-3}\ cm^2/sec$	E7
$D = \lambda T/\rho_o C_p$	$D_{\perp} = 1.2 \times 10^{-3}\ cm^2/sec$	E7
Thermal Index Gradients		
$\frac{dn_{\perp}}{dT}$	$10^{-3}/^{\circ}K$	5CB at 25°C
	$10^{-2}/^{\circ}K$	5CB near T_c
	$10^{-4}/^{\circ}K$	$T > T_c$ [isotropic]
	[visible - IR]	
$\frac{dn_{\parallel}}{d7}$	$-2 \times 10^{-3}/^{\circ}K$	5CB at 25°C
	$-10^{-2}/^{\circ}K$	5CB near T_c
	$10^{-4}/^{\circ}K$	$T > T_c$ [isotropic]
	[visible - IR]	

4. Liquid Crystal Nonlinearities - Summary

As the dynamical equations (4) - (8) clearly show, the magnitude of liquid crystal nonlinearities are highly dependent on various time scales involved, and the specific interaction geometry. In references [3b], the optical nonlinearities associated with these processes for wave mixing (i.e., grating) type geometries are explicitly derived by solving these equations and comparison with experimental observations. In table 2, these nonlinearities as well as electronic (resonant) optical nonlinearity (c.f., reference [3b]) are summarized.

From these data, one can see that nematic liquid crystals possess two large optical nonlinearities in the millisecond regime. One is the laser induced collective molecular reorientation effect characterized by a nonlinear index coefficient n_2 on the order of 10^{-4} cm^2/watt ($\chi^{(3)}$ of about 10^{-3} esu). The other is the thermal effect which is about an order of magnitude smaller. We have quoted these values for the millisecond time scale as it is also the typical response time of most photorefractive crystals for the following nonlinear optical phenomena.

It is important to note that some of these nonlinearities listed in table 2 are very broadband. For example, the orientational and thermal optical nonlinearities of nematic liquid crystals span over the entire visible-near infrared-infrared spectrum owing to the fact that the large birefringence of liquid crystal spans over the same spectral regime, c.f., figure 5. Consequently, almost all kinds of nonlinear optical phenomena have been observed including two- and multi-photon absorption, self-phase modulation, -focusing, -defocusing, wave mixing and beam amplification, optical phase conjugation, stimulated Brillouin, thermal and orientational scattering, bistability and instability, nonlinear fabry-perot action, harmonic generation, nonlinear interface switching polarization switching, optical limiting, nonlinear waveguiding, etc. In the following sections, we will review recent studies and "discoveries." Most of the other phenomena, and the liquid crystal physics involved, have been treated in the review articles and book [3,4].

Table 2 Optical Nonlinearities of Liquid Crystals

Nonlinear Mechanism [for both isotropic and nematic phases unless specified	Response Times [quoted here characteristic relaxation time]	Nonlinearity $\tau_{laser} > \tau_{response}$ $\chi^{(3)}$ in esu unit
Individual molecular reorientation	$\tau_{off} \sim$ ps	$\lesssim 10^{-12}$
Electronic hyper-polarizability (see section 3.8)	$\tau_{off} <$ ps	$\gtrsim 10^{-12}$ (may be as large as 10^{-9}
Collective molecular reorientation		
- Nematic	$\tau_{off} \sim$ ms to s	$\gtrsim 10^{-4}$; typical $\approx 10^{-3}$
- Isotropic	$\tau_{off} \lesssim 10^2$ ns	$\gtrsim 10^{-9}$
Density Effect (both phases)	$\tau_{off} \lesssim 10^2$ ns	$> 10^{-7}$
Thermal Effect		
- Nematic	$\tau_{off} \gtrsim 10^2$ µs	$\gtrsim 10^{-5}$ ~typically $\approx 10^{-4}$ (positive for n_0) (negative for n_e)
- Isotropic	$\tau_{off} \lesssim 10^2$ µs	$\gtrsim 10^{-7}$

Comment: For conversion of $\chi^{(3)}$ in esu to nonlinear index coefficient, use n_2 (in unit of cm^2/watt) $= (9.54\ n^2)^{-1} \chi^{(3)}$ (in esu) [c.f., reference 3b].

All values are critically dependent on laser pulse duration, absorption, characteristic diffusion length, geometry, etc., as explained in the text. For a discussion of $\chi^{(3)}$ in the transient regime [$\tau_{laser} < \tau_{response}$], see reference 3b.

5. Nonlinear Optical Phenomena

Most applications of photorefractive crystals, discussed in other articles contained in this review volume, are based on optical wave mixing process. In these processes, the basic ingredient for beam coupling effect leading to exchange of optical power, mutual influences of phase and polarization states of the optical waves present are the laser induced index gratings. In photorefractive crystals, a naturally occurring phase shift between the imparted optical intensity grating and the resulting refractive index grating mediates the energy coupling between the two interacting waves [7].

In Kerr-like media, similar phase shift may be generated by several ways, such as the coupling of the input beams to the generated side-diffraction, frequency difference between the interacting waves [8] and transient wave mixings [9]. Some of these effects have been demonstrated in recent studies with liquid crystals, using either laser induced molecular reorientational effect [10] or thermal index changes [11]. In liquid crystals, special wave mixing processes could also be mediated by the birefringence or bipolarity of the nonlinearities [12] of the nematic phase.

5.1 Self-Starting Phase Conjugation and Transient Beam Amplification

Transient wave mixing processes do not require the generation of side-diffraction, or externally imparted frequency shifting of one of the interacting waves. Consequently, longer interaction length and/or larger wave mixing angles could be employed, allowing larger gain of the signal beam, and "oscillations" to occur. In a recent study, we have demonstrated for the first time that such strong two beam coupling effect could lead to self-starting phase conjugation effect [13] in liquid crystals.

Figure 6 depicts the experimental setup; insert shows the wave vector matching condition. In this process, the input beam transverses the medium twice, once as the input pump wave $\vec{E}_1$ (along $\vec{K}_1$), and then as a reflected wave $\vec{E}_2$ (along $\vec{K}_2$). Because of the path difference, $\vec{E}_1$ and $\vec{E}_2$ are not necessarily coherent with respect to each other at the liquid crystal sample. However, $\vec{E}_1$ could coherently interact with its own scattered noise (E_3), just as $\vec{E}_2$ with is coherent noise $\vec{E}_4$, via transient two wave mixing effect.

Figures 7a and 7b depicts the results at the observation plane. For this experiment, the following parameters are used: sample is 150μm planar aligned 5CB (Pentyl- Cyano- Biphenyl) nematic liquid crystal doped with dichroic dye D16 (from EM Chemicals, NY); the nonlinearity is due to laser induced thermal index (extraordinary wave) change; laser pulses are of 2ms in duration; beam intensities of E_1 and E_2 at the sample are about 160 watts/cm^2 (power is 0.4 watt; beam diameter is 0.5 mm); sample temperature is at $T_c - 2^\circ$; interaction length within the liquid crystal sample is about 1mm, owing to the glazing angle of incidence in the prism-cladded sample; wave mixing angle used is 1 . It is observed that a very well defined phase conjugation signal is formed at this intensity level and its occurrence is attributable to the amplification of the coherent noise sources E_3 and E_4 by the input $\vec{E}_1$ and the reflection $\vec{E}_2$. In spite of the aberrations and large divergence (c.f., reflected beam in figure 7) imparted by the input lens and other optics, the phase conjugated signal is observed to be spatially of the same quality as the input beam, and has the same (small) divergence. Figure 7c shows that the self-starting phase conjugation buildup time is shortened as the input laser power is increased from 200 mW to 600 mW. We had observed establishment time of 1 ms or less at input power above 600 mW. We have also observed similar shortening of the establishment (or rise) time as the temperature of the nematic cell is brought closer to T_c, where the thermal index gradients are greatly enhanced.

Such strong beam coupling effect has also been observed in a simple forward propagating pump-probe beam experiments, c.f., figure 8. A typical result obtained in a similar 200μm planar sample of dye doped 5CB liquid crystal is shown in figure 9 obtained at $T = T_c - 2°$; crossing angle is 0.2°; pump power = 100 mWatt; probe power = 1.25 mWatt; laser pulse width = 2.5 ms; beam diameter = 2 mm. In this case, the interaction length is 200μm, as the lasers are incident at normal to the glass-slide enclosed liquid crystal cell. Because of the shorter interaction length, and the small wave mixing angle, side-diffractions are also generated. Nevertheless, as shown in figure 9, a probe amplification factor of about 10 is observed. The side diffraction (E_{-3}) on the probe side, as shown on the upper trace, follows the dynamics of the amplified probe beam; its magnitude, and that of the pump-side diffraction E_{+2}, depend on several factors, such as input pump/probe beam ratio, crossing angle, beam power, temperature, etc.

Such beam amplification effects have also been observed with millisecond infrared (10.6 μm CO_2) laser pulses [3b, 11] where self-oscillations of the phase conjugation signal in a four-wave mixing geometry depicted in figure 6) are also observed. These wave mixing effects in the infrared are typically performed with undoped liquid crystals (e.g., 5CB-Pentyl-Cyano-Biphenyl) whose absorption constant in this spectral range is on the order of 40-100 cm^{-1}.

Some of these preliminary experimental studies, along with theoretical modelling of the dynamics of these processes, are described in references [13], and [14]. Since the phase conjugation effect originates from the speckle noise source, the theory dealing with the effect of speckle noise in the dynamics of self-starting phase conjugation process as reported in reference 15 is particularly relevant. The theory given in reference 15 could be modified to describe the dynamics of these self-starting processes in liquid crystals.

5.2 Birefringent Nonlinearities, Stimulated Scattering and Mixing of Extraordinary and Ordinary Waves

Birefringence of liquid crystals may be utilized for special wave mixing phenomena. These include: beam amplification and polarization switching effects using the bipolar thermal nonlinearities of nematic liquid crystals [14], and stimulated extraordinary-ordinary ray switching via laser induced reorientation effects [16]. Both effects made use of the unusually large birefringence of nematics, this large birefringence holds throughout the visible-infrared regime, nematic liquid crystal are potentially competitive materials for such so-called cross-polarized beam coupling effect compared to photorefractive materials [17].

Detailed theories for the bipolar beam amplification (c.f., figure 10) and polarization switching processes are described in references 12. These considerations of the birefringent nature of the nonlinear material have also been extended to photorefractive materials [18]. Experimental results [18] for such e-o ray mixing and amplification effects in a photorefractive crystal are in good agreement with theoretical predictions.

Recently [19], we have succeeded in observing steady-state e-o wave scattering involving laser induced director-axis reorientational effects, which is in principle non-resonant, i.e., not dependent on the wavelength. The process of e-o ray mixing via stimulated molecular reorientations in nematic is analogous to the noise initiated beam amplification and phase conjugation effects described in the preceding section. In this case, the scattered noise from the incident beams originate from orientational fluctuations which, in conjunction with the birefringence of the nematics, give rise to orthogonally polarized scattered waves. Consider, for example, the interaction geometry as depicted in figure 11a in which a linearly input laser is incident at normal to a planar aligned nematic film, i.e., an incident o-wave. Owing to the orientational fluctuation $\Delta\vec{n}$, the scattering "noise" will contain some extraordinary wave component. Since these input and scattered waves are phased coherently, they could interfere and thus produce a moving grating along the direction of propagation of the waves. The moving grating is caused by the frequency difference $\Omega = \omega_o - \omega_e$ between the generated e-wave (w_e) and the input o-wave (w_o). The grating constant is given by $|q| = (\frac{\omega}{c})(n_e - n_o)$.
Depending on the duration of the input laser compared to the orientational grating relaxation time $\tau \sim \eta/Kq^2$, (K:elastic constant; η:viscosity) this process falls into the transient or steady state regime. Transient effects have been investigated by Zeldovich et al. [16], where fairly high power pulsed Ruby laser is required to generate

observable signals. The more interesting and practically useful one is the steady state case, corresponding to input laser pulse duration longer than τ. Typically, for the grating constant $\lambda_q = \frac{2\pi}{|q|} = \lambda/\Delta n \sim 3\mu m$, the corresponding orientational relaxation time constant is on the order of about $\tau \approx 6$ ms, using typical liquid crystal parameters ($K \approx 5 \times 10^{-7}$ dyne, $\eta \sim 1.2$ Poise). Using laser pulses of duration much longer than this will allow one to observe the steady state version of the o-e ray stimulated scattering effects.

Figure 11b depicts the experimental setup. A linearly polarized low power Ar^+ laser pulse of 100ms duration is focused onto a planar aligned 200µm thick nematic liquid crystal (E7, from EM Chemicals) either as an e- or an o- wave. A crossed polarizer at the exit monitors the generated orthogonally polarized wave. Figure 11c shows the experimental observations; indeed, above a certain threshold, there is significant stimulated scattering of the input o-wave into the e-wave component. For this unoptimized setup, the threshold power for observing the effect is only 400 mW. In some recent experiments [22], we have observed even lower threshold for these stimulated scatterings and polarization switching effects. Compared to typical stimulated scattering threshold power in the kilowatt and megawatt range, this lower threshold phenomena certainly deserves more detailed investigation. An even more important point is that this phenomena is based on the non-wavelength selective molecular reorientation process. We expect, therefore, that the process is observable in a large spectral regime, e.g., diode laser wavelength, near IR (1.3-1.5µm), mid-IR (3-6µm), etc., where nematic liquid crystals are quite transparent and possess large birefringence, c.f., figure 5.

5.3 Dye-Doped Nematic Liquid Crystals and Isotropic Liquid Crystal Cored Fibers

Recently, we have observed some rather unusual characteristics in liquid crystalline materials that are doped with dichronic dye molecules.

As reported in reference 20 and 21, the presence of a

small amount ($\lesssim$ 1%) of anthraquinone dyes or dichroic dyes such as D4, D16, etc. (EM Chemicals)] dramatically increases the orienting action of an optical field on the liquid crystal. A reduction by a factor of several hundred times reduction in the optical threshold for reorienting nematic liquid crystal director is observed in homeotropic LC film (doped with anthraquinone dyes [20]) and also in planar film (doped with dicroic dyes [21]), c.f., figure 12. The dominant nonlinear mechanism is due to an enhanced reorientational effect (from planar to homeotropic), caused by the dye molecules in their photoexcited states. In terms of nonlinear coefficient, this effect gives a n_2 on the order of -10^{-2} cm^2/watt, thus putting such dye doped liquid crystal among the most nonlinear materials.

The preceding discussions are centered on the nematic liquid crystals. Owing to the large orientational fluctuation induced scattering loss (typically about 20 db/cm) [22] in this phase, they are usually assembled in thin film forms for nonlinear optical applications, with the laser-LC interaction region limited to less than 10^2 μms. On the other hand, orientational and density fluctuations are considerably reduced in the isotropic phase [much less than 1 cm^{-1}], and thus it permits interaction geometries of much longer optical path length. Isotropic liquid crystals also possess sizable optical nonlinearities, c.f., Table 2.

Figure 13 shows the experimental set up used in the experimental study of nanosecond laser pulse propagation and backward stimulated scattering effect in liquid crystalline (LC) fiber. The LC fiber is made by filling the hollow core of a glass capillary (refractive index of 1.45) with an isotropic liquid crystal mixture (TM74A from EM Chemicals; refractive index of 1.54) which has previously been shown to exhibit stimulated back scattering and phase conjugation effect [23]. The length of the fiber is 3 cm, and the core diameter is 50 μm. (We have also experimented with fibers of other core diameters, ranging from 2 μm to 200 μm.) The laser used is the second harmonic of a Nd:Yag at 0.53 mm, with a FWHM of 20 ns. The transverse mode structure of the laser is nearly Gaussian. It is focused into the fiber by means of a microscope objective of focal length 1 cm. The linear transmission loss of the fiber is very small; most of the coupling loss is due to the interface reflections. The input, back reflected, and the transmitted pulses, are recorded by a fast storage oscilloscope.

The photo insert (a) shows a typical input laser pulse temporal behavior, which contains some envelope modulations because of the double longitudinal modes present in the pulse. At small input laser energy, the transmitted pulse is identical in form to the input. Above an input energy of about 20 μJ, the transmitted pulse begins to exhibit reduced transmission of the later portion of the pulse, as shown in the photo insert (b). This is attributed to self-defocusing effect associated with the negative thermal/density optical nonlinearities induced by the focused Gaussian laser pulse, similar to the defocusing optical limiting effect observed previously in a thin film of the same LC mixture. By measuring the output energy versus the input, we obtained a typical optical limiting behavior, c.f., Fig. 14. The threshold fluence for this unoptimized set up is measured to be about 0.4 J/cm^2, which ranks among the lowest in nonlinear optical material based optical limiting devices.

Above an input laser pulse energy of about 0.1 mJ, we begin to observe a stimulated backscattered signal, attributable to Stimulated Brillouin scattering. The backscattered pulse is of much shorter duration compared to the input, as shown in the photo insert (c). The temporally "clipped" form of the signal is most likely due to the defocusing effect responsible for the depressed tail portion of the forward propagating pulse. Nevertheless, the backscattered signal appeared to possess phase conjugation properties, similar to those reported in reference [23]. Similar stimulated backscattering and intensity dependent transmission effects are observed in LC fibers of different core diameters [24].

6. Conclusion

As demonstrated by the myriad of nonlinear optical phenomena discussed above and in the literature, pure or dye-doped liquid crystals, in thin film or optical fiber (or waveguide) structures, possess several unusual physical characteristics that enable the observation of these effects. The main advantages offered by liquid crystal are the broadband nonlinearities, low cost, and great flexibility in physical configurations and molecular engineering of new liquid crystalline compounds. Many more novel nonlinear optical phenomena are anticipated.

7. Acknowledgement

This work is supported in part by the National Science Foundation, Air Force Phillips Laboratory and Army Research Office.

References:

1. P. G. deGennes, "The Physics of Liquid Crystals," (Oxford University Press, London, 1974).

2. G. W. Gray and J. W. Goodby, "Smectic Liquid Crystal," (Leonard Hill, Glasgow, 1984).

3. a) I. C. Khoo, "Nonlinear Optics of Liquid Crystal," in Progress in Optics," ed. E. Wolf, North-Holland, Amsterdam (1988).

   b) I. C. Khoo and S. T. Wu, "Optics and Nonlinear Optics of Liquid Crystals," (World Scientific, Singapore 1993).

   c) I. C. Khoo and Y. R. Shen, "Liquid Crystals: Nonlinear Optical Properties and Processes," Optical Engineering 24, 579 (1985).

4. N. V. Tabiryan and B. Ya. Zel'dovich, in a series of articles published in Mol. Cryst. Liq. Cryst., 62, p. 637, 69, p. 19, and 69, p. 31 [1981] on nematic, smectic and cholesteric liquid crystals.

5. I. C. Khoo, R. G. Lindquist, R. R. Michael, R. J. Mansfield, and P. LoPresti, "Dynamics of Picosecond Laser-induced Density, Temperature, and Flow-Reorientation Effects in the Mesophases of Liquid Crystals," J. Appl. Phys., 69, 3853 (1991); "Flow" effect, where the liquid crystal molecules are displaced by the intense optical field, has also been observed in the smectic phase. I. C. Khoo and R. Normandin, "Nanosecond Laser-induced Transient and Erasable Permanent Grating Diffractions and Ultrasonic Waves in a Nematic Film," J. Appl. Phys.,

55, 1416 (1984). See also, K. Miyano and J. B. Ketterson, "Ultrasonic study of liquid crystals," Phys. Rev. A12, p. 615 (1975).

6. H. J. Eichler and R. Macdonald, "Flow Alignment and Inertial Effects in Picosecond Laser-induced Reorientation Phenomena of Nematic Liquid Crystals," Phys. Rev. Lett., 67, p. 2666 (1991). For polarization selective transient grating experiment, see also, F. W. Deeg and M. D. Fayer, "Analysis of Complex Molecular Dynamics in an Organic Liquid by Polarization Selective Subpicosecond Transient Grating Experiments," J. Chem. Phys. 91, p. 2269 (1989) and references therein.

7. J. Feinberg, Opt. Lett. 7, 486 (1982); J. O. White, M. Cronin-Golomb, B. Fisher and A. Yariv, Appl. Phys. Lett. 40, 450 (1982); M. Cronin-Golomb, B. Fisher, J. O. White and A. Yariv, IEEE J. Quant. Electron. QE20, 12 (1984); Appl. Phys. Lett. 42, 919 (1983); Ian McMichael and P. Yeh, Opt. Lett. 12, 48 (1987). See also P. Gunter and J. P. Huignard, "Photorefractive Materials and Their Applications," Vol. 1 and Vol. II (Springer-Verlag, Berlin, 1988). See also reference __ here on self-pumped phase conjugation and P. Yeh, IEEE J. Quant. Electron. QE25, p. 484 (1989).

8. For atomic media, see for example, C. J. Gaeta, J. F. Lam and R. C. Lind, Opt. Lett. 14, 245 (1989). See also G. Grynberg et al. Opt. Lett. 16, 1071 (1991).

9. V. L. Vinetskii et al., Sov. J. Quant. Electron. 7, p. 230 (1977); Sov. Phys. Usp. 22, p. 742 (1979) on transient two-wave mixing. For transient multiwave mixing, see I. C. Khoo et al., Phys. Rev. A41, 408 (1990); A41, 1544 (1990). See also I. C. Khoo, et al., Technical Digest '92 International Quant. Electron. Conference, Vienna, Austria, p. 238.

10. I. C. Khoo et al., J. Opt. Soc. Am. B8, 1433 (1991). I. C. Khoo and T. H. Liu, Phys. Rev. A39, p. 4036 (1989); T. H. Liu and I. C. Khoo, IEEE J. Quant. Electron. QE23, 2020 (1987).

11. L. Richard, J. Maurin and J. P. Huignard, Opt. Comm. 57, 365 (1986). I. C. Khoo, P. Y. Yan, G. M. Finn,

T. H. Liu and R. R. Michael, J. Opt. Soc. Am. B5, p. 202 (1988); see also F. Sanchez, P. H. Kayoun and J. P. Huignard, J. Appl. Phys. 64, p. 26 (1988).

12. I. C. Khoo, Phys. Rev. Lett. 64, 2273 (1990); I. C. Khoo and N. V. Tabiryan, Phys. Rev. A41, 5528 (1990). See also N. V. Tabiryan, A. V. Sukhov and B. Ya. Zeldovich, Mol. Cryst. Liq. Cryst. 136, 1 (1986).

13. I. C. Khoo, H. Li and Y. Liang, Conference on Lasers and Electro Optics '93, Baltimore, MD. '93 CLEO Technical Digest, p. 428. Also submitted for publication in Opt. Letts. (Accepted for publication June 1993).

14. I. C. Khoo, H. Li and Y. Liang, accepted for publication in J. Quant. Electron (1993). See also I. C. Khoo et al., Technical Digest, 1992 International Quant. Electron. Conference, Vienna, Austria.

15. I. C. Khoo, N. Beldyugina, H. Li, A. V. Mamaev and V. V. Shkunov, Opt. Lett. 18, p. 473 (1993).

16. The effect briefly described here is the steady-state low power laser induced counterpart of the transient high-power laser induced effect reported by B. Ya Zeldovich et al., JETP Lett. 41, p. 514 (1985).

17. See, for example, Frederick Vachss and Tallis Y. Chang, J. Opt. Soc. Am. B6, p. 1683 (1989).

18. I. C. Khoo, Y. Liang and H. Li, IEEE J. Quant. Electron. QE28, p. 1816 (1992).

19. I. C. Khoo et al. To be presented at the 1993 IEEE-LEOS Annual Meeting, San Jose, CA.

20. I. Janossy, L. Csillag and A. D. Lloyd, Phys. Rev. A44, p. 8410 (1991); I. Janossy and T. Kosa, Opt. Letts. 17, p. 1183 (1992).

21. I. C. Khoo, H. Li and Y. Liang, IEEE J. Quant. Electron. (May 1993). In press.

22. J. R. Whinnery, Chenming Hu and Y. S. Kwon, IEEE J. Quant. Electron. QE13, p. 262 (1977); T. G. Giallorenzi and J. P. Sheridan, J. Appl. Phys. 46, p. 1271 (1975). Note: Mino Green and S. J. Madden, Appl. Opt. 28, p. 5202 (1989), however, reported that very small core diameter (<10 μm) nematic liquid crystal fibers suffer much less scattering loss. The effect, however, remains to be quantitatively documented.

23. I. C. Khoo R. R. Michael and P. Y. Yan, IEEE J. Quantum Electronics QE23, 1344 (1987).

24. I. C. Khoo, H. Li, Y. Liang and P. LoPresti, to be presented at the '93 Annual IEEE-LEOS Technical Meeting.

Figure Captions

Fig. 1 Chemical structure of the liquid crystal Pentyl-Cyano-Biphenyl [5CB].

Fig. 2 a) Nematic phase, b) Smectic phase, c) Cholesteric phase.

Fig. 3 a) Typical temperature dependence of the refractive indices of SCB (pentyl-cyano biphenyl) in the visible spectral region.

b) Typical temperature dependence of the refractive indices of SCB (pentyl-cyano biphenyl) in the infrared spectral region.

Fig. 4 Interaction of a linearly polarized (extraordinary ray) laser with a homeotropically aligned nematic liquid crystal film.

Fig. 5 Measured birefringence of three typical nematic liquid crystals (after reference 3b).

Fig. 6 Experimental setup for self-starting phase conjugation. Insert shows the wave vector phase matching diagram.

Fig. 7 Photograph of the phase conjugation signal (as indicated by arrow) and ordinary reflection (left): (a) at low power (< 0.2 mWatt), no signal; (b) at a higher power (> 0.2 mWatt), signal appears. Double reflections are due to the glass-slide beam splitter used. (c) Oscilloscope traces of the time-evolution of the phase conjugated signal for an input square pulse, showing the shortening of the build up time and the intensification of the phase conjugation signal as the input power is increased from 200 mW to 600 mW.

Fig. 8 Schematic depiction of a forward pump-probe beam amplification experimental set up. In thin sample, side-diffractions are also generated.

Fig. 9 Lower trace: lower, square pulse is the probe pulse when the pump pulse is absent. Amplified (upper) signal occurs when the pump beam is on.

Upper trace: detected diffracted beam (E_{-3}) power when both pump and probe pulses are on. Small "bumps" recorded at the end of the pulse in both upper and lower trace is the noise scattered off the pump beam when the probe beam is blocked.

Fig. 10 Field and wave vector configurations of two mixed-polarization beams incident from the left side [air] on a birefringent nonlinear optical crystal drawn for the case $n_o > n_e$ (e.g., photorefractive crystal). In liquid crystals, usually $n_e > n_o$.

Fig. 11 a) Schematic depiction of the optical field propagating as an o-wave in a planar nematic liquid crystal cell. $\mathbf{\Delta}\vec{n}$ is the orientational fluctuation of the director axis $\hat{n}'$ is the perturbed director axis.

b) Experimental setup for observing o-e wave stimulated scattering. Sample is located at the focal plane of the input lens.

c) Observed e-wave power as a function of the input o-wave power, showing a nonlinear dependence above a certain threshold input value.

Fig. 12 Optical fields configuration and propagation through a planar dye doped liquid crystal. $\vec{E}_1$ and $\vec{E}_2$ are excitation or pump beam, the He-Ne laser probe the induced grating; (a)(b), (c), (d) show the four polarization orientations used to study the polarization dependence of the effect.

Fig. 13 Schematic depiction of the experimental setup for studying forward transmitted and backward propagating signals from a liquid crystal cored fiber. (a) Input pulse shape; (b) transmitted pulse shape; (c) "backscattered" pulse shape. Input pulse is 20 ns in duration, derived from a Q-switched Nd:Yag second-harmonic (0.53 μm); pulse energy ≈ 5 μJ.

Fig. 14 Observed transmitted (output) laser pulse energy dependence on the input, showing a typical self-limiting effect.

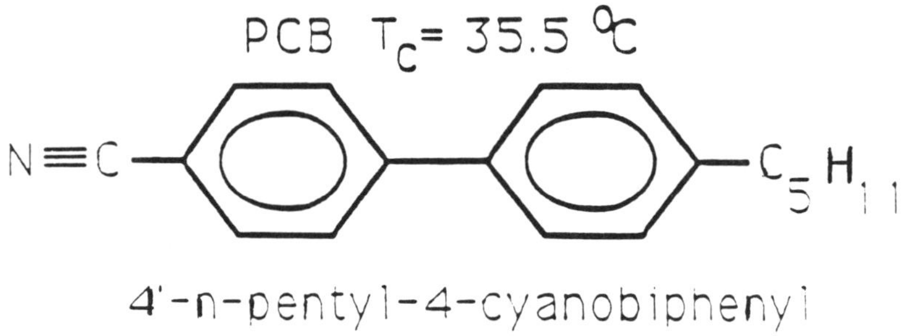

Fig. 1

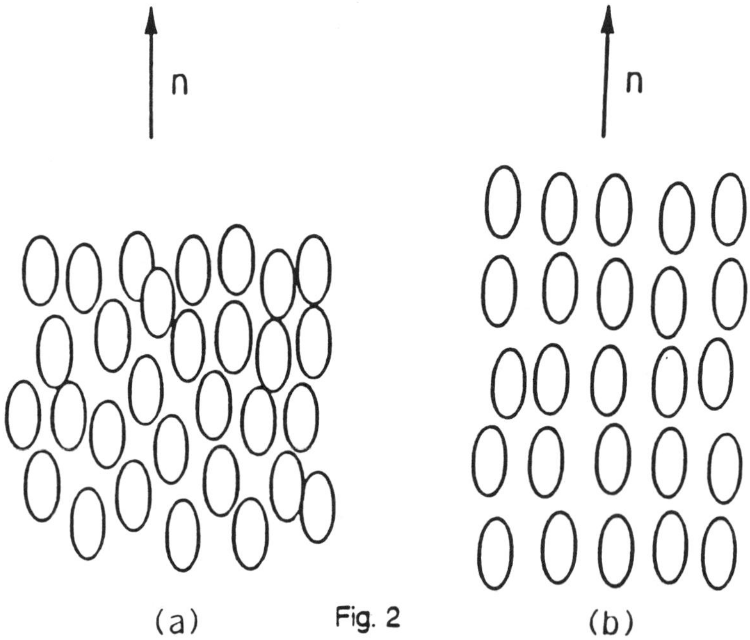

Fig. 2

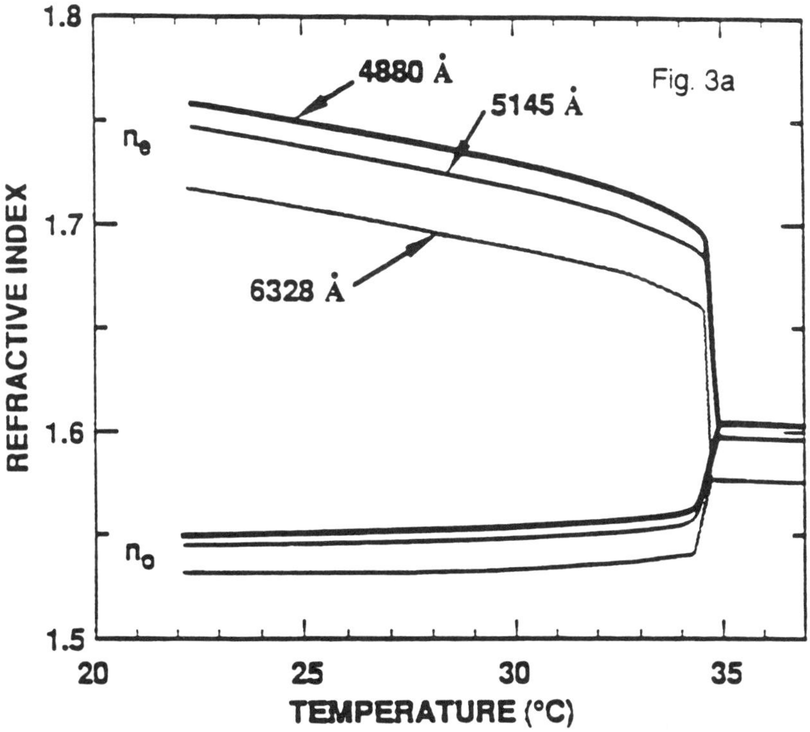
1.8
1.7
1.6
1.5
REFRACTIVE INDEX
20
25
30
35
TEMPERATURE (°C)
4880 Å
5145 Å
6328 Å
Fig. 3a
n_e
n_o

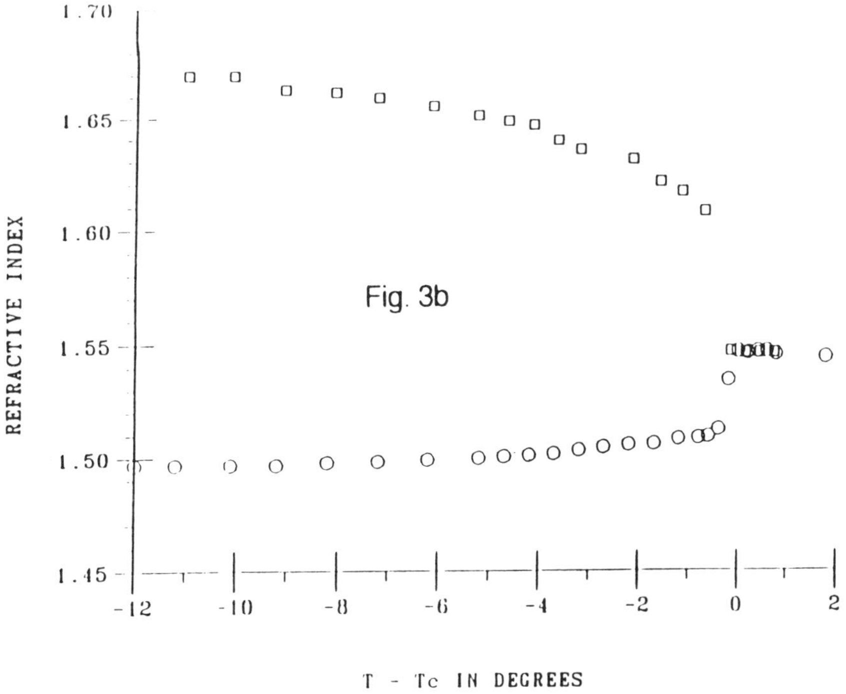
1.70
1.65
1.60
1.55
1.50
1.45
REFRACTIVE INDEX
-12
-10
-8
-6
-4
-2
0
2
T - Tc IN DEGREES
Fig. 3b

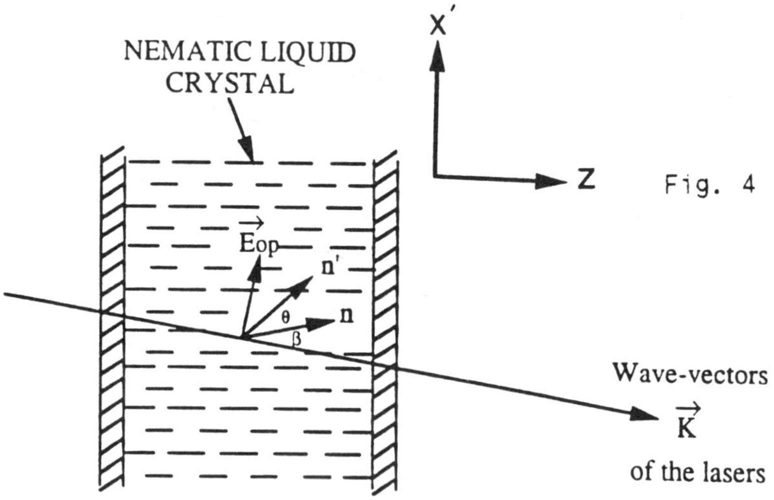
NEMATIC LIQUID
CRYSTAL
X'
Z
Eop
n'
θ
n
β
Wave-vectors
K
of the lasers

Fig. 4

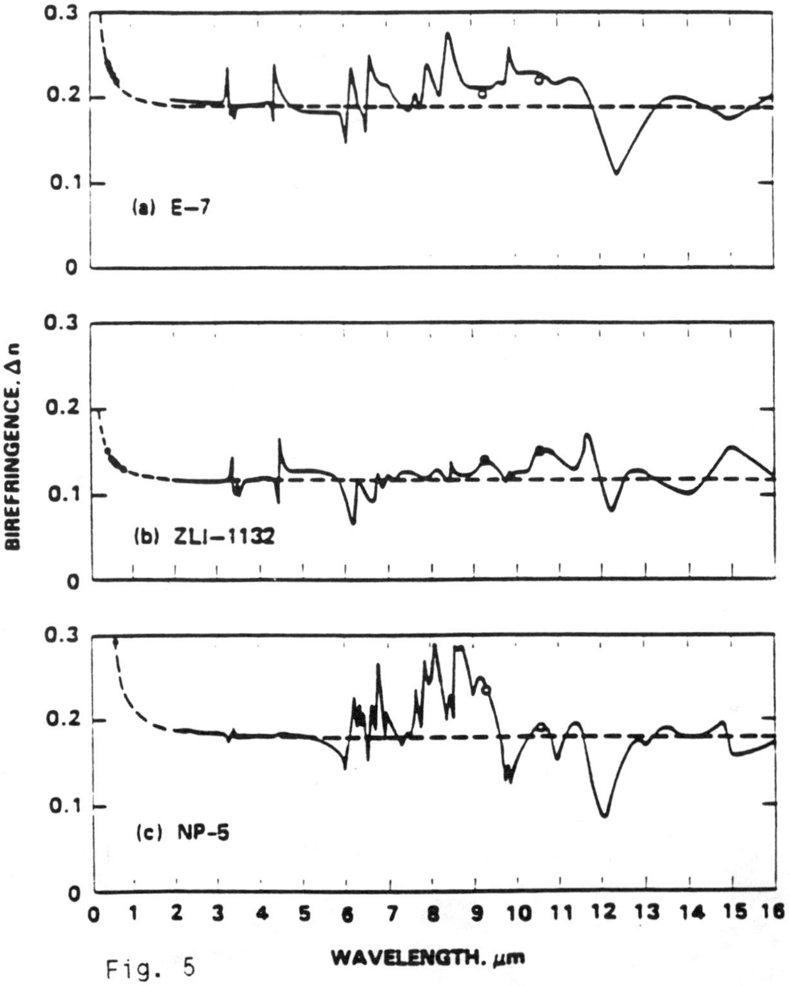
0.3
0.2
0.1
0
(a) E–7
BIREFRINGENCE, Δn
(b) ZLI–1132
(c) NP–5
0 1 2 3 4 5 6 7 8 9 10 11 12 13 14 15 16
WAVELENGTH, μm

Fig. 5

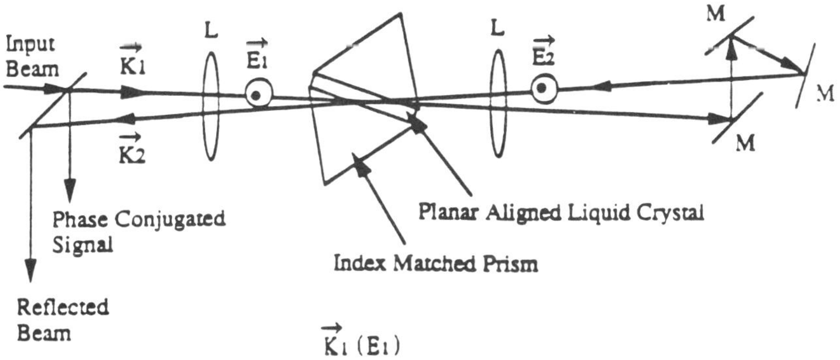

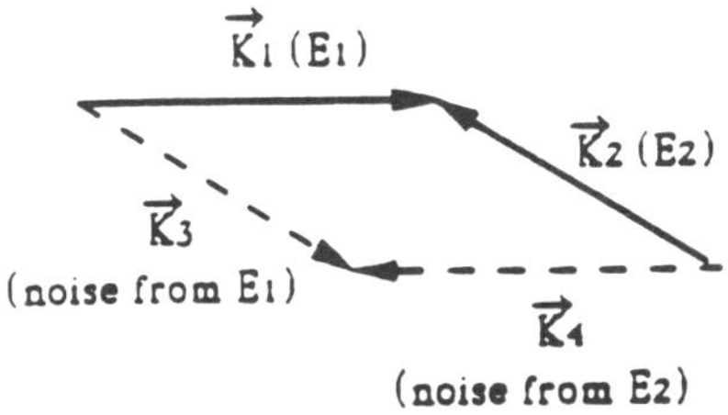

Fig. 6

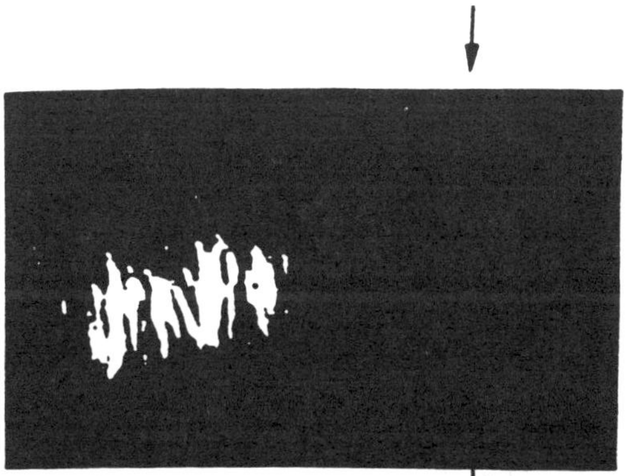

Fig. 7a

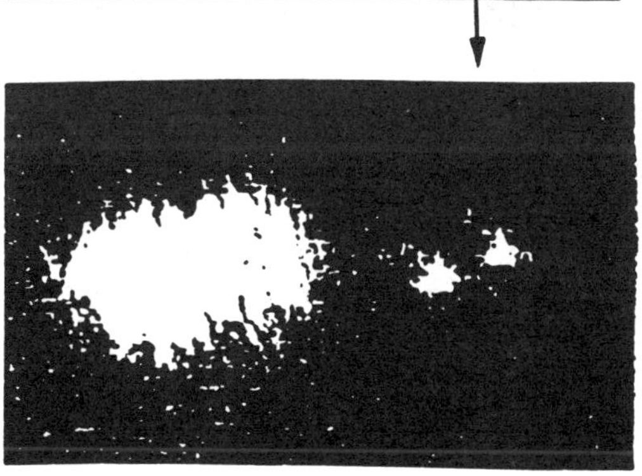

Fig. 7b

Fig. 8

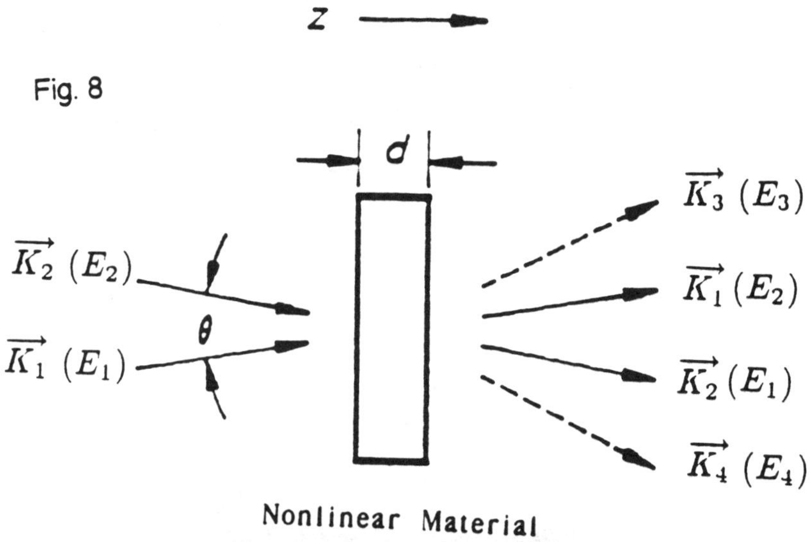
z
d
$\vec{K_2}$ (E_2)
θ
$\vec{K_1}$ (E_1)
$\vec{K_3}$ (E_3)
$\vec{K_1}$ (E_2)
$\vec{K_2}$ (E_1)
$\vec{K_4}$ (E_4)
Nonlinear Material

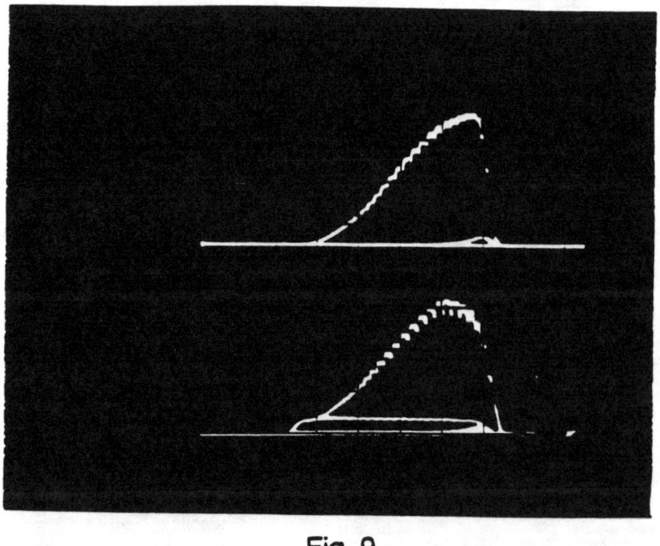

Fig. 9

Fig. 10

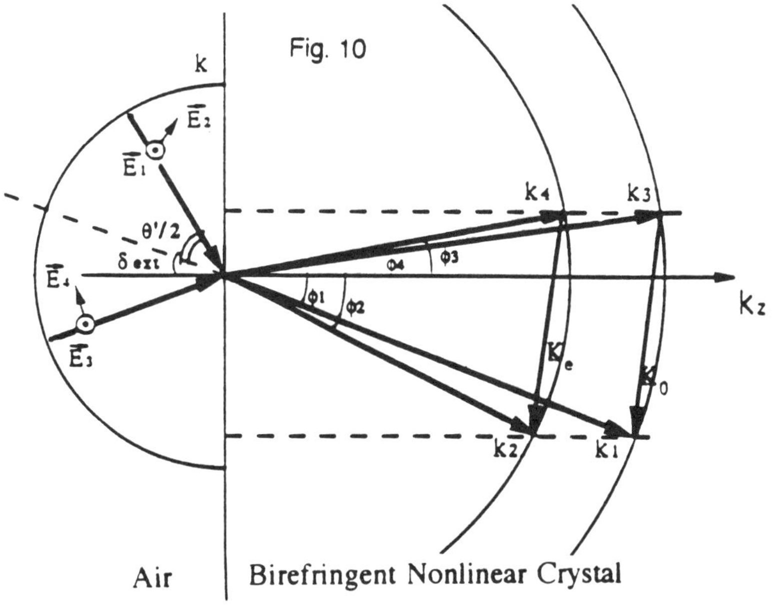

Fig.11a

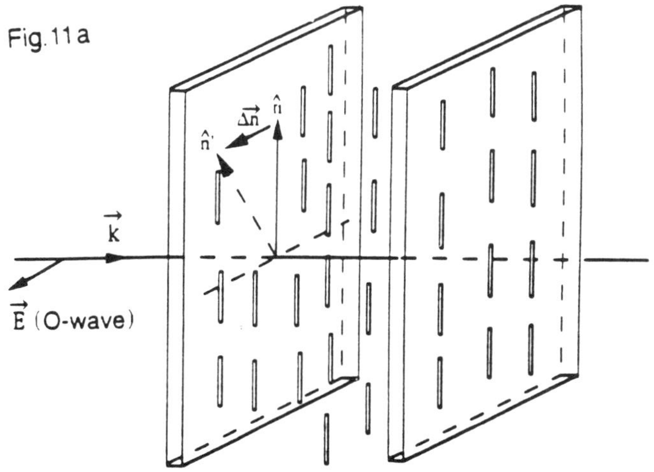

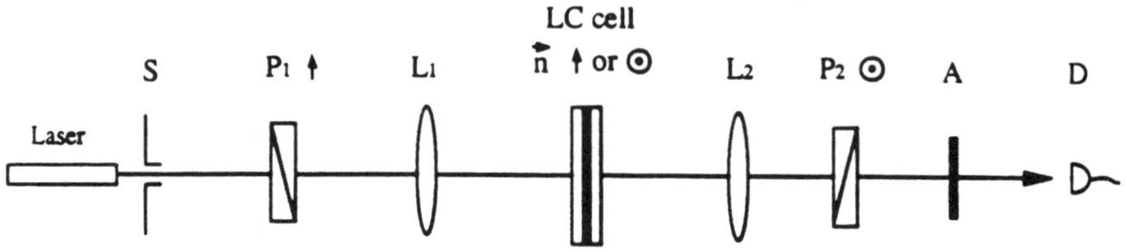

Laser : CW Ar laser, 5145 Å

LC cell: Planar, E7, 200 μm

S : Shutter, 100 ms

P1, P2 : Polarizers

L1, L2 : Lens, f = 5 cm

A : Attenuator

D : Photo-detector, connected to oscilloscope

Fig. 11b

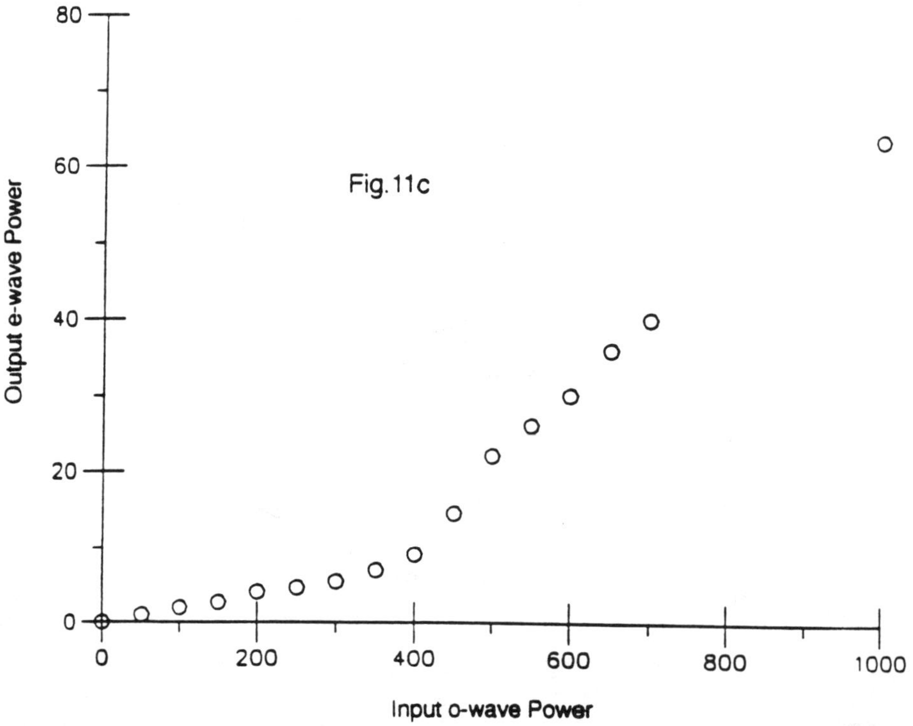

Fig. 11c

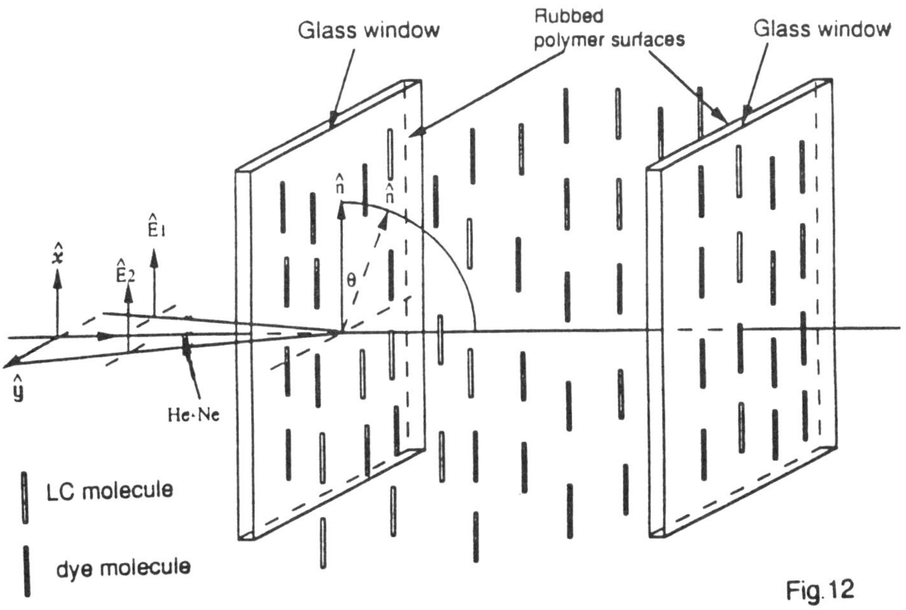
Glass window
Rubbed
polymer surfaces
Glass window
n̂
n̂
Ê1
Ê2
x̂
θ
ŷ
He-Ne
LC molecule
dye molecule

Fig.12

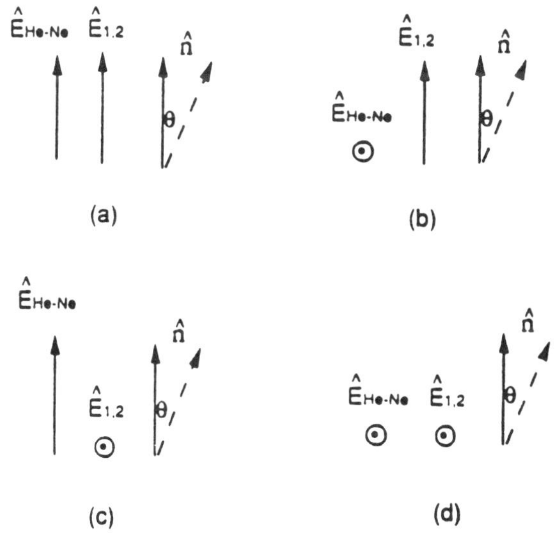
ÊHe-Ne
Ê1,2
n̂
θ
(a)
Ê1,2
n̂
ÊHe-Ne
θ
(b)
ÊHe-Ne
n̂
Ê1,2
θ
(c)
n̂
ÊHe-Ne
Ê1,2
θ
(d)

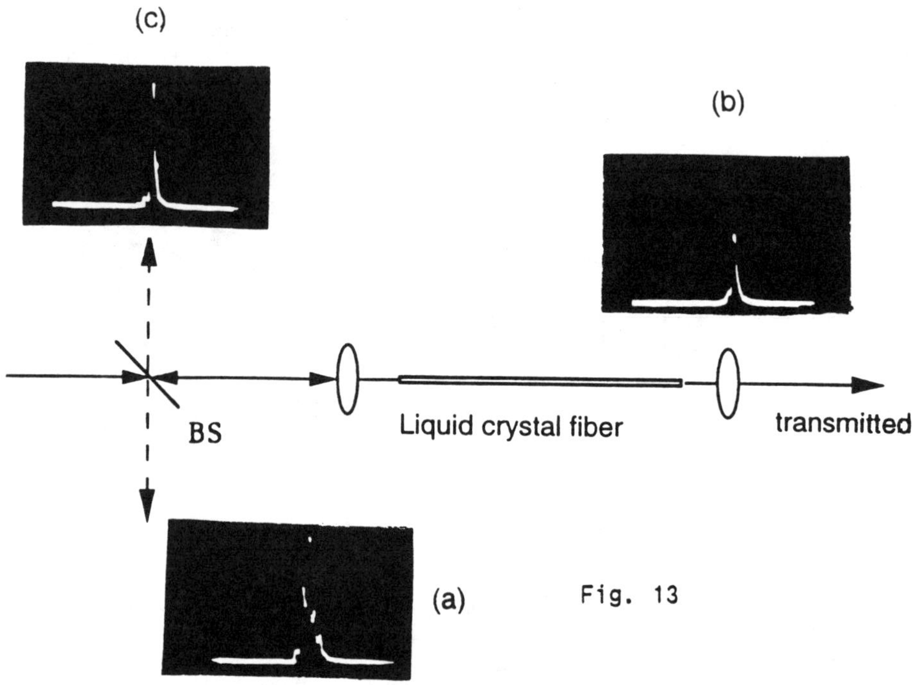
(c)
(b)
BS
Liquid crystal fiber
transmitted
(a)

Fig. 13

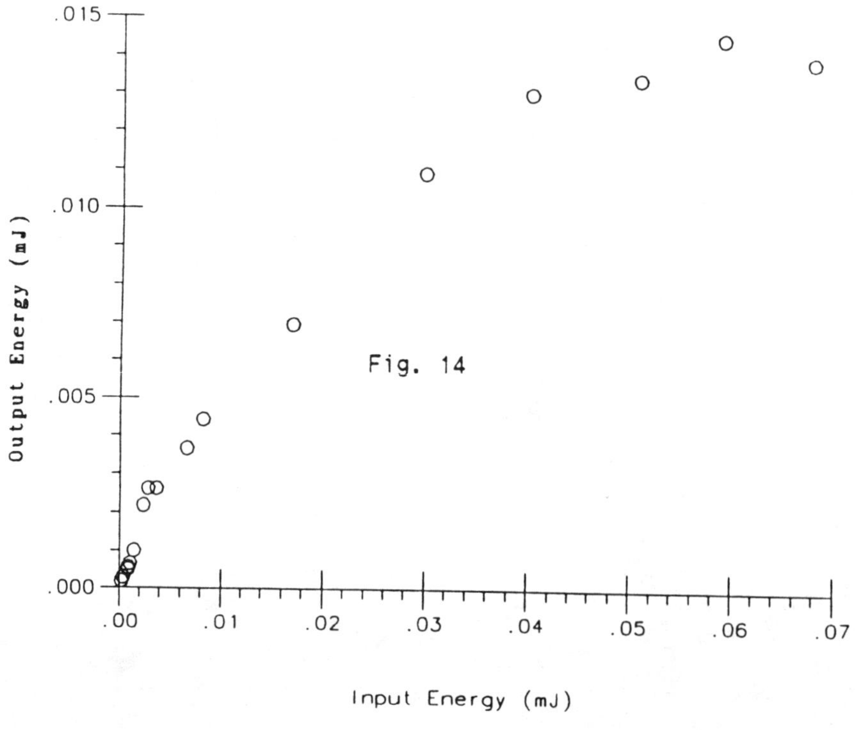
.015
.010
.005
.000
Output Energy (mJ)
.00
.01
.02
.03
.04
.05
.06
.07
Input Energy (mJ)

Fig. 14

Diffraction properties of photorefractive gratings

Ragini Saxena
Rockwell International Science Center
1049 Camino dos Rios, Thousand Oaks, CA 91360
(805) 373 - 4157

ABSTRACT

Volume phase gratings in photorefractive crystals have nonuniform amplitude and phase due to the energy exchanged by the writing beams within the material. Analytic expression is obtained for the diffraction efficiency of a weak reading beam that does not perturb the grating and has a different polarization from the write beams. For a read beam of arbitrary intensity, the diffraction efficiency is a nonlinear function of the read beam intensity and is nonreciprocal with respect to readout from the two input ports. These properties of photorefractive gratings are studied for arbitrary phase shifts of the index grating from the intensity pattern.

I. INTRODUCTION

The photorefractive crystal has proven to be an almost ideal holographic medium, demonstrating its ability to write efficient volume holograms at low levels of light without the need for special development procedures. The amplitude and phase of such holograms are nonuniform[1] within the crystal due to two-beam coupling effects[2] between the pair of coherent writing beams. Further, the read beam and its diffracted beam induce a new photorefractive grating that may be in or out of phase with the original grating being read.[3] The diffraction is therefore quite different from that of a simple Bragg scattering off a fixed thick grating, and the coupled wave theory of Kogelnik[4] must be modified to describe the diffraction efficiency η of such gratings. A recent study[5] examined diffraction efficiency of photorefractive gratings when polarization asymmetry occurs during writing and readout. The new photorefractive grating formed by the read beam and its diffracted component is ignored for a reading beam that is weak compared to the writing beams. For this case, η is independent of the read beam intensity, and the argument of the sine function in η is a nonlinear function of grating thickness, coupling constants during writing and readout, and input intensity ratio of writing beams. Inclusion of the effects arising from a read beam having an intensity comparable to that of the writing beams has shown η to be a nonlinear function of the read beam intensity and nonreciprocal with respect to readout from the two input ports.[6] Similar nonreciprocity with respect to direction of readout has been observed for fixed volume gratings in photorefractive media.[7] A theoretical explanation[8] of this effect was obtained by taking into account the new photoinduced grating formed by the read beam and its diffracted component, besides the fixed grating present initially. The nonlinear and nonreciprocal behavior of η has been examined for arbitrary phase shifts of the photorefractive grating from the intensity pattern.[9] Analytic expressions for the diffraction efficiency are obtained for limiting values of the phase shift between index change and fringe pattern.

II. THEORY

[A] Diffraction Efficiency of Fixed, Uniform Gratings

Periodic media with spatially varying conductivity σ are known as absorption gratings, while periodic media with spatially varying dielectric constant ε are known as phase gratings; when both σ and ε are periodically varying functions in space, then we have a "mixed" grating. An absorption grating is necessarily lossy because the nonzero conductivity of the medium leads to the decay of an incident light beam. An example of a 1-d phase grating with a sinusoidal index variation is given by

$$n(x) = n_0 + n_1 \cos Kx \quad , \tag{1}$$

where n_0 is the average index of refraction of the medium, n_1 is the uniform amplitude of the fixed index grating, K is the grating wave number equal to $2\pi/\Lambda$ (Λ being the grating spacing), and x is the spatial coordinate along the grating wave vector. The "thickness" of a grating is determined by the dimensionless parameter $Q = 2\pi\lambda L/n_0\Lambda^2$, where λ is the wavelength of light and L is the thickness of the grating in the direction of light propagation. A grating is planar or thin if $Q < 1$, and it is a thick volume grating if $Q \geq 10$. For grating spacing comparable to wavelength ($\Lambda \approx \lambda$), Q is less than or greater than unity depending on whether the grating thickness L is smaller or greater than $\Lambda \approx \lambda$. If the grating planes are perpendicular (parallel) to the grating surface, then we have unslanted transmission (reflection) gratings. The diffracted beam appears on the same side of the grating as the transmitted (reflected) read beam for a transmission (reflection) grating (Fig. 1). Table I by Kogelnik[10] summarizes the various diffraction orders and the maximum diffraction efficiency that can be obtained theoretically from uniform fixed gratings. Note that the diffraction efficiency of thin phase gratings is limited to about 34% in the first orders in the Raman–Nath regime. For volume phase gratings, maximum diffraction efficiency is 100% to the Bragg-matched first order, all higher orders being suppressed.

The fixed volume phase grating of Eq. (1) is formed by recording the interference pattern of two coherent writing beams I_{AW} and I_{BW}

$$I_W(x) = I_{0W}\,(1+m \cos Kx) \quad , \tag{2a}$$

where $I_{0W} = I_{AW} + I_{BW}$ is the sum of the intensities of the two writing beams, $m = 2\sqrt{I_{AW}I_{BW}}\,/\,(I_{AW}+I_{BW})$ is the modulation depth of the intensity pattern, and the subscript W indicates writing beams. If $\beta = I_{BW}/I_{AW}$ is the intensity ratio of the writing beams, then $m = 2\sqrt{\beta}\,/\,(1+\beta)$. The recording medium (a photographic plate, for example) is exposed to the intensity pattern of Eq. (2a). After processing with a developer and a fixer, the plate is bleached to convert the recorded amplitude modulation into a phase modulation. Assuming a linear response of the photographic plate to the incident exposure, the refractive index is given by

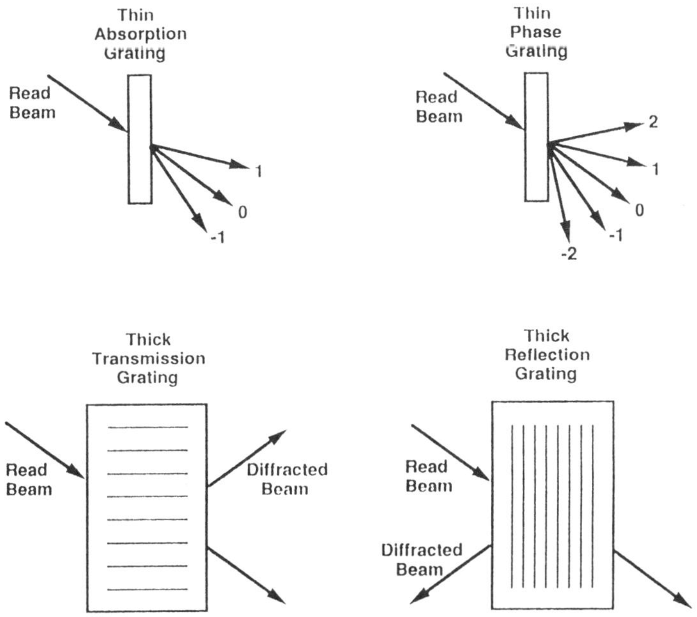

Fig. 1. Diffraction of a read beam from various types of fixed, uniform gratings.

Table I

Diffraction Properties of Various Gratings

$Q = \frac{2\pi\lambda L}{n_0\Lambda^2}$	**Type**	**Geometry**	**Diffraction Orders**	η_{max}
$Q < 1$ Thin	Absorption $\alpha(x)=\alpha(x+\Lambda)$	Transmission	Two: ± 1	6.25%
	Phase $n(x)=n(x+\Lambda)$	Transmission	Many	33.9% 1st-order
$Q > 1$ Thick	Absorption $\alpha(x)=\alpha(x+\Lambda)$	Transmission	One Bragg-Matched Order	3.7%
		Reflection		7.2%
	Phase $n(x)=n(x+\Lambda)$	Transmission		100%
		Reflection		100%

$$n(x) = n_o + n_{10}\, m \cos Kx = n_o + 2n_{10} \frac{\sqrt{\beta}}{1+\beta} \cos Kx \qquad (2b)$$

where n_{10} is a constant of proportionality that depends on the incident exposure and the recording medium.

To derive Kogelnik's formula[4] for diffraction efficiency of volume phase gratings with uniform amplitude, consider the transmission grating shown in Fig. 2 with an index of refraction given by Eq. (1). The electric field of the incident read beam and its diffracted component are described as plane waves

$$E_{AR}(\mathbf{r},t) = \frac{1}{2}\, A_R(z) \exp[i(\mathbf{k}_{AR}\bullet\mathbf{r} - \omega t)] + \text{c.c.} \;, \qquad (3)$$

where A_R is the complex amplitude of the read beam at steady-state, ω is the frequency and $\mathbf{k}_{AR}$ is the corresponding wave vector in the medium at an angle θ_{AR} with respect to the z-axis. To describe the diffracted beam, A is replaced by B. The subscript R denotes the read beam and its coherently generated diffracted beam. When the electric field of the

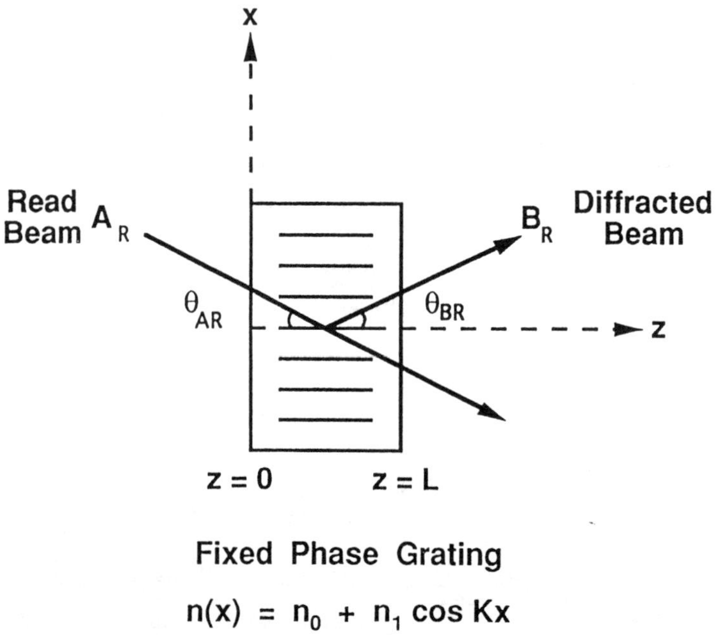

Fig. 2. The schematic representation of diffraction from a uniform phase grating.

optical wave is described by Eq. (3), then the intensity is equal to $\varepsilon_o n_o c|A_R(z)|^2/2$ in mks units,[11] where ε_o is the electric permittivity of free space and c is the velocity of light. Ignoring the multiplicative constant factor, $I_{AR}(z) \equiv |A_R(z)|^2/2$ is taken to represent the intensity of the beam in the rest of this paper. The z axis is taken normal to the surface of the medium, and the complex amplitudes are assumed to be functions of z due to absorption and diffraction. The amplitudes are independent of the transverse coordinate x because the grating is assumed to be infinite in that direction. To describe the propagation and coupling of the two waves at steady state in the thick transmission phase grating, we start with the scalar wave equation

$$\nabla^2 E + \frac{n^2\omega^2}{c^2} E = 0 \quad , \tag{4}$$

where $E = E_{AR} + E_{BR}$ is the total electric field. Substituting for n and E in the wave equation, using the slowly varying envelope approximation and ignoring phase-mismatched terms and terms that are quadratic in the index change, we arrive at the following coupled wave equations

$$\cos\theta_{AR} \frac{dA_R}{dz} = -\frac{\alpha_R}{2} A_R + i \frac{\pi n_1}{\lambda_R} B_R e^{i\Delta k_z z}$$

$$\cos\theta_{BR} \frac{dB_R}{dz} = -\frac{\alpha_R}{2} B_R + i \frac{\pi n_1}{\lambda_R} A_R e^{-i\Delta k_z z} , \tag{5a}$$

where

$$\Delta k_z = k(\cos\theta_{BR} - \cos\theta_{AR}) \quad , \quad k \sin\theta_{BR} + K = k \sin\theta_{AR} \quad , \tag{5b}$$

α_R is the intensity absorption coefficient of the grating at the reading wavelength that has been introduced phenomenologically to the coupled wave equations and Δk_z is the phase-mismatch along the z-direction. Integration of the coupled wave equations (5a) over the transverse coordinate x lead to the last relation expressed in Eq. (5b). For phase-matched readout, $\Delta k_z = 0$ (Fig. 3). Ignoring the trivial solution of $\theta_{BR} = \theta_{AR}$ which corresponds to the diffracted beam propagating along the read beam, the relevant solution is $\theta_{AR} = -\theta_{BR} \equiv \theta_R$ for $\Delta k_z = 0$ (Bragg-matched readout), with $\theta_R = \sin^{-1}(\lambda_R/2n_0\Lambda)$. Hence the Bragg-matched diffracted beam propagates at an angle to the grating planes that is equal and opposite to the reading beam angle. Using boundary conditions $A_R(0) = A_{R0}$ and $B_R(0) = 0$, the solutions for the beam amplitudes are given by

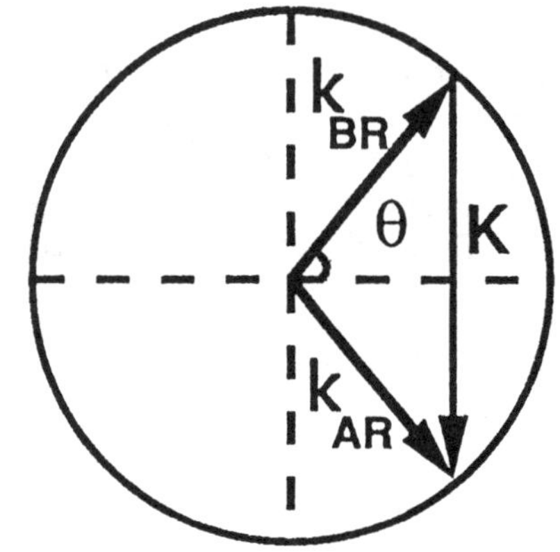

$$K = k_{AR} - k_{BR}$$

Fig. 3. Bragg–matched readout with $\Delta k_z = 0$.

$$A_R(z) = A_{R0} \exp(-\frac{\alpha_R z}{2\cos\theta_R}) \cos\kappa z ,$$

$$B_R(z) = i\, A_{R0} \exp(-\frac{\alpha_R z}{2\cos\theta_R}) \sin\kappa z, \tag{6}$$

where $\kappa = \pi n_1/\lambda_R \cos\theta_R$, θ_R being the Bragg angle for the reading beam. Figure 4 shows the normalized beam intensities $|A_R(z)|^2/|A_{R0}|^2$ (solid curve) and $|B_R(z)|^2/|A_{R0}|^2$ (dashed curve) as a function of distance z into a lossless grating ($\alpha_R = 0$). At the beginning of the thick phase grating, there is no diffracted beam, and the normalized read beam intensity is unity. Periodic exchange of energy takes place between the read beam and its diffracted component. The ratio of the diffracted beam intensity at the output of the grating to the read beam intensity at the input of the grating determines the diffraction efficiency: $\eta = |B_R(L)|^2 / |A_{R0}|^2$. Substituting for κ, we arrive at Kogelnik"s formula[4]

$$\eta = \exp(-\frac{\alpha_R L}{\cos\theta_R}) \sin^2 (\frac{\pi n_1 L}{\lambda_R \cos\theta_R}) \quad . \tag{7}$$

The argument of the sine function depends on the product of grating amplitude times its thickness and on the wavelength and Bragg angle at readout, but the argument is independent of the read beam intensity. For a lossless grating ($\alpha_R = 0$) and Bragg–matched readout ($\Delta k_z = 0$), diffraction efficiency is equal to unity whenever the argument of the sine function is an integral multiple of $\pi/2$, implying total transfer of the read beam energy to the diffracted beam. Substituting Eq. (2b) into Eq. (7), we obtain

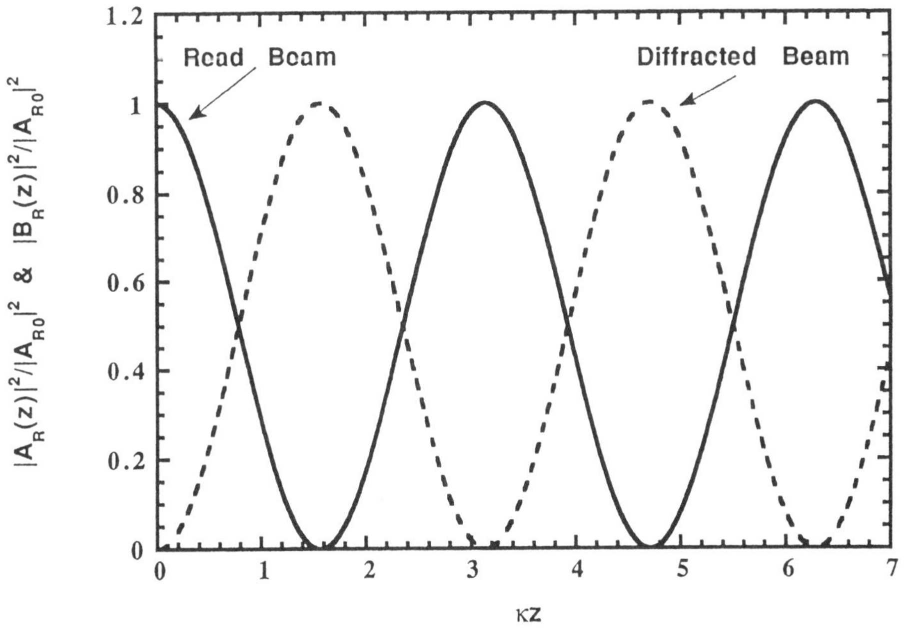

Fig. 4. Intensities of the read beam (solid curve) and the diffracted beam (dashed curve) as a function of normalized depth κz within the medium.

$$\eta = \exp(-\frac{\alpha_R L}{\cos\theta_R}) \sin^2 [\frac{\pi n_{10} L}{\lambda_R \cos\theta_R} \frac{2\sqrt{\beta}}{1+\beta}] . \tag{8}$$

The above equation illustrates the variation of the diffraction efficiency of uniform phase grating with the intensity ratio of the writing beams.

[B]. Photorefractive Gratings with Nonuniform Amplitude and Phase

Two coherent light beams intersect in a photorefractive crystal when writing a grating (Fig. 5); due to photorefractive two-beam coupling, one writing beam gains energy at the expense of the other writing beam. The nonreciprocal energy transfer between the two beams is due to Bragg scattering off the grating the beams themselves wrote. This phenomenon is known as "Self-Diffraction"[12], and the direction of energy transfer is determined by the orientation of the crystal and the nature of charge carriers (electrons or holes). Such an energy exchange is not found in most nonlinear materials because they respond locally to optical beams, with no phase shift between index grating and intensity pattern. We will now show that the amplitude and phase of the index modulation are no longer uniform due to the exchange of energy between the two writing beams in the photorefractive medium. Similar to the previous section, we describe the two writing beams as plane waves with complex amplitudes A_W and B_W. These two beams interfere to produce a sinusoidal intensity pattern given by

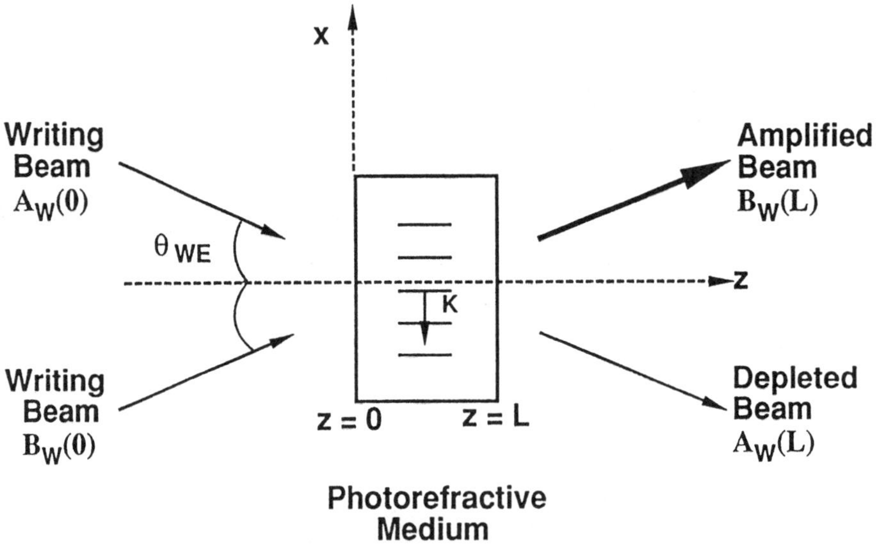

Fig. 5. Schematics of photorefractive two-beam coupling.

$$I_W(x,z) = I_{0W}(z)\left(1+\frac{m(z)}{2}e^{iKx}+\frac{m^*(z)}{2}e^{-iKx}\right) , \qquad (9)$$

where $I_{0W}(z)$ is the local sum of the two beam intensities: $I_{0W}(z) = I_{AW}(z)+I_{BW}(z) = [\,|A_W(z)|^2 + |B_W(z)|^2\,]/2$. K is the magnitude of the grating wavevector: $|\mathbf{K}| = |\mathbf{k}_{AW} - \mathbf{k}_{BW}| = 2\pi/\Lambda$, Λ being the fringe separation equal to $\lambda_W/(2\sin\theta_{WE})$ when the two beams are symmetrically incident at an external angle of $2\theta_{WE}$ about the crystal face normal. λ_W is the wavelength of the two writing beams and x is the direction along the grating wave vector. m(z) is the local modulation depth of the incident intensity pattern: $m(z) = A_W(z)B_W^*(z)/I_0(z)$; it depends on the polarizations and intensity ratio of the two writing beams. The high intensity regions of the spatially periodic intensity pattern cause photoionization of the impurity levels in the bandgap. The mobile charges drift and diffuse before recombining and being trapped in regions of low intensity, leading to a separation of ionic charges inside the crystal. The resulting space-charge field produces a photorefractive index grating via the linear electro-optic effect.[12] Assuming electrons to be the charge carriers, the photorefractive index grating can be written as

$$n = n_b + \frac{n_{1W}}{2}\left[\, e^{i\phi}\,\frac{A_W B_W^*}{|A_W|^2 + |B_W|^2}\, e^{iKx} + c.c.\right] , \qquad (10a)$$

where

$$n_{1W} = n_b^3 \, r_{eff,W} \, E_q \sqrt{\frac{E_0^2+E_d^2}{E_0^2+(E_d+E_q)^2}} \quad ,$$

$$\tan\phi = \frac{E_0^2+E_d(E_d+E_q)}{E_0E_q} \quad . \tag{10b}$$

Here n_b is the background refractive index of the photorefractive medium in the absence of light, n_{1W} is the amplitude of the index grating at the fundamental harmonic of the grating vector and ϕ represents the spatial phase shift of the index gratings with respect to the light intensity pattern. $r_{eff,W}$ is an effective electro-optic coefficient that depends on the polarization of the writing beams, the orientation of the grating wave vector that is written, and the non-zero Pockels coefficients of the photorefractive medium. E_0 is the dc electric field applied to the photorefractive crystal, E_d is the characteristic electric field due to diffusion, and E_q is the saturation value of the space charge field. Substituting for n and E in the scalar wave Eq. (4), using the slowly varying envelope approximation and ignoring phase-mismatched terms and terms that are quadratic in the index change, we arrive at the following coupled wave equations

$$\frac{dA_W}{dz} = -\frac{\alpha_W}{2\cos\theta_{WI}} A_W + i\, e^{i\phi} \frac{\pi n_{1W}}{\lambda_W \cos\theta_{WI}} \frac{A_W B_W^*}{|A_W|^2 + |B_W|^2} B_W \quad ,$$

$$\frac{dB_W}{dz} = -\frac{\alpha_W}{2\cos\theta_{WI}} B_W + i\, e^{-i\phi} \frac{\pi n_{1W}}{\lambda_W \cos\theta_{WI}} \frac{A_W^* B_W}{|A_W|^2 + |B_W|^2} A_W \quad , \tag{11}$$

where α_W is the intensity absorption coefficient at the writing wavelegth and θ_{WI} is the half angle between the beams inside the crystal. Expressing the complex beam amplitudes in polar form: $A_W = \sqrt{2I_{AW}}\exp(i\psi_{AW})$, we obtain the following equations for beam intensities and phases

$$\frac{dI_{AW}}{dz} = -\frac{\alpha_W}{\cos\theta_{WI}} I_{AW} - \gamma \frac{I_{AW}I_{BW}}{I_{AW}+I_{BW}} \quad , \qquad \frac{d\psi_{AW}}{dz} = \zeta \frac{I_{BW}}{I_{AW}+I_{BW}} \quad ,$$

$$\frac{dI_{BW}}{dz} = -\frac{\alpha_W}{\cos\theta_{WI}} I_{BW} + \gamma \frac{I_{AW}I_{BW}}{I_{AW}+I_{BW}} \quad , \qquad \frac{d\psi_{BW}}{dz} = \zeta \frac{I_{AW}}{I_{AW}+I_{BW}} \quad , \tag{12a}$$

where

$$\gamma = \frac{2\pi n_{1W}\sin\phi}{\lambda_W\cos\theta_{WI}} \quad , \qquad \zeta = \frac{\pi n_{1W}\cos\phi}{\lambda_W\cos\theta_{WI}} \quad . \tag{12b}$$

The intensity equations and subsequently the phase equations can be solved as functions of distance z in the crystal

$$I_{AW}(z) = I_{AW}(0)\,\frac{1+\beta}{1+\beta\,e^{\gamma z}}\,\exp(-\frac{\alpha_W z}{\cos\theta_{WI}}) \quad ,$$

$$I_{BW}(z) = I_{BW}(0)\,\frac{1+\beta}{\beta+e^{-\gamma z}}\,\exp(-\frac{\alpha_W z}{\cos\theta_{WI}}) \quad , \tag{13a}$$

$$\psi_{AW}(z) - \psi_{BW}(z) = \psi_{AW}(0) - \psi_{BW}(0) - \zeta z + \frac{2\zeta}{\gamma}\ln[\frac{1+\beta e^{\gamma z}}{1+\beta}] \quad , \tag{13b}$$

where $\beta = I_{BW}(0)/I_{AW}(0)$ is the intensity ratio of the writing beams at the input. We find that the intensity of each beam at a distance z depends on the input intensities of **both** the beams, and there is an exponential dependence on the product γz. If the beam being amplified (≡ probe or writing beam B) is initially much weaker than the beam being depleted (≡ pump or writing beam A) such that $\beta \ll 1$, and $\beta \ll \exp(-\gamma L)$, we obtain an exponential buildup for the probe; while the pump is undepleted and remains constant at its input value in a lossless photorefractive medium: $I_{BW}(L) \approx I_{BW}(0)\,e^{\gamma L}$ and $I_{AW}(L) \approx I_{AW}(0)$. Substituting the solutions for the beam intensities and phases given by Eqs. (13) into the expression for the photorefractive index grating given by Eq. (10a), we arrive at the following expression

$$n(x,z) = n_b + \frac{n_{1W}}{\sqrt{\beta}\,e^{\gamma z/2} + \frac{e^{-\gamma z/2}}{\sqrt{\beta}}}\ \cos[Kx + \phi + \Delta\psi(0) - \zeta z + \frac{2\zeta}{\gamma}\ln(\frac{1+\beta e^{\gamma z}}{1+\beta})]$$

$$\equiv n_b + n_{1eff}(z)\ \cos[Kx+\phi+\Delta\psi_{eff}(z)] \tag{14}$$

where $\Delta\psi = \psi_{AW} - \psi_{BW}$. This is the general expression for a photorefractive index grating using transmission geometry that is valid for arbitrary phase shift ϕ of index grating from the intensity pattern. In contrast to the fixed, volume phase grating with uniform amplitude and phase of Eq. (1), the photorefractive index grating in the above equation has an amplitude and phase that vary with the depth in the grating. Bending of the photorefractive hologram through the crystal thickness is caused by $\Delta\psi_{eff}(z)$ - there is no bending for $\phi = \pi/2$, and bending increases as ϕ shifts from $\pi/2$. Note that Eq. (14) is *independent of absorption losses* ($\propto \alpha_W$) in the medium at the writing wavelength. Without loss of generality, we arbitrarily set the relative phase between the two writing beams equal to zero at the input ($\Delta\psi(0) = 0$). Figure 6a is a plot of the nonuniform grating amplitude $2n_{1eff}(z)/n_{1W}$ and Fig. 6b is a plot of the nonuniform phase of the grating $\Delta\psi_{eff}(z)$ as a function of depth z into the grating. The phase shift ϕ of the index grating from the intensity pattern is 0 (dashed curve), $\pi/6$ (dotted-dashed curve) and $\pi/2$ (solid curve), while $\beta = 0.01$ and $2\pi n_{1W} / (\lambda_W \cos\theta_{WI}) = 20\ cm^{-1}$. Note that $\Delta\psi_{eff}(z)$ changes sign in the bulk of the grating for $\phi = \pi/6$, so that the surfaces of equal index change are curved and tilted.[12]

For $\phi = 0$, there is no phase shift of the index grating from the intensity pattern, so that the intensity coupling constant γ between the writing beams is equal to zero, and the phase coupling constant ζ is equal to its maximum value of $\zeta_m = \pi n_1/\lambda_W \cos\theta_{WI}$. In this case, Eq. (14) simplifies to

$$n(x,z) = n_b + n_{1W} \frac{\sqrt{\beta}}{1+\beta} \cos(Kx - \zeta_m z) \quad , \tag{15a}$$

which is a grating with uniform amplitude and a spatial frequency that is modified by the phase coupling between the writing beams. The phase of the grating varies linearly with depth into the crystal, so that the grating planes are rotated to form slanted gratings. Maximum diffraction efficiency will occur at a read angle different from that during writing. Comparing with Eq. (2b), the amplitude n_{1W} of the photorefractive index grating given in Eq. (15a) must be replaced by $2n_{10}$ to yield the results of the uniform phase grating.

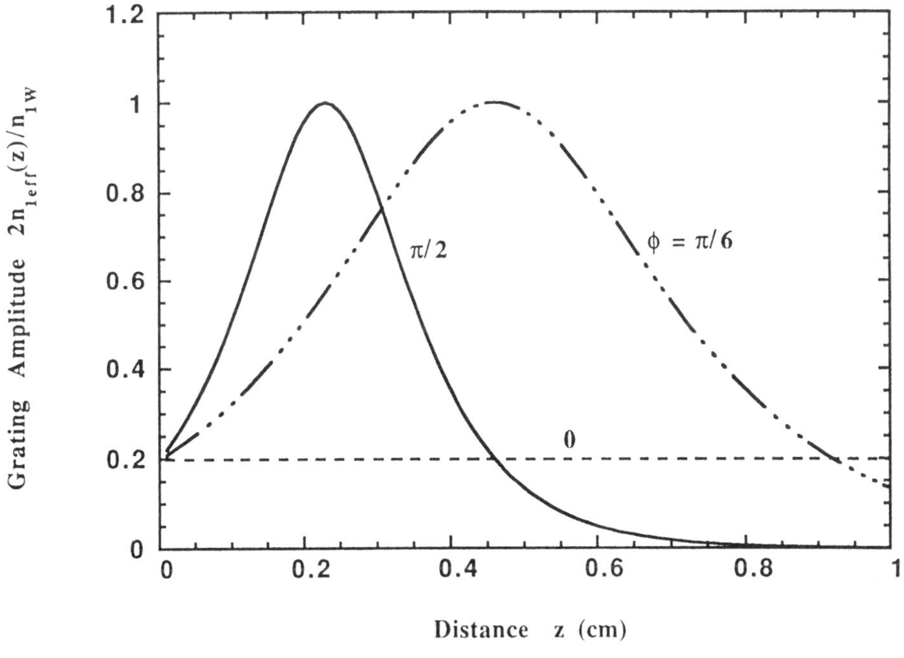

Fig. 6a. Nonuniform amplitude $2n_{1eff}(z)/n_{1W}$ of photorefractive grating for different values of phase shift ϕ of index grating from intensity pattern. The beam ratio β during writing is equal to 0.01, and $2\pi n_{1W}/\lambda_W \cos\theta_{WI}) = 20\ cm^{-1}$.

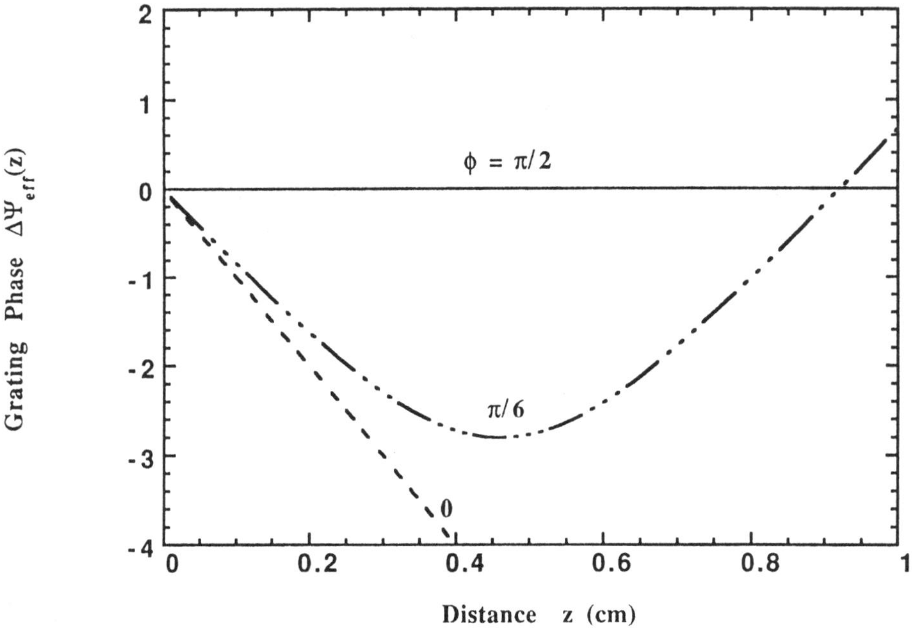

Fig. 6b. Nonuniform phase $\Delta\psi_{eff}(z)$ of photorefractive grating for different values of phase shift ϕ of index grating from intensity pattern. The other parameters are identical to Fig. 6a.

Similarly, if $\phi = \pi/2$, the phase coupling constant ζ is equal to zero and γ is equal to its maximum value of $\Gamma = 2\pi n_1/\lambda_W \cos\theta_{WI}$. In this case, Eq. (14) simplifies to

$$n(x,z) = n_b - n_{1W} \frac{1}{\sqrt{\beta}\, e^{\Gamma z/2} + \frac{e^{-\Gamma z/2}}{\sqrt{\beta}}} \sin Kx\,. \tag{15b}$$

Hence the photorefractive index grating in the above equation has a uniform phase and an amplitude that varies with the depth in the grating and is spatially shifted by $\pi/2$ from the intensity pattern of Eq. (9). The nonuniformity $2n_{1eff}(z)/n_{1W}$ of the grating amplitude is a function of the input beam ratio β and the strength of the beam coupling Γ and is illustrated in Fig. 7 for $\Gamma=20\text{cm}^{-1}$, a coupling value easily achieved in single crystal $BaTiO_3$. The grating amplitude is maximum when the two writing beams have equal intensity; for $\beta = 1$ case, this occurs at the input plane itself, while for the $\beta = 0.01$ case, the two writing beams have equal intensity only after they have propagated some distance into the crystal and the weak probe has been amplified at the expense of the strong pump beam via photorefractive two-beam coupling. For $\beta = 100$, the probe is stronger than the pump to begin with, and two-beam coupling effects serve to further amplify the probe in the medium. Hence the optimum beam ratio is never achieved and the grating amplitude decreases monotonically in the medium. It is important to study the effects of this departure from uniformity of photorefractive index gratings on the diffraction efficiency of a weak reading beam. As we shall show in the next section, maximum diffraction efficiency occurs for $\beta = 0.01$, when the grating peaks near the middle of the interaction length.[12]

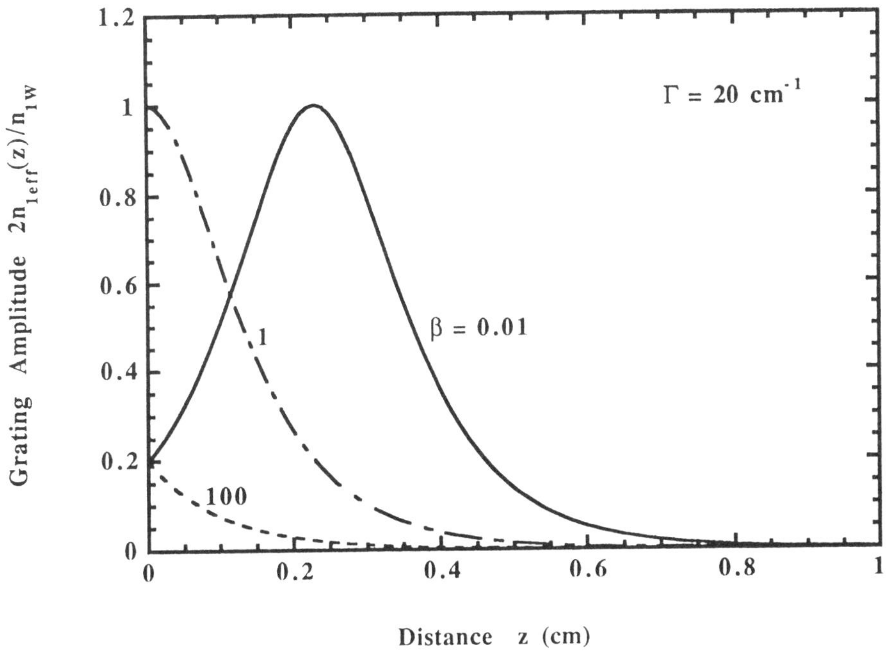

Fig. 7. Nonuniform amplitude $2n_{1eff}(z)/n_{1W}$ of photorefractive grating for different values of beam ratio β during writing. The two-beam coupling constant Γ is equal to 20 cm^{-1}.

[C]. Readout of π/2 Shifted Photorefractive Gratings with a Weak Incoherent Beam

The basic interaction geometry is illustrated in Fig. 8. A_W and B_W are the two coherent writing beams of wavelength λ_W that intersect within a photorefractive crystal of thickness L. The angle between the two beams is $2\theta_{WI}$ in the medium. The photorefractive grating induced by this pair of beams is read by an incoherent third beam A_R of wavelength λ_R that is incident at a different angle (θ_{RI} in the medium) for Bragg matched readout. A fourth beam B_R is generated by the diffraction of the read beam off this grating. The theoretical problem of interest is to find an expression for the diffraction efficiency of the photorefractive index grating when $\phi = \pi/2$, as given by Eq. (15b). The intensity of the read beam I_{AR} is assumed to be weaker than the writing beam intensities I_{AW}, I_{BW}, so that the readout process does not perturb the grating being read. Also, Bragg condition is satisfied during readout, so that $\mathbf{k}_{AR} - \mathbf{k}_{BR} = \mathbf{k}_{AW} - \mathbf{k}_{BW} = \mathbf{K}$. By following the methods used in the previous sections, we arrive at the coupled wave equations for the amplitudes of the read beam A_R and the diffracted beam B_R

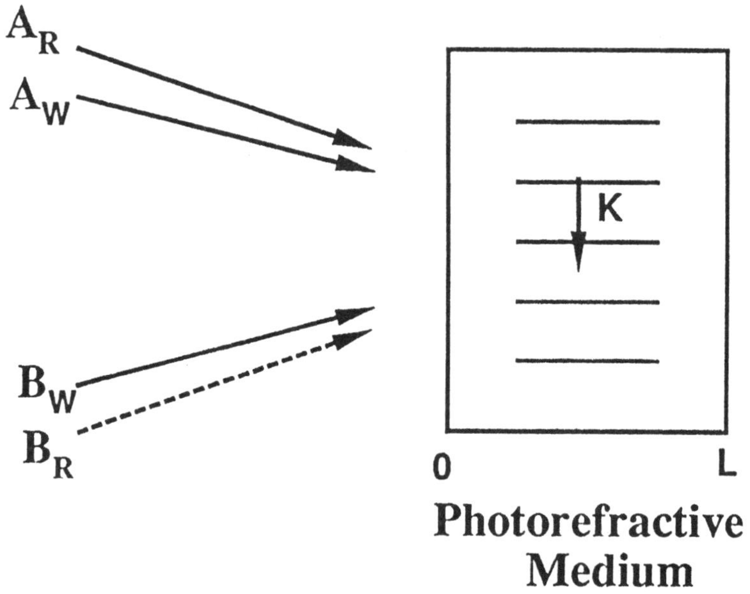

Fig. 8. Schematics of simultaneous writing and readout of photorefractive index grating.

$$\frac{dA_R}{dz} = -\frac{\alpha_R}{2\cos\theta_{RI}} A_R - \frac{\pi n_{1R}}{\lambda_R \cos\theta_{RI}} \frac{1}{\left(\sqrt{\beta}e^{\Gamma z/2} + \frac{e^{-\Gamma z/2}}{\sqrt{\beta}}\right)} B_R$$

$$\frac{dB_R}{dz} = -\frac{\alpha_R}{2\cos\theta_{RI}} B_R + \frac{\pi n_{1R}}{\lambda_R \cos\theta_{RI}} \frac{1}{\left(\sqrt{\beta}e^{\Gamma z/2} + \frac{e^{-\Gamma z/2}}{\sqrt{\beta}}\right)} A_R \quad . \tag{16}$$

Since the read beam and its diffracted component can have a polarization different from that of the writing beams, r_{eff} at readout can be different from that during writing. Such a possibility is allowed for by distinguishing between the grating amplitudes n_1 during writing and readout, even though both the processes involve the same space charge grating with wave vector **K**. Using boundary conditions $A_R(0) = A_{R0}$, $B_R(0) = 0$, we obtain the following expression for the diffraction efficiency $(= |B_R(L)/A_{R0}|^2)$ of the hologram[5]

$$\eta = \exp\left(-\frac{\alpha_R L}{\cos\theta_{RI}}\right) \sin^2\left[\frac{2\kappa}{\Gamma}\left(\tan^{-1}(e^{\Gamma L/2}\sqrt{\beta}) - \tan^{-1}(\sqrt{\beta})\right)\right] , \tag{17}$$

where $\kappa = \pi n_{1R}/\lambda_R \cos\theta_{RI}$ is the coupling constant during readout. κ and $\Gamma = 2\pi n_{1W}/\lambda_W \cos\theta_{WI}$ are related but can be quite different in value depending on the polarization states of the writing beams and that of the reading beam. Note that κ (Γ) is the amplitude (intensity) coupling constant, so that $2\kappa = \Gamma$ for identical polarization and wavelength during writing and readout. Then the expression (17) for diffraction efficiency simplifies to

$$\eta = \exp\left(-\frac{\alpha_R L}{\cos\theta_{RI}}\right)\frac{\beta}{1+\beta}\,\frac{\left[\exp(\Gamma L/2) - 1\right]^2}{1 + \beta\exp(\Gamma L)}\ , \tag{18}$$

which is identical to the result obtained by Kukhtarev et al[12] for a lossless medium ($\alpha_R = 0$). In the limit of no coupling during writing ($\Gamma \rightarrow 0$, κ remains finite), Eq. (17) reduces to the Kogelnik result[4] given by Eq. (8) if we use the fact that n_{1R} is equal to $2n_{10}$ of the fixed, uniform grating; this limit can be achieved in noncubic photorefractive crystals by writing a grating that can be read out only by anisotropic diffraction. In particular, we examine the diffraction efficiency of holograms written in $BaTiO_3$ for the following two cases. In the first case, the writing beams at λ_W = 514.5 nm are extraordinary-polarized, resulting in strong two-beam coupling (Γ_e=20cm^{-1}). In such a case, the index modulation varies significantly over the thickness of the crystal because one beam is strongly amplified as the other strongly depletes. For the second case, the writing beams are ordinary-polarized for which the beam coupling constant is much smaller (Γ_o=1cm^{-1}). In both cases, the grating is read out by a weak beam at λ_R = 633 nm that is extraordinary-polarized ($\kappa=\Gamma_e/2$=10cm^{-1}). Plots of the diffraction efficiency η as a function of medium thickness L, as given by Eq. (17), are shown in Fig. 9, where the beam ratio is taken to be 0.1 (absorption is neglected, for simplicity). Also plotted for comparison is the diffraction efficiency of a uniform hologram (with no coupling, corresponding to Eq. (8)), with the same reading beam coupling constant κ. Note that for the case of polarization asymmetry during writing/reading, the argument of the sine function in Eq. (17) can be quite large, allowing η to exhibit one complete cycle. Figure 10 is a plot of the diffraction efficiency η as a function of beam ratio β for L = 0.5 cm. The values of the coupling constants are the same as in Fig. 9. Note that writing with ordinary-polarized beams gives results similar to the Kogelnik case due to the smaller two-beam coupling effects. Corresponding to this choice of parameter values, the maximum diffraction efficiency occurs at an input intensity ratio of 0.01 for extraordinary-polarized writing beams, while the maxima occur at $\beta \sim 0.1$ and 8 for ordinary-polarized writing beams.

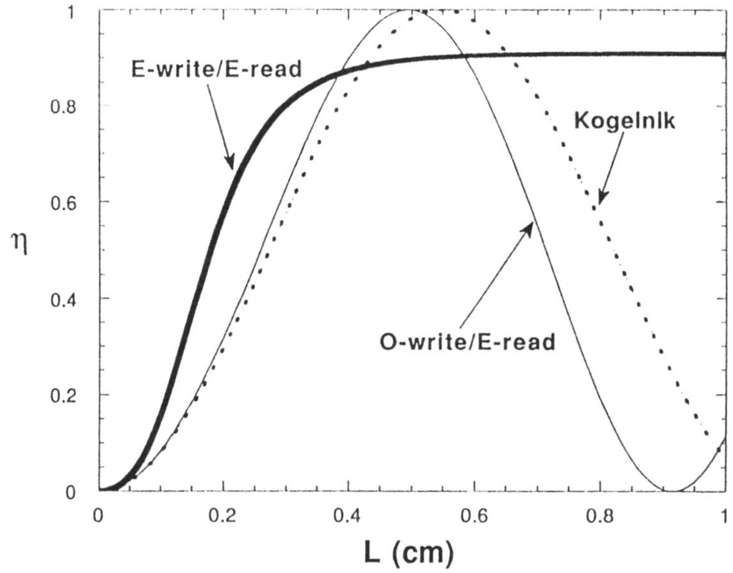

Fig. 9. Diffraction efficiency η as a function of medium thickness L. Input beam ratio β is equal to 0.1, and absorption effects are neglected (α = 0).

(i) $\Gamma = 2\kappa = 20\ cm^{-1}$ (E-write / E-read curve)
(ii) $\Gamma = 1\ cm^{-1}$, $2\kappa = 20\ cm^{-1}$ (O-write / E-read curve)
(iii) $\Gamma = 0$, $2\kappa = 20\ cm^{-1}$ (Kogelnik case)

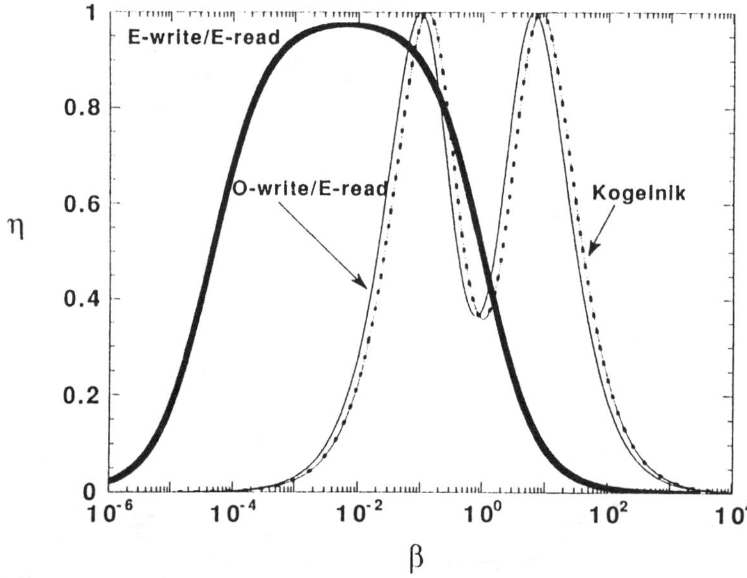

Fig. 10. Diffraction efficiency η as a function of beam ratio β during writing. Grating thickness L is equal to 0.5 cm. All other parameters are the same as in Fig. 9.

[D]. Readout of Photorefractive Gratings with a Strong Incoherent Beam

The basic interaction geometry is identical to that shown in Fig. 8, and the incoherent read beam intensity I_{AR} can be comparable to or larger than the intensities of the writing beams I_{AW}, I_{BW}. Hence the new grating formed by the read beam and its coherently generated diffracted beam must be taken into account, besides the grating being read. In this section, the diffraction efficiency of a photorefractive grating is examined for arbitrary phase shifts of the index grating from the intensity pattern. All the four waves are assumed to be plane waves with identical linear polarization, so that the grating amplitude n_1 is equal for the writing and readout beams. The total intensity is the sum of the intensity of all the four beams and the interference terms arising from the two pairs of coherent beams

$$I(\mathbf{r}) = \frac{1}{2}\sum_j \{|A_j|^2 + |B_j|^2 + [A_j B_j^* \exp(i\mathbf{K}_j\cdot\mathbf{r}) + c.c.]\} \quad , \tag{19}$$

where the grating vectors are defined by $\mathbf{K}_j \equiv \mathbf{k}_{Aj} - \mathbf{k}_{Bj}$, $j = W, R$. If Bragg condition is satisfied for the two coherent beam pairs: $\mathbf{K}_W = \mathbf{K}_R \equiv \mathbf{K}$, then the gratings formed by the two beam pairs combine to form a single grating that couples all the beams

$$n = n_b + \frac{n_1}{2}\left[e^{i\phi}\frac{A_W B_W^* + A_R B_R^*}{|A_W|^2 + |B_W|^2 + |A_R|^2 + |B_R|^2}\exp(i\mathbf{K}\cdot\mathbf{r}) + c.c.\right] . \tag{20}$$

We can derive the following coupled wave equations by using the methods developed in the previous sections

$$\frac{dA_j}{dz} = \gamma \frac{A_W B_W^* + A_R B_R^*}{|A_W|^2 + |B_W|^2 + |A_R|^2 + |B_R|^2} B_j$$

$$\frac{dB_j}{dz} = -\gamma^* \frac{A_W^* B_W + A_R^* B_R}{|A_W|^2 + |B_W|^2 + |A_R|^2 + |B_R|^2} A_j, \tag{21}$$

where $j = W$, R, and γ is the complex field coupling coefficient defined as $\gamma = i\pi n_1 e^{i\phi}/(\lambda\cos\theta_I)$. In arriving at these equations, we have ignored the difference between the wavelengths ($\lambda_W \approx \lambda_R \equiv \lambda$) and angles ($\theta_{WI} \approx \theta_{RI} \equiv \theta_I$) of the writing and readout beams, so that the coupling constant is the same for both pairs of beams. For Bragg-matched readout, the frequencies of the various beam pairs can differ by a few tens of gigahertz and their angles differ by a few milliradians for typical values of 1 μm grating period and 1mm thickness of the photorefractive grating. The coupled wave equations (21) can be solved by using methods developed in Refs. 13–15: the constants of integration associated with these equations are identified; two new variables $f \equiv A_R/A_W$ and $g \equiv B_R/B_W$ are defined and their equations are solved. The intensities of the four beams can then be expressed as the following functions of z[9]

$$|A_R(z)|^2 = \frac{c_W\,|f(z)g(z)|^2 - c_R\,|f(z)|^2}{|g(z)|^2 - |f(z)|^2}$$

$$|B_R(z)|^2 = \frac{c_R\,|g(z)|^2 - c_W\,|g(z)f(z)|^2}{|g(z)|^2 - |f(z)|^2}$$

$$|A_W(z)|^2 = \frac{c_W\,|g(z)|^2 - c_R}{|g(z)|^2 - |f(z)|^2}$$

$$|B_W(z)|^2 = \frac{c_R - c_W\,|f(z)|^2}{|g(z)|^2 - |f(z)|^2} \quad , \qquad (22)$$

where

$$f(z) = \frac{sf_0 + (\sigma f_0 + 2\overset{*}{b})\tanh \kappa z}{s - (\sigma - 2bf_0)\tanh \kappa z}$$

$$g(z) = \frac{sg_0 - (\sigma g_0 + 2\overset{*}{b})\tanh \kappa^* z}{s + (\sigma - 2bg_0)\tanh \kappa^* z} \quad , \qquad (23)$$

and $c_R = |A_R(0)|^2 + |B_R(0)|^2$, $c_W = |A_W(0)|^2 + |B_W(0)|^2$, $\sigma = c_R - c_W$, $b = \overset{*}{A}_R(0)A_W(0) + \overset{*}{B}_R(0)B_W(0)$, $s^2 = \sigma^2 + 4|b|^2$, $\kappa = s\gamma/4I_0$, $f_0 = A_R(0)/A_W(0)$, $g_0 = B_R(0)/B_W(0)$, $I_0 = [|A_R(0)|^2 + |B_R(0)|^2 + |A_W(0)|^2 + |B_W(0)|^2]/2$. The above equations are general solutions for forward four-wave mixing. For the problem at hand, we have the boundary conditions $B_R(0) = 0$, and $A_R(0)$, $A_W(0)$, and $B_W(0)$ are known quantities. The diffraction efficiency is then defined as $\eta \equiv |B_R(L)|^2 / |A_R(0)|^2$, where $|B_R(L)|^2$ can be calculated by evaluating the second equation of (22) at z = L. From the expressions for f(L) and g(L), it is obvious that η will be a function of $\tanh \kappa L$, where κ is a complex quantity for any general value of the spatial phase shift ϕ. Using mathematical identities, $\tanh \kappa L$ can be expressed in terms of $\tanh \kappa_r L$ and $\tan \kappa_i L$, where the subscripts r and i refer to the real and imaginary parts of a complex quantity, respectively. Hence we can expect η to exhibit both oscillatory and exponential behavior as a function of $|\kappa|L$.

[i] $\phi = \pi / 2$

γ is a real, negative quantity equal to $-\gamma_r$, with $\gamma_r = \pi n_1/(\lambda \cos\theta_1)$. The expression for η simplifies to[9]

$$\eta = \frac{16\, I_{AW}(0)\, I_{BW}(0)\, \tanh^2\kappa_r L}{s^2(1 + \tanh^2\kappa_r L) - 4\, s \tanh \kappa_r L\, [I_{AR}(0) + I_{AW}(0) - I_{BW}(0)]} \quad , \tag{24}$$

where $\kappa_r = s\gamma_r/4I_0$, $I_{Aj}(0) = |A_j(0)|^2/2$, etc. The expression for η involves only hyperbolic functions of $\kappa_r L$, and is asymmetric with respect to change in sign of κ_r, which implies unequal diffraction efficiencies for readout from the two input ports. A change in the sign of the coupling constant is equivalent to readout from the other input port. This is true for both equal and different intensity writing beams, except that the latter case would involve an interchange of the unequal intensities of the writing beams besides the change in read beam intensity to describe readout from the other input port.

For a weak reading beam $[I_{AR}(0) << I_{AW}(0), I_{BW}(0)]$, the input intensities dependent factor $s/(2I_0)$ that is present in κ_r is equal to unity, so that $\kappa_r = \gamma_r/2$. Hence Eq. (24) simplifies to

$$\eta = \frac{\beta}{1+\beta}\, \frac{[\exp(\gamma_r L) - 1]^2}{1 + \beta \exp(2\gamma_r L)} \quad , \tag{25}$$

where β is the input beam ratio of the writing beams defined as: $\beta \equiv I_{BW}(0) / I_{AW}(0)$. This result is identical to the diffraction efficiency obtained in Eq. (18) if we neglect absorption and use $\Gamma = 2\gamma_r$. The same expression for η was obtained by Kukhtarev *et al.*[12] for a weak read beam and $\pi/2$ shifted photorefractive grating. For small coupling strengths, $\gamma_r L << 1$, and Eq. (25) reduces to $\eta \approx \beta\gamma_r^2 L^2 / [(1+\beta)^2]$. This expression is identical to Kogelnik's result (Eq. (8)) for small arguments of the sine function with $n_1 \equiv 2n_{10}$.

Figure 11 is a plot of the diffraction efficiency as a function of the read beam intensity, as given by Eq. (24). The equal intensity of the write beams is taken to be unity: $I_{AW}(0) = I_{BW}(0) = 1$. Note that the modulation depth can still be smaller than unity due to the read beam intensity in the denominator of the expression for m. For small coupling strengths ($\gamma_r L \approx 1$) and weak read beam ($I_{AR}(0) < I_{AW}(0), I_{BW}(0)$), the diffraction efficiency is independent of the read beam intensity. Increase of the read beam intensity above the write beam intensities serves to erase the grating being read, and the diffraction efficiency falls to zero. For strong coupling, the behavior of η is different—the diffraction efficiency is ~50% for weak read beams, and increases as the read beam becomes stronger than the write beams, reaches a broad maximum and eventually falls to zero when $I_{AR}(0) >> I_{AW}(0), I_{BW}(0)$. Enhancement of the diffraction efficiency occurs

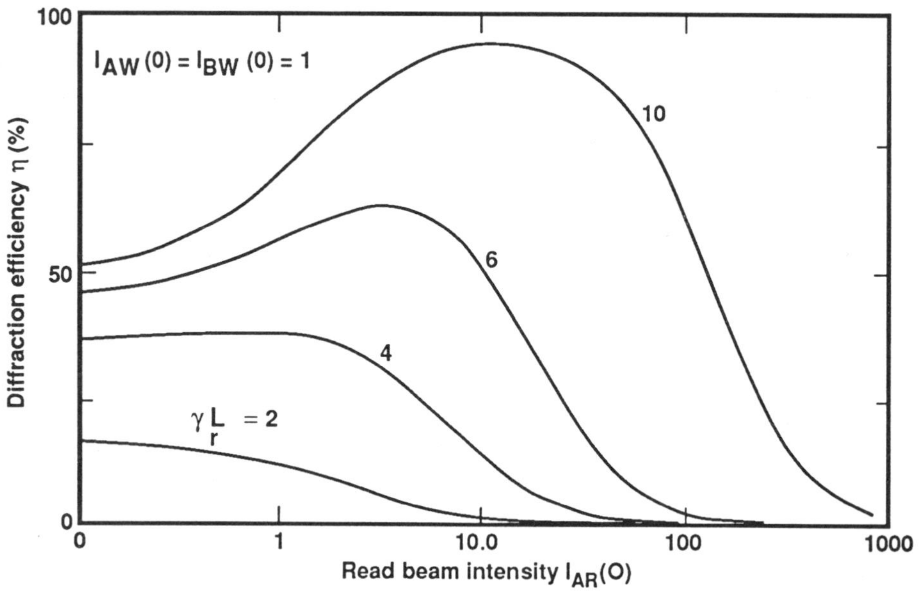

Fig. 11. Diffraction efficiency η as a function of read beam intensity $I_{AR}(0)$ for various coupling strengths $\gamma_r L$. The normalized, equal intensities of the write beams are unity: $I_{AW}(0) = I_{BW}(0) = 1$.

when the new grating formed by the read beam and its diffracted component is in phase with the grating being read. Figure 12 is the same set of curves as in Fig. 11 but for negative coupling constant γ_r, which corresponds to readout from the other input port. In this case, the two gratings are out of phase, so that there is no enhancement of the diffraction efficiency for any intensity of the read beam. As predicted by Eq. (24), the maximum diffraction efficiency is only 50% at large coupling strengths. Figure 13 is a plot of the diffraction efficiency η versus the normalized coupling strength $\gamma_r L$ for negligible absorption and equal intensity of write beams, $I_{AW}(0) = I_{BW}(0) = 1$. The asymmetry in η is negligible for a weak read beam and becomes appreciable only when $I_{AR}(0) > I_{AW}(0), I_{BW}(0)$.

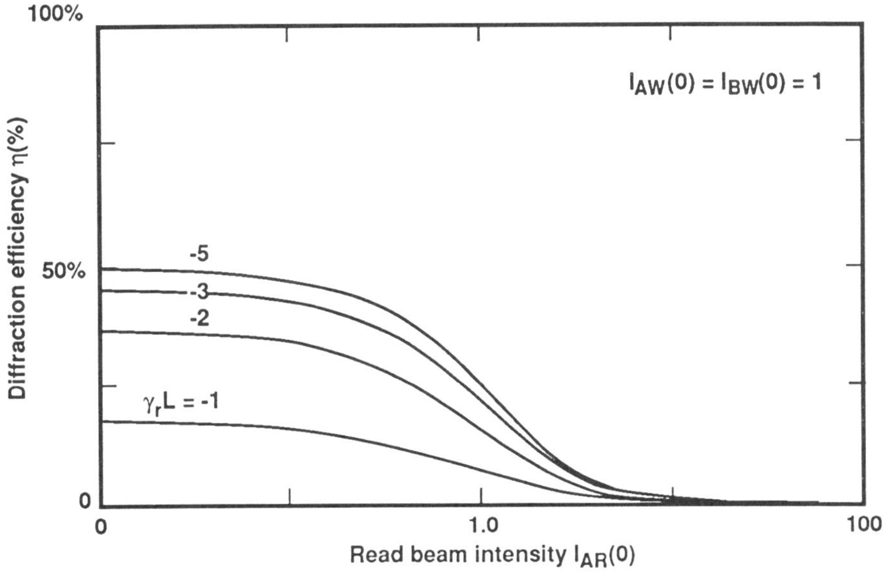

Fig. 12. Diffraction efficiency η as a function of read beam intensity $I_{AR}(0)$ for various negative coupling strengths $\gamma_r L$. The normalized, equal intensities of the write beams are unity: $I_{AW}(0) = I_{BW}(0) = 1$.

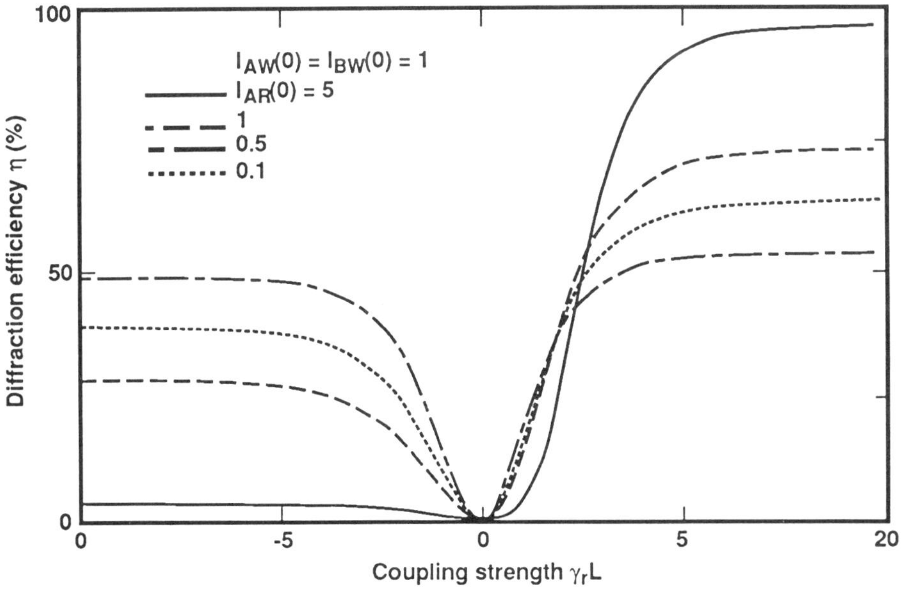

Fig. 13. Diffraction efficiency η as a function of coupling strength $\gamma_r L$ for various read beam intensities $I_{AR}(0)$. The normalized, equal intensities of the write beams are unity: $I_{AW}(0) = I_{BW}(0) = 1$.

[ii] $\phi = 0$

For the case of $\phi = 0$, which corresponds to a local index grating that follows the intensity pattern, γ (and hence κ) is a purely imaginary number that we take to be equal to $i\gamma_i(i\kappa_i)$. The expression for diffraction efficiency simplifies to[9]

$$\eta = \frac{4I_{AW}(0)I_{BW}(0)}{[I_{AR}(0) + I_{AW}(0) - I_{BW}(0)]^2 + 4I_{AW}(0)I_{BW}(0)} \sin^2 \kappa_i L \quad . \tag{26}$$

Apart from the multiplicative factor before the trignometric function and the read beam intensity dependence of κ_i, this result is reminiscent of Kogelnik's expression (Eq. (7)) for diffraction efficiency of uniform gratings. For a weak reading beam $[I_{AR}(0) << I_{AW}(0), I_{BW}(0)]$ and writing beams of comparable intensity $[I_{AW}(0) \approx I_{BW}(0)]$, the multiplicative factor and the input intensities dependent factor $s/(2I_0)$ that is present in κ_i are both equal to unity. Hence Eq. (26) reduces to: $\eta = \sin^2(\gamma_i L/2)$, where $\gamma_i \equiv \pi n_1/(\lambda \cos\theta_I)$ is the amplitude coupling constant in Eqs. (21); this expression is identical to Kogelnik's formula (Eq. (7)) with $n_1 \equiv 2n_{10}$. If the writing beams have different intensities, then Eq. (26) simplifies to

$$\eta = \frac{4\beta}{(1 + \beta)^2} \sin^2 \left(\frac{\gamma_i L}{2}\right) \quad , \tag{27}$$

which is identical to the expression obtained by Kukhtarev *et al.*[12] for replay of a local photorefractive grating with a weak read beam. For small coupling strengths, $\gamma_i L << 1$, and Eq. (27) reduces to $\eta \approx \beta\gamma_i^2 L^2/[(1+\beta)^2]$, which is identical to Kogelnik's expression[4] for small arguments of the trignometric function. Increasing the read beam intensity serves to reduce the multiplicative factor in Eq. (26) until it is zero for a strong read beam $[I_{AR}(0) >> I_{AW}(0), I_{BW}(0)]$, implying erasure of the grating from readout. The $s/2I_0$ factor that is present in κ_i is not very sensitive to changes in read beam intensity and decreases slightly from unity as the intensity of the weak read beam is increased. Once $I_{AR}(0) > I_{AW}(0), I_{BW}(0)$, the factor increases and is again equal to unity for a strong read. Note that η is symmetric with respect to change in sign of κ_i, which implies that there is no nonreciprocity in η for local photoinduced gratings, even when the read beam is intense.

For the general case of a nonzero spatial phase shift between photorefractive grating and fringe pattern and a read beam intensity comparable to that of the writing beams, the parameter dependence of diffraction efficiency is studied numerically. Figure 14 shows the nonlinear behavior of η as a function of the read beam intensity $I_{AR}(0)$ for various phase shifts ϕ. Note that the nonlinearity is more pronounced and that higher values of η are possible when $\phi = \pi / 6$. However, photorefractive gratings formed at $\phi = \pi / 2$ are more difficult to erase. The curve for $\phi = \pi/2$ is identical to that shown in

Fig. 11. For $\phi = 0$ case, both the multiplicative factor and $s/2I_0$ in κ_i decrease as the reading beam intensity is increased, and the decrease of the first function is more rapid. The coupling strength $\gamma_i L$ is equal to 10 in the calculations, so that for an initial read beam intensity of 0.1, $\kappa_i L$ is equal to 2.38. This is considerably more than $\pi/2$ at which a sine function peaks, so even though the argument of the sine function is decreasing from the initial value of 2.38 with increasing read beam intensity, η still increases with $I_{AR}(0)$ until the factor $s/2I_0$ in κ_i reduces to $2\pi/(\gamma_i L) = 0.63$. Beyond that value, η starts decreasing mainly due to the rapidly falling values of the multiplicative factor with $I_{AR}(0)$. The curve for $\phi = \pi/6$ in Fig. 14 represents a combination of effects present in $\phi = \pi/2$ and $\phi = 0$ cases. Figure 15 shows the nonreciprocal behavior of η with respect to direction of readout for various phase shifts ϕ and a read beam intensity equal to that of the writing beams. For equal intensity writing beams, a change in the sign of the coupling constant is equivalent to readout from the other input port. Note that $\phi = 0$ ($\phi = \pi/2$) gives oscillatory (exponential) behavior of η as a function of $|\gamma|L$, while $\phi = \pi/6$ exhibits a combination of both behaviors, with decaying oscillations. The nonreciprocity is enhanced for $\phi = \pi/6$, while $\phi = 0$ case is symmetric with respect to readout from either input port even when the read beam is as intense as the write beams. Figure 16 is a plot of η versus $|\gamma|L$ for a weak reading beam ($I_{AR}(0) = 0.01$). The

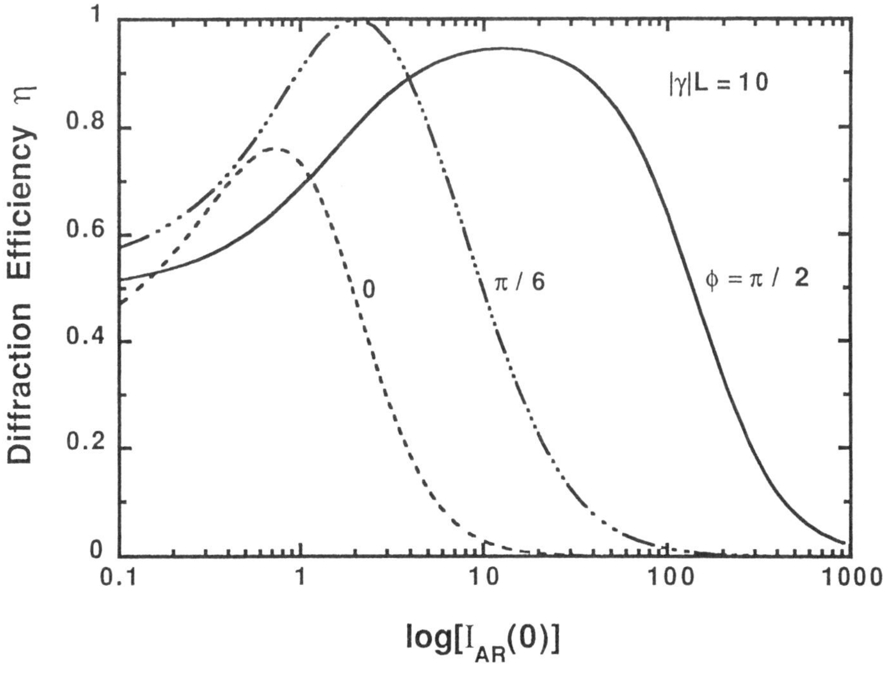

Fig. 14. Diffraction efficiency η as a function of read beam intensity $I_{AR}(0)$ for various values of spatial phase shift ϕ. The modulus of the coupling strength, $|\gamma|L$, is chosen to be 10, and the writing beams are taken to be equal; i.e., $I_{AW}(0) = I_{BW}(0) = 1$.

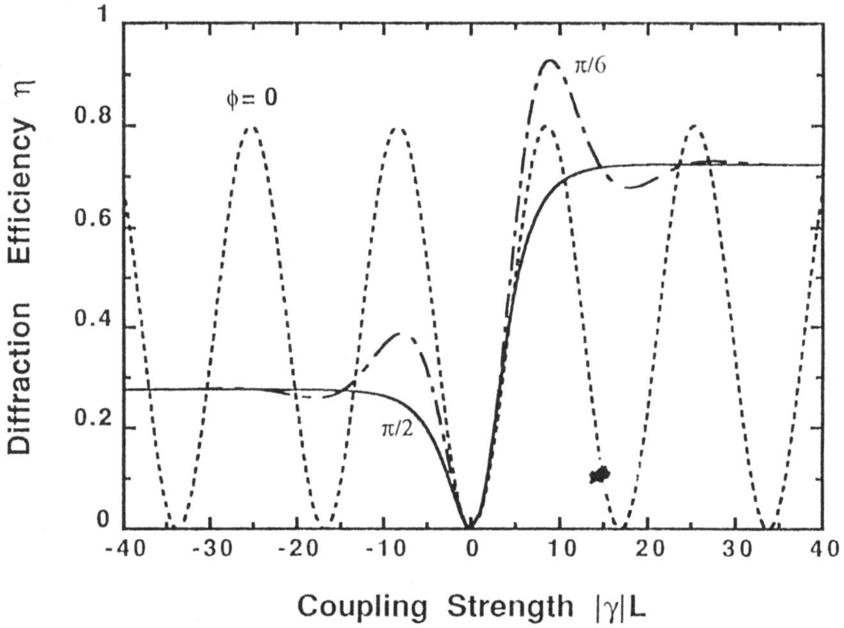

Fig. 15. Diffraction efficiency η as a function of coupling strength |γ|L for various values of spatial phase shift φ. The read beam is taken to be equal to the writing beams; i.e., $I_{AR}(0) = I_{AW}(0) = I_{BW}(0) = 1$.

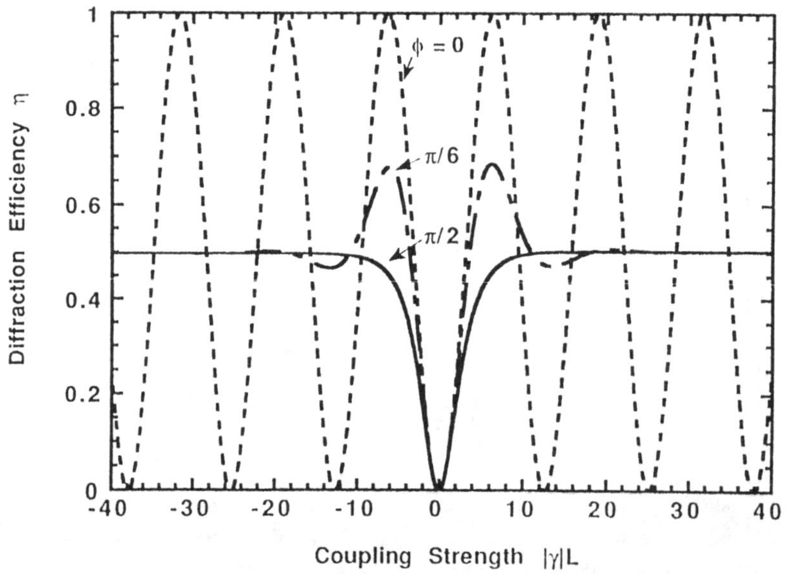

Fig. 16. Same as in Fig. 15, but for $I_{AR}(0) = 0.01$.

diffraction efficiency is now a symmetric function of coupling constant for all values of ϕ, and the oscillatory behavior for $\phi = 0$ reaches a peak value of unity, as expected from Eq. (26)—for the choice of parameters in this figure, the slant of the gratings is zero, leading to a Bragg matched readout with 100% maximum diffraction efficiency.

III. SUMMARY

Kogelnik's formula for diffraction efficiency of a volume phase grating is derived for Bragg matched readout. The differences between a photorefractive index grating and a thick phase grating of uniform amplitude and phase are outlined. For a weak read beam and polarization asymmetry during writing and readout, the diffraction efficiency of a photorefractive grating is independent of the read beam intensity; it is a function of grating thickness, coupling constants during writing and readout, and input intensity ratio of writing beams. For a strong read beam, the diffraction efficiency is a nonlinear function of the read beam intensity. The diffraction efficiency is also nonreciprocal with respect to readout from the two input ports. The nonlinear and nonreciprocal properties of diffraction efficiency are enhanced when the phase shift has a value lying between 0 and $\pi/2$.

ACKNOWLEDGEMENT

This research is supported in part by the U. S. Air Force Office of Scientific Research under contract F49620-92-C-0023.

REFERENCES

1. J.M. Heaton, P.A. Mills, E.G.S. Paige, L. Solymar and T. Wilson, *Optica Acta* **31**, 885 (1984).

2. P. Yeh, *IEEE J. Quant. Electron.* **QE- 25**, 484 (1989).

3. D.L. Staebler and J.J. Amodei, *J. Appl. Phys.* **43**, 1042 (1972).

4. H. Kogelnik, *Bell Syst. Tech. J.* **48**, 2909 (1969).

5. J. Hong and R. Saxena, *Opt. Lett.* **16**, 180 (1991).

6. R. Saxena, F. Vachss, I. McMichael and P. Yeh, *J. Opt. Soc. of Am.* B **7**, 1210 (1990).

7. N.A. Vaines, S.L. Chapman and R.W. Eason, *Appl. Opt.* **28**, 4381 (1989).

8. Claire Gu and Pochi Yeh, *J. Opt. Soc. of Am.* B **7**, 2339 (1990).

9. R. Saxena, C. Gu and P. Yeh, J. Opt. Soc. of Am. B **8**, 1047 (1991).

10. H. Kogelnik, *Laser Technology—November issue*, 68 (1967).

11. A. Yariv and R.A. Fisher, "Optical Phase Conjugation," ed., R.A. Fisher (Academic, New York, 1983), Chap. 1

12. N.V. Kukhtarev, V.B. Markov, S.G. Odulov, M.S. Soskin, and V.L. Vinetskii, *Ferroelectrics* **22**, 949 (1979).

13. B. Fischer, J.O. White, M. Cronin-Golomb, and A. Yariv, *Opt. Lett.* **11**, 239 (1986).

14. P. Yeh, *J. Opt. Soc. Am.* **B4**, 1382 (1987).

15. K.H. Ringhofer and L. Solymar, *Appl. Phys.* **B48**, 395 (1989); D.C. Jones and L. Solymar, *Appl. Phys.* **B50**, 355 (1990).

Photorefractive Mutually Pumped Phase Conjugators

Q. Byron He
Department of Electrical and Computer Engineering
University of California at Santa Barbara
Santa Barbara, CA 93106
Tel (805) 893-8376
Fax (805)893-3262

ABSTRACT

In this paper, we review the theoretical development, various experimental configurations and applications of mutually pumped phase conjugators in photorefractive materials.

1. Introduction

Since the discovery of the photorefractive nonlinearity in the late 1970's, nonlinear optics, especially nonlinear optical phase conjugation, has made significant progress in the last fifteen years [1, 2]. Due to the unique characteristics of photorefractive nonlinearity, optical phase conjugation can be easily demonstrated with a low power cw laser (even in miliwatts) and a photorefractive crystal without using high power laser sources. Various novel photorefractive phase conjugators and applications of phase conjugation have been discovered and developed. One of the amazing discoveries of photorefractive phase conjugators is the mutually pumped phase conjugation (MPPC).

Mutually pumped phase conjugation is a nonlinear optical process in which two mutually incoherent laser beams interact each other in the nonlinear medium and generate two phase conjugate waves simultaneously. It differs from two individual self-pumped phase conjugation processes in the following way. In the mutually pumped phase conjugation process, the photons of one phase conjugate beam come from the other input (pump) beam. In the self-pumped phase conjugation, however, the photons of the phase conjugate beam come from its own input beam. Therefore, the reflectivity of a self-pumped phase conjugator is always less than one, but the reflectivity of a mutually pumped phase conjugator can be higher than 100%. In the mutually pumped phase conjugation process, one pump beam reads out the phase conjugate spatial wavefront of the other without significant spatial and temporal crosstalk. These unique characteristics led to many potential applications of photorefractive mutually pumped phase conjugators.

Mutually pumped phase conjugation in photorefractive media was first proposed theoretically by M. Cronin-Golomb et al. in 1983 [3]. It was first demonstrated experimentally by S. Weiss et al. in photorefractive $BaTiO_3$ crystals in 1987 [4]. Since then, a variety of different geometry have been discovered and invented in many kinds of photorefractive crystals, such as SBN, $KNbO_3$, GaAs, InP, etc. [5-7] In this paper, we review the basic configurations of mutually pumped phase conjugators and discuss the characteristics of each configurations. Then we review the theoretical development of mutually pumped phase conjugation, especially the multiple-grating effects in MPPC. We also present some applications of mutually pumped phase conjugators in optical communications, optical image processing and optical interconnections.

2. Photorefractive Mutually Pumped Phase Conjugators

In the late 70's, it was realized that photorefractive crystals are excellent materials not only for volume holographic storage but also for nonlinear optics. Due to the nonlocal effect of the photorefractive nonlinearity, strong beam coupling (energy transfer) can be easily observed among coherent beams from low power cw lasers. As a result, a variety of optical parametric oscillators based on the photorefractive nonlinearity were demonstrated. Self pumped phase conjugators and mutually pumped phase conjugators were discovered or invented [8-9]. So far more than 10 different geometry and configurations of mutually pumped phase conjugators have been demonstrated in a variety of photorefractive crystals.

As shown in Figure 1, commonly observed photorefractive mutually pumped phase conjugators may be divided into three classes, one-interactive region, two-interactive region, and external cavity-assisted. Double Phase Conjugate Mirror (DPCM) is an one-interactive region mutually pumped phase conjugator. It was the first MPPC demonstrated in photorefractive $BaTiO_3$ by S. Weiss et al. in the experiment [4]. In this geometry, two incident beams are directly overlapped in the photorefractive crystal in such a way that each incident beam is incident in the fanning region of the other. Usually two conical scattering rings are observed first crossing through both of the input beams. If the photorefractive coupling is strong enough or if the input beams carry the spatial modulation, the conical rings would gradually collapse into the phase conjugate beams. Double phase conjugate mirrors have high phase conjugate reflectivity, large input dynamic range. They can even generate two quasi-phase conjugate spatial wavefront when the incident beams have different wavelength. This is because the Bragg diffraction condition can be approximately fulfilled for two different wavelength in the one-interactive region of four wave mixing process. Since DPCM geometry requires low coupling threshold for the oscillation, it has been observed in many other photorefractive crystals, such as SBN, $KNbO_3$, BGO, GaAs, InP etc.. However, the phase conjugate fidelity of DPCM could be undermined by the slow decay of the conical scattering and the poor angular selectivity in the perpendicular direction to the incident plane. The second class of MPPC is the two-interactive region phase conjugators. Depending on the number of the total internal reflection (TIR) of the beam path, there are Bridge Conjugator (0 TIR) [10], Bird Wing Phase Conjugator (1 TIR) [11], Mutually Incoherent Beam Coupler (2 TIR) [12], and Frog-leg Phase Conjugator (3 TIR) [13]. The common characteristic of the two-interactive region mutually pumped phase conjugator is that the two input beams do not directly overlap each other, or it is the overlap (counter-propagating) of the fanning beams that interconnect the two regions of four wave mixing. The fanning is usually collapse into kinky beam filament that chooses its optimized direction for the phase conjugator. Therefore the beam path inside the crystal that connects the two incident beams may look quite complicated depending on the incident angles and the geometrical size of the crystal. Due to the presence of multiple interactive regions, the wavelength of the two input beams have to be nominally the same in order to satisfy the Bragg condition for all regions. It is also for this reason that the conical scattering rings do not appear in this class of conjugators so that they usually have good phase conjugate fidelities. The third class of MPPC includes the external cavity-assisted mutually pumped phase conjugators, as shown in Figure 1(f, g, h) [14-16]. Using additional mirror (or mirrors) that forms either open or closed cavity, mutually pumped phase conjugation process can be fully optimized and the competing self pumped phase conjugation process can be minimized.

3. Theoretical Development

The first theoretical model of mutually pumped phase conjugation process, proposed by M. Cronin-Golomb, was based on the self-oscillation theory of four wave mixing in one interactive region. It explained the coupling threshold and the finite dynamic range in the double phase conjugate mirror geometry. This theory was extended to two interactive regions to study the multiple-region mutually pumped phase conjugators [17]. Comparing with the one region MPPC, the theory showed that with the same input beam ratio, the oscillation threshold for the one-region MPPC is lower than the two-region's and the phase conjugate reflectivity of one-region configuration is also higher than the two-region's. In explaining the phase conjugate properties, P. Yeh proposed the hologram sharing theory and showed that only when the output beams are phase conjugate of their inputs, the shared hologram is enhanced without suffering mutual erasure [18]. Recently, A. Zozulya proposed the two dimensional theory of double phase conjugate mirror. The result from two dimensional theory indicates that the photorefractive double phase conjugate mirror behaves more like a convective stimulate Brillouin scattering amplifier than an oscillator [19]. Here we discuss the multiple grating competition effects in photorefractive mutually pumped phase conjugators. In the standard four wave mixing theory of MPPC, such as double phase conjugate mirror, only one transmission was considered. However, when the two input beams are partially coherent, for example in the applications of laser phase-locking, the generation of reflection grating and 2K reflection gratings is usually inevitable. It is important to consider the effect of reflection gratings in the process of mutually pumped phase conjugation.

Referring to Figure 2, we consider a photorefractive double phase conjugate mirror that has not only the transmission grating, but also reflection gratings. The grating wave vectors are expressed as

$$
\begin{aligned}
KT &= k1 - k2 = k3 - k4 \\
KR &= k1 - k3 = k2 - k4 \\
K2k &= k3 - k2 \\
K2k' &= k1 - k4
\end{aligned}
\tag{1}
$$

where $\mathbf{k}_1$, $\mathbf{k}_2$, $\mathbf{k}_3$, $\mathbf{k}_4$ are wave vectors of plane waves A_1, A_2, A_3, A_4, respectively. The coupled wave equations for the phase conjugate beams A_2 and A_4 are

$$
\begin{aligned}
\frac{\partial A_2}{\partial z} &= \frac{1}{2}\Gamma_T G_T A_1 - \frac{1}{2}\Gamma_R G_R A_4 - \frac{1}{2}\Gamma_{2k} G_{2k} A_3 \\
\frac{\partial A_4}{\partial z} &= -\frac{1}{2}\Gamma_T G_T A_3 - \frac{1}{2}\Gamma_R G_R A_2 - \frac{1}{2}\Gamma_{2k'} G_{2k'} A_1 \\
\tau_T \frac{\partial G_T}{\partial t} + G_T &= \frac{A_1 A_2 + A_3 A_4}{I_0} \\
\tau_R \frac{\partial G_R}{\partial t} + G_R &= \frac{A_1 A_3 + A_2 A_4}{I_0} \\
\tau_{2k} \frac{\partial G_{2k}}{\partial t} + G_{2k} &= \frac{A_3 A_2}{I_0} \\
\tau_{2k'} \frac{\partial G_{2k'}}{\partial t} + G_{2k'} &= \frac{A_1 A_4}{I_0}
\end{aligned}
\tag{2}
$$

where we assume all the coupling constants are real. Γ_T, Γ_R, Γ_{2k}, and $\Gamma_{2k'}$ are the intensity coupling constants for the transmission, reflection and 2K reflection gratings, respectively. τ_T, τ_R, τ_{2k}, and $\tau_{2k'}$ are the time constants for the respective gratings. I_0 is the total intensity. The boundary condition may be written as

$$
\begin{aligned}
A_4(z = L) &= 0 \\
A_2(z = 0) &= A_{20} H(t)
\end{aligned}
\tag{3}
$$

where H(t) is a unit-step function. Under the nondepleted pump approximation, we can assume that two pump beams A_1 and A_3 are constants and $I_0 = I_1 + I_3$. Taking the Laplacian transform with respect to time of the Eq. (2), the coupled wave equations in the Laplacian domain are

$$
\begin{aligned}
\frac{d\overline{A}_2}{dz} &= \frac{1}{2}\Gamma_T \frac{I_1\overline{A}_2 + A_1 A_3 \overline{A}_4}{(\tau_T s + 1) I_0} - \frac{1}{2}\Gamma_R \frac{A_1 A_3 \overline{A}_4}{(\tau_R s + 1) I_0} - \frac{1}{2}\Gamma_{2K} \frac{I_2 \overline{A}_2}{(\tau_{2K} s + 1) I_0} \\
\frac{d\overline{A}_4}{dz} &= -\frac{1}{2}\Gamma_T \frac{I_3\overline{A}_4 + A_3 A_1 \overline{A}_2}{(\tau_T s + 1) I_0} - \frac{1}{2}\Gamma_R \frac{A_3 A_1 \overline{A}_2}{(\tau_R s + 1) I_0} - \frac{1}{2}\Gamma_{2K'} \frac{I_1 \overline{A}_4}{(\tau_{2K'} s + 1) I_0}
\end{aligned}
\tag{4}
$$

This equation can be easily solved with the boundary condition (3). Although it is difficult to obtain the inverse Laplacian transform of the solutions for the above equation, the reflection grating effects on the photorefractive mutually pumped phase conjugators can still be analyzed by studying the poles of the solution. For example, when the coupling is reaching the threshold, the phase conjugate beams start to oscillate, i.e., the solution should have a pole that becomes real positive at the threshold conditions. Figure 3 shows the coupling threshold for the transmission gratings as a function of the coupling of the reflection gratings, assuming $\Gamma_{2k} = -\Gamma_{2k'}$, and the beam intensity ratio q=1. It is clear that the presence of reflection grating always raises the oscillation threshold of the mutually pumped phase conjugator. The effect of the 2K grating depends on the sign of the coupling constant. It can either raise or reduce the oscillation threshold. As shown in Figure 4, the dynamic range of the mutually pumped phase conjugator is also investigated in presence of the reflection gratings. Similar to the oscillation threshold, the formation of reflection grating always reduce the dynamic range. The effect of 2K grating on the dynamic range also depends on the sign of the 2K grating coupling constant. From these analysis, it is expected that the reflectivity will decrease as a result of reflection grating since the reflection grating always channels the energy away from the phase conjugate beams in the configuration of MPPC. Complicated dynamic effects may also arise because of the high nonlinearity.

Another type of multiple grating competition effect in mutually pumped phase conjugator is the build-up and the decay of the conical scattering [20]. This is the competition among the shared transmission gratings that are all Bragg phase matched. Assuming the conical scattering wave as a plane wave, Figure 5 shows the vector diagram of optical waves and the shared gratings. Two mutually incoherent pump beams A_2 and A_4 are incident on the photorefractive crystal that generates the phase conjugate beams A_1, A_3, and the conical scattering beams A_1' and A_3'. Three transmission gratings $\mathbf{K}_1$, $\mathbf{K}_2$, $\mathbf{K}_3$, are formed where

$$
\begin{aligned}
K_1 &= k_1 - k_4 = k_3 - k_2 \\
K_2 &= k_1' - k_4 = k_3' - k_2 \\
K_3 &= k_1 - k_1' = k_3 - k_3'
\end{aligned}
\qquad (5)
$$

We can see that the grating $\mathbf{K}_1$ is responsible for the generation of phase conjugate beams, $\mathbf{K}_2$ is responsible for the conical scattering, $\mathbf{K}_3$ describes the coupling between the phase conjugate beam and the conical scattering beam. Assuming Γ_1, Γ_2, and Γ_3 are the coupling constants for the respective gratings, we can solve the steady state and the dynamic coupled wave equations with the presence of all three transmission gratings. The results of the steady state solution show that when $\Gamma_1>\Gamma_2>\Gamma_{th}$, and $\Gamma_3 \sim 0$, the phase conjugate wins the competition and the conical scattering beams decay to zero. When $\Gamma_1=\Gamma_2>\Gamma_{th}$, and $\Gamma_3 >0$, the output is still phase conjugate. However, when $\Gamma_2>\Gamma_1>\Gamma_{th}$, and $\Gamma_3 \sim 0$, the conical scattering beams will win the competition and the conical ring will collapse to direction where the gain is the highest, thus lead to the noncoplanar oscillation. If Γ_3 is not negligible in this case, the solution will depend on how strong the mutual coupling Γ_3 is. Figure 5 shows the typical transient process of a photorefractive mutually phase conjugator where $\Gamma_1>\Gamma_2>\Gamma_{th}$, and $\Gamma_3 \sim 0$. Clearly, two distinct time evolution phases exist in the transient process. The first phase involves the build-up of phase conjugate and the conical scattering beams as a result of coherent amplification of noise from noise scattering through shared gratings. The second phase shows strong competition among generated beams which leads to steady growth of the phase conjugate and the slow decay of the conical scattering.

4. Applications of Photorefractive MPPC

Since the discovery of photorefractive mutually pumped phase conjugators, various applications of MPPC have been proposed and demonstrated. Due to the unique characteristics of photorefractive MPPC such as low power, mutually incoherent laser sources, high phase conjugate reflectivity and good phase conjugate fidelity, it attracted a lot of attention of the researchers in different areas of optics. We present a brief review of some of the basic applications suggested in recent years.

One of the applications is in the field of optical image processing. Many image processing tasks, for example image thresholding [21], edge enhancement [21], image addition and subtraction [22], novelty filtering etc., can be performed in a simple configuration as shown in Figure 7. Since the photorefractive mutually phase conjugator oscillates only when the intensity ratio of the two input beams is within certain value or finite dynamic range, it is possible to adjust the beam ratio such that only the high spatial frequency components of the input image I_1, oscillate. The output at S_1 will be an edge-enhanced image of I_1. The principle of image subtraction and novelty filtering is based on the Stokes' principle and the time reversal property of optical phase conjugation. The image in one arm I_1 (amplitude modulation), is subtracted from the image in the other arm I_2 at the output port S_3 because two wavefronts are 180° out of phase. At screen S_4, the output will be addition of the two images. If these two images are replaced by a spatial light modulator(SLM) operated at the phase only mode, at the steady state the phase change at SLM induced by the input scene will be corrected by optical phase conjugation. Therefore the output at S_3 should be a null if the intensity ratio of two arms are properly adjusted. However, when the object in the input scene is moving suddenly, the novel scene will appear at the port S_3 because the phase disturbance cannot be canceled due to the finite response time of the MPPC.

One of the unique characteristics of the photorefractive mutually pumped phase conjugator is that phase conjugate waves can be generated using two independent laser sources. This is especially important in the applications of long distance optical communications since mutual coherence are hard to achieve between remote stations [23-25]. Using MPPC, a two-way communication link may be established between two independently modulated lasers and there is no temporal crosstalk between the two communication channels, as illustrated in Figure 8. The temporal modulation can be amplitude, frequency or phase modulations. As long as the modulation is faster than the response time of the MPPC, no crosstalk occurs between the two channels. Figure 9 shows the signals simultaneously received at two detectors, D_1 and D_2. Interestingly, when the two lasers are amplitude-modulated at the same frequency but out of phase, i.e., two laser beams are never present in the crystal at the same time, phase conjugate beams can still be generated, as shown in Figure 10 [25]. This is due to the fact that photorefractive material has finite response times of writing and erasure. As long as the modulation rate is faster than the photorefractive response time, the crystal will always respond to the integral as if they were continuous. Since the spatial wavefront aberration is corrected through phase conjugation, the received signal can be readily combined with the local oscillator to achieve homodyne or heterodyne detection with a wide field of view [24, 25].

Photorefractive mutually pumped phase conjugator can also be applied in optical reconfigurable interconnection [26, 27], laser-phase locking [28], laser beam clean-up [29], optical neural networks [30] et al..

5. Conclusions

In this paper, we have reviewed the basic properties, theories and applications of photorefractive mutually pumped phase conjugation. It is shown that this extraordinary nonlinear optical process not only has its fundamental theoretical significance but also its great potential in various applications. The main obstacle that limits the practical application of MPPC devices is the slow response time of the photorefractive material. As the photorefractive material performance improves or novel nonlinear material develops, mutually pumped phase conjugation will play an important role in optical communication and optical computing.

6. Acknowledgment

Author would like to acknowledge the encouragement and support from Drs. Pochi Yeh, John Caulfield, Graeme Duthie, Don Gregory and Joseph Shamir in various period of the research on this subject. Author thanks Dr. Claire Gu for many helpful discussions.

7. References

[1] J. Feinberg, D. Heiman, A. R. Tanguay, Jr., R. W. Hellwarth, "Photorefractive effects and light induced charge migration in barium titanate", J. Appl. Phys. 51, pp1297-1305, 1980.

[2] M. Cronin-Golomb, B. Fischer, J. O. White, and A. Yariv, "Theory and applications of four-wave mixing in photorefractive media," IEEE J. Quantum. Electron., QE-20, pp12-30, 1984.

[3] M. Cronin-Golomb, "Large nonlinearities in four-wave mixing in photorefractive crystals in passive optical phase conjugation", Ph.D dissertation, California Institute of Technology, 1983.

[4] S. Weiss, S. Sternklar, and B. Fischer, "Double phase conjugate mirror: analysis, demonstration and applications", Opt. Lett. 12, pp114-116, 1987.

[5] E. J. Sharp, W. W. Clark III, M. J. Miller, G. L. Wood, B. Monson, G. L. Salamo, and R. R. Neurogaonkar, "Double phase conjugation in tungstern bronze crystals," Appl. Opt., 29, pp743-749, 1990.

[6] P. L. Chua, D. T. H. Liu, and L. J. Cheng, "Self-pumped and double phase conjugation in GaAs with applied DC electric field," Appl. Phys. Lett., 57, pp858-860, 1990.

[7] V. Vieux, P. Gravey, N. Wolffer, and G. Plcoli, "Double phase conjugated mirror and double color pumped oscillator in photorefractive InP-Fe," Appl. Phys. Lett., 58, pp2880-2882, 1991.

[8] J. O. White, M. Cronin-Golomb, B. Fischer, and A. Yariv, "Coherent oscillation by self-induced gratings in the photorefractive $BaTiO_3$", Appl. Phys. Lett.40, pp450-452, 1982.

[9] J. Feinberg, "Self-pumped, continuous-wave phase conjugator using internal reflection," Opt. Lett., 7, pp486-488, 1982.

[10] D. Wang, Z. Zhang, Y. Zhu, S. Zhang, and P. Ye, "Observations on the coupling channel of two mutually incoherent beams without internal reflection in $BaTiO_3$," Opt. Comm., 73, pp495-500, 1989.

[11] M. D. Ewbank, "Mechanism for photorefractive phase conjugation using incoherent beams," Opt. Lett., 13, pp47-49, 1988.

[12] A. M. Smout and R. W. Eason, "Analysis of mutually incoherent beam coupling in $BaTiO_3$," Opt. Lett., 12, pp498-500, 1988.

[13] M. D. Ewbank, R. A. Vazquez, R. R. Neurgaonkar, J. Feinberg, "Mutually pumped phase conjugation in photorefractive strontium barium niobate: theory and experiment," J. Opt. Soc. Am. B, 7, pp2306-2316, 1990.

[14] P. Yeh, T. Y. Chang, and M. D. Ewbank, "Model for mutually pumped phase conjugation," J. Opt. Soc. Am. B, 5, pp1743-1749, 1988.

[15] Q. C. He, and J. G. Duthie, "Semilinear double phase conjugate mirror," Opt. Comm., 75, pp311-314, 1990.

[16] A. A. Zozulya, and A. V. Mamaev, "Phase conjugation of mutually incoherent light beams in a photorefractive medium," Sov. Phys. JETP, 70, pp56-62, 1990.

[17] Q. C. He, "Theory of photorefractive phase conjugators with mutually incoherent beams," IEEE J. Quantum Electron. 24, pp2507-2513, 1988.

[18] P. Yeh, "Coupled mode theory of hologram sharing model of MPPC's," Appl. Opt., 28, pp1961-1964, 1989.

[19] A. A. Zozulya, "Double phase conjugate mirror is not an oscillator," Opt. Lett., 16, pp545-547, 1991.

[20] Q. B. He, P. Yeh, C. Gu, and R. R Neurgaonkar, "Multigrating competition effects in photorefractive mutually pumped phase conjugation," J. Opt. Soc. Am. B, 9, pp114-120, 1992.

[21] S. Sternklar, S. Weiss, and B. Fischer, "Optical information processing with the double phase conjugate mirror," Opt. Eng., 26, pp423-427, 1987.

[22] Y. Tomita, R. Yahalom, and A. Yariv, "Real-time image subtraction with the use of wave polarization and phase conjugation," Appl. Phys. Lett., 52, pp425-427, 1988.

[23] Q. C. He, J. G. Duthie, J. Shamir, and H. J. Caulfield, "Optical parallel communication with photorefractive materials," in Technical Digest, Topical meeting on photorefractive materials, effects, and devices, 1987.

[24] Q. C. He, J. Shamir, and J. G. Duthie, "Optical communication via phase conjugation through distortion media: experiment," Appl. Opt., 28, pp306-310, 1989.

[25] Q. C. He, "Mutually pumped phase conjugation in photorefractive materials," Ph.D dissertation, University of Alabama in Huntsville, 1990.

[26] H. J. Caulfield, J. Shamir, and Q. C. He, "Flexible two-way optical interconnects in layered computers," Appl. Opt., 26, pp2291-2292, 1987.

[27] S. Weiss, M. Segev, S. Sternklar, and B. Fischer, "Photorefractive dynamic optical interconnects," Appl. Opt., 27, pp3422-3428, 1988.

[28] S. Sternklar, S. Weiss, M. Segev, and B. Fischer, "Beam coupling and locking of lasers using photorefractive four-wave mixing," Opt. Lett., 11, pp528-530, 1986.

[29] T. Y. Chang, "Spatial-mode cleanup of a pulsed laser beam through mutually pumped phase conjugation with a cw reference," Opt, Lett. 15, pp1342-1344, 1990.

[30] G. Dunning, Y. Owechko, and B. H. Soffer, "Hybrid optoelectric neural networks using a mutually pumped phase conjugate mirror," Opt. Lett., 16, pp928-930, 1991.

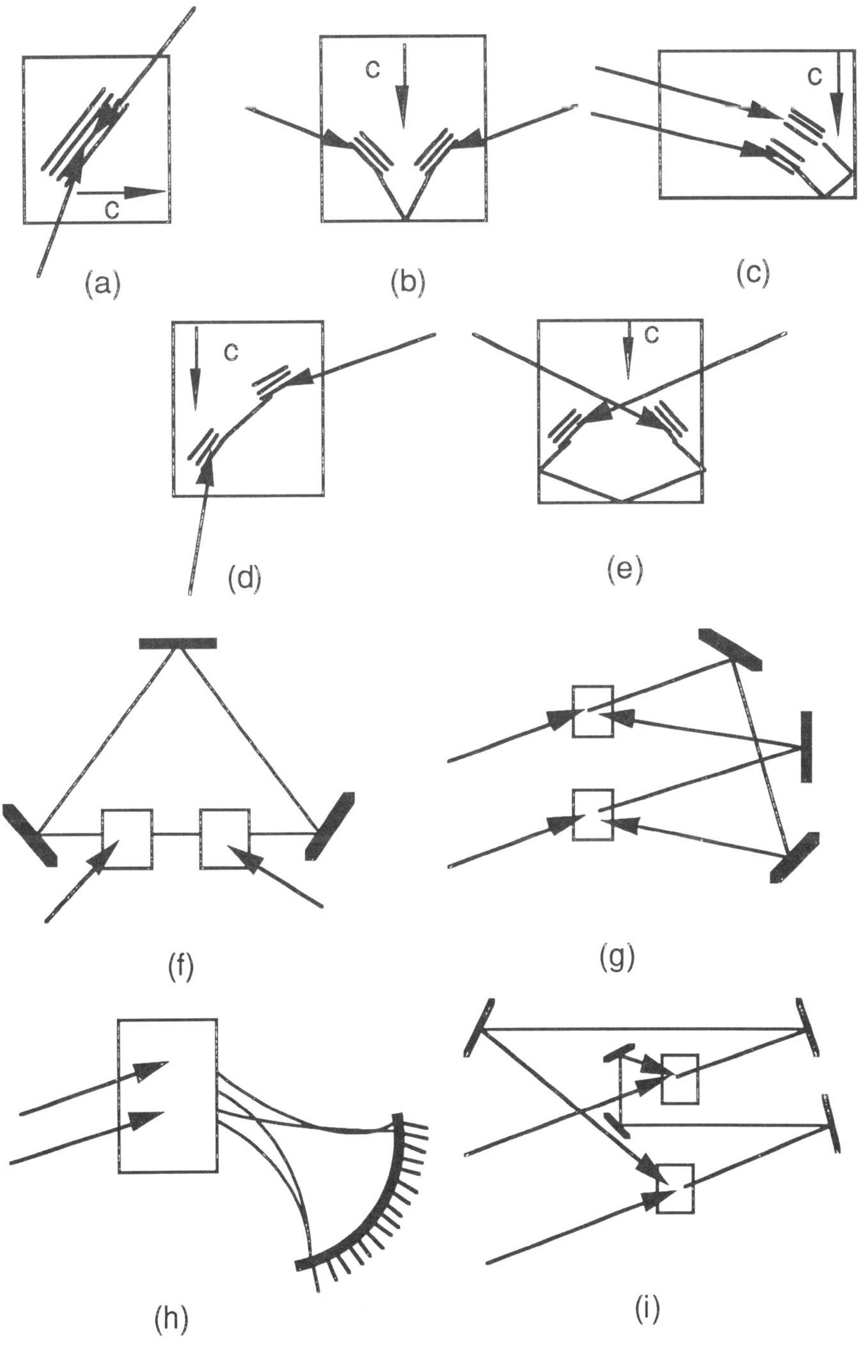

Figure 1 Configurations of mutually pumped phase conjugators. (a), Double Phase Conjugate Mirror [4]. (b), Bird-Wing Phase Conjugator [11]. (c), Mutually Incoherent Beam Coupler [12]. (d), Bridge Phase Conjugator [10]. (e), Frog-Leg Phase Conjugator [13]. (e-i), External cavity-assisted mutually pumped phase conjugators [14-16]

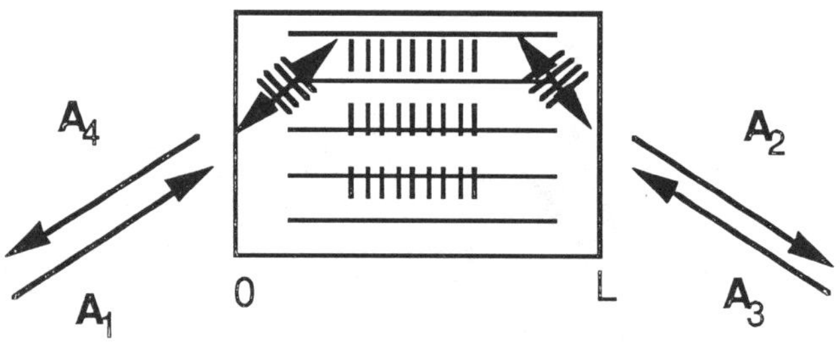

Figure 2 Reflection gratings in photorefractive MPPC.

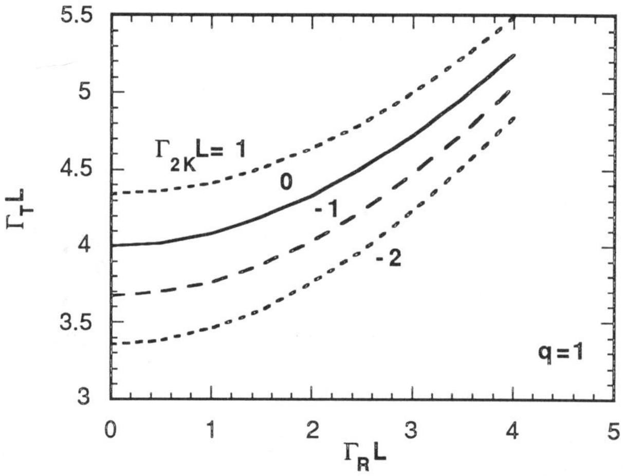

Figure 3 Coupling threshold of the MPPC as functions of reflection grating coupling constants.

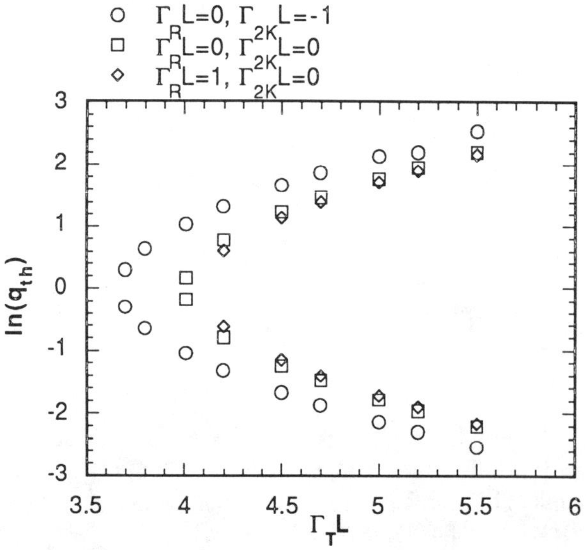

Figure 4 Dynamic range of the MPPC as functions of coupling constants.

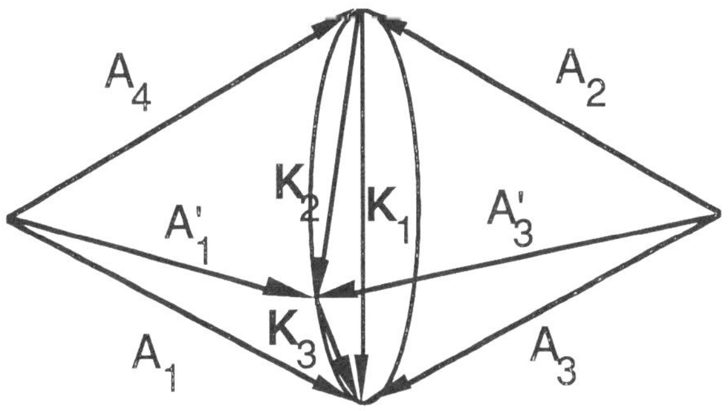

Figure 5 Vector diagram of the optical waves and the shared gratings.

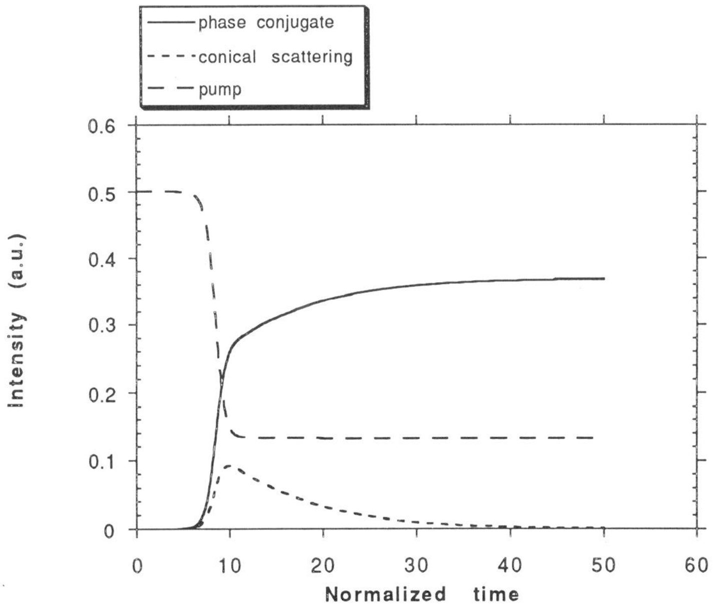

Figure 6 Typical transient process of photorefractive MPPC

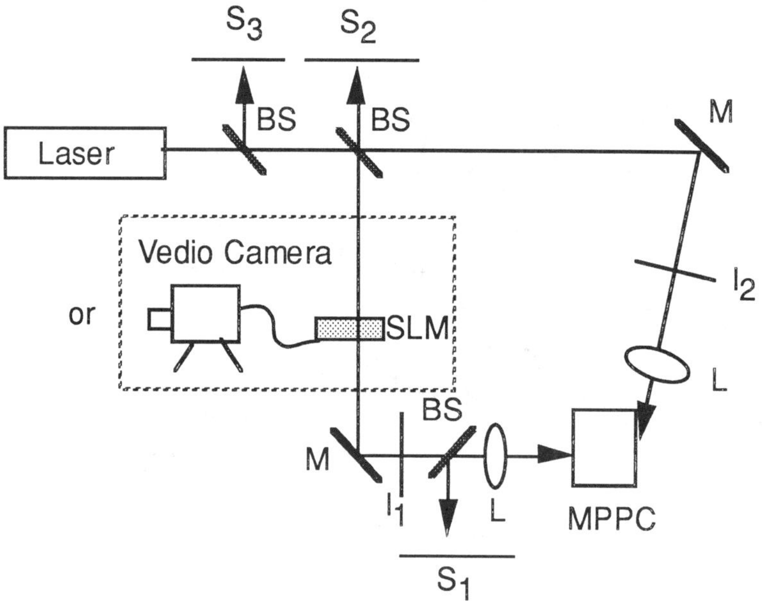

Figure 7 Optical image processing using photorefractive MPPCs.

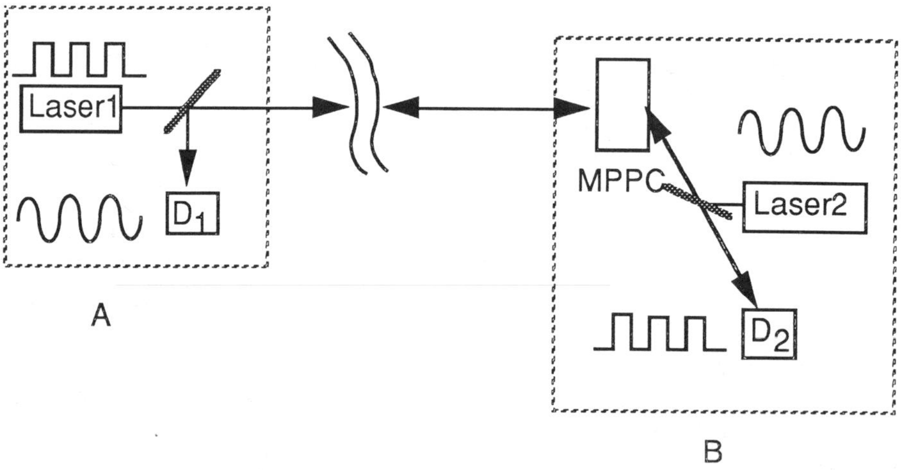

Figure 8 Two-way optical communication using photorefractive MPPCs.

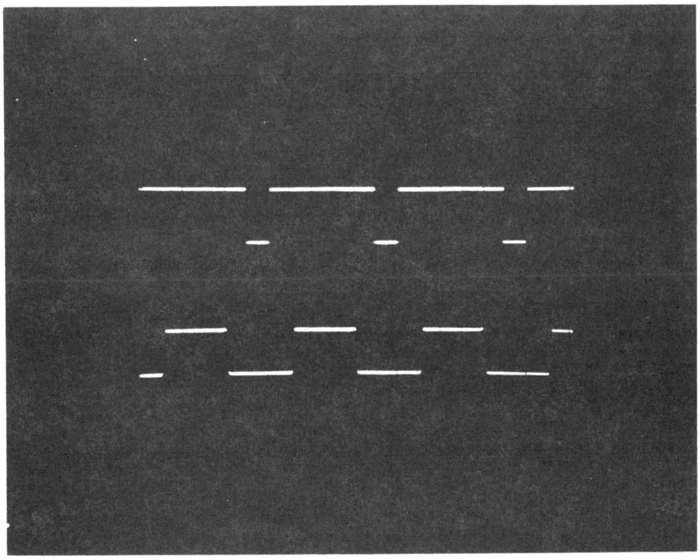

Figure 9 Two-way simultaneous transmission. The upper trace is the signal received at B transmitted from A, while lower trace is the signal received at A transmitted from B.

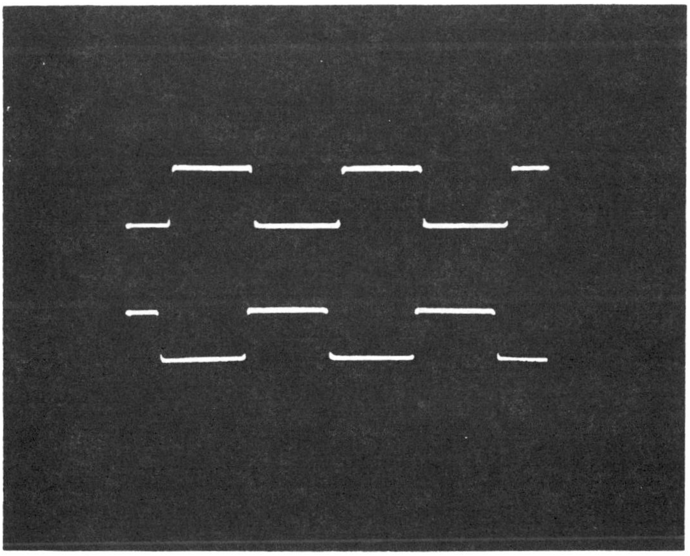

Figure 10 Two-way simultaneous transmission with the same modulation frequency but opposite phase.

Photorefractive Materials II

Photorefractive tungsten bronze materials for data storage applications

R.R. Neurgaonkar, W.K. Cory and J.R. Oliver
Rockwell International Science Center
Thousand Oaks, CA 91360

and

Demetri Psaltis
California Institute of Technology
Pasadena, CA 91125

ABSTRACT

The photorefractive and electrical properties of ferroelectric tungsten bronze (T.B.) materials are reviewed with respect to optical data storage applications. Our work indicates that both the photorefractive properties (speed and coupling) and dark conductivity are sensitive to the type of T.B. host, dopants, and structural defects. The filled-lattice T.B. crystals exhibit faster response and lower dark conductivities, which are needed for data storage applications, as compared to unfilled T.B. crystals. Concepts for 3–D holographic memory applications using these materials are discussed.

INTRODUCTION

The strong photorefractive effects that have been achieved in organic and inorganic optical materials are being explored for applications in optical data storage, optical computing and pattern recognition. Among these materials, tungsten bronze ferroelectrics have demonstrated exceptional coupling coefficients and speed of response, and various groups have also achieved holographic fixation[1–5] under the influence of external fields.

Figure 1 lists a number of parameters that depend strongly on host/dopant interaction and are important for these applications. The tungsten bronze materials grown at Rockwell have been used for various applications including sensor protection, phase conjugation, and image processing.[6–12] In the course of this study, we have systematically established the relationship between the host crystals and dopants to optimize the speed of response and diffraction efficiency. However, we need to have a better control over the dark conductivity and the conditions for holographic fixation which determine the retention of holograms as well as their nondestructive readout. At present, various groups are investigating these problems in the tungsten bronzes as well as other photorefractive materials. Our progress in this area is reviewed in the present paper with respect to:

- High optical figure of merit (n^3r_{ij}/ε) T.B. hosts
- Dark conductivity: dark retention time
- Role of dopants: speed of response, coupling and diffraction efficiency
- Effects of external electric field: Erase/write time ratio, and fixation
- 3–D holographic disks

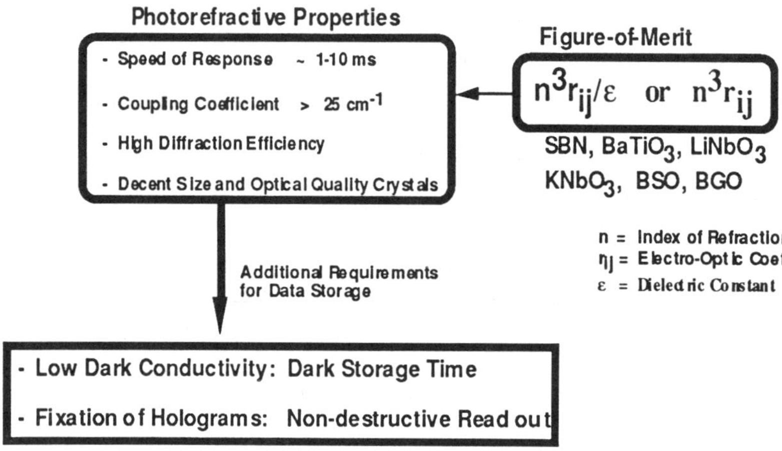

Fig. 1. Important material properties for data storage applications.

EXPERIMENTAL RESULTS AND DISCUSSION

Tungsten Bronze Materials

The T.B. structure is typified by oxygen octahedra linked at the corners in a complex way to yield three types of openings, two of which normally contain the A_1 and A_2 cations.

The B cations, typically niobium (Nb^{5+}) or tantalum (Ta^{5+}), are inside the oxygen octahedra. The tungsten bronze compositions are characterized by the chemical formulae $(A_1)_4(A_2)_2C_4B_{10}O_{30}$ or $(A_1)_4(A_2)_2B_{10}O_{30}$, in which the A_1 cations are in the 15–fold coordinated site, the A_2 cations are in the 12–fold coordinated site, the C cations are in the 9–fold coordinated site and the B cations are in two different 6–fold coordinated sites.[13–15]

At Rockwell, we have grown and investigated the ferroelectric and optical properties of both tetragonal and orthorhombic tungsten bronze crystals[16-21] and have proposed the following classification for these materials[22]:

1. Crystals exhibiting strong transverse effects with 4 mm symmetry (e.g., SBN $Sr_2KNb_5O_{15}$, $Ba_6Ti_2Nb_8O_{30}$) : <001> polar axis.

2. Crystals exhibiting strong longitudinal effects with 4 mm symmetry (e.g., BSKNN, $K_3Li_2Nb_5O_{15}$): <001> polar axis.

3. Crystals exhibiting strong transverse and longitudinal effects with mm2 symmetry (e.g., $Sr_{2-x}Ca_xNaNb_5O_{15}$): <001> polar axis.

4. Crystals exhibiting strong longitudinal effects with mm2 symmetry (e.g., $Pb_2KNb_5O_{15}$, $Pb_{1-x}Ba_xNb_2O_{15}$): <100> or <010> polar axis.

Although a large number of T.B. crystals are now available for data storage applications, we are concentrating on a few bronze compositions that are easy to grow and that exhibit high optical figures of merit (n^3r_{ij}/ε), e.g., $Sr_{1-x}Ba_xNb_5O_{15}$ (SBN), $Ba_{2-x}Sr_xK_{1-y}Na_yNb_5O_{15}$ (BSKNN) and $(Ba,Sr)_6Ti_2Nb_8O_{30}$ (BSTN) solid solution crystals as summarized in Fig. 2.

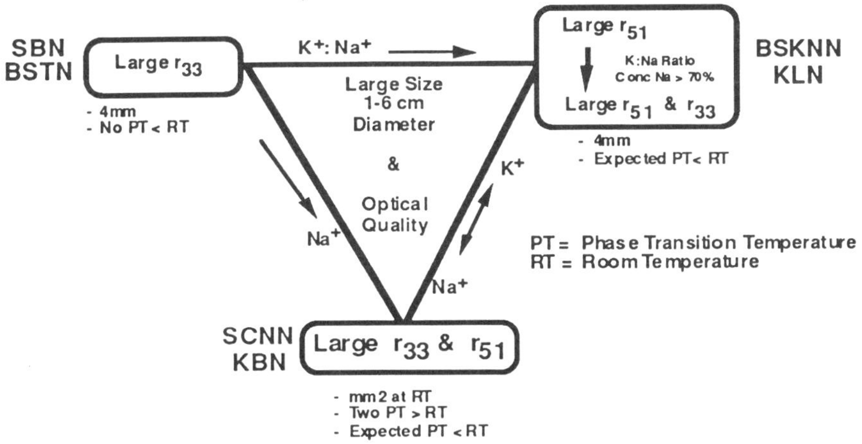

Fig. 2. High figure-of-merit T.B. Crystals of Interest.

The optical figures of merit (FOM) for BSTN and BSKNN are significantly higher than for SBN, since the former have higher spontaneous polarization and lower dielectric constants (Fig. 3). This behavior is usually observed in the compositions that exhibit a “filled” bronze structure.

Photorefractive measurements indicate that both BSTN and BSKNN have a faster response with two-beam coupling compared to SBN:60. For these crystals, we also expect much lower dark conductivities ($< 10^{-16}\ \Omega^{-1}\text{-cm}^{-1}$) and thus longer dark storage times.

The Czochralski pulling technique is well established for the growth of these tungsten bronze crystals,[16–21] and we have succeeded in growing optical quality SBN:60 crystals up to 6 cm diameter (Fig. 4). We do not anticipate any major problems in scaling up the crystal diameter in the range of 9–10 cm. Table 1 summarizes the crystal size and the key properties of selected bronze crystals.

<u>Dark Conductivity in T.B. Crystals</u>

Besides the photorefractive properties, the dark conductivity in these crystals must be controlled since this dictates the dark storage time. Figure 5 schematically illustrates the dark conductivity for tungsten bronze crystals compared to $LiNbO_3$. The dark retention time for $LiNbO_3$ is known to be significantly higher (> 1 year) than other photorefractive materials because of the low dark conductivity, which is due almost entirely to ionic motion.

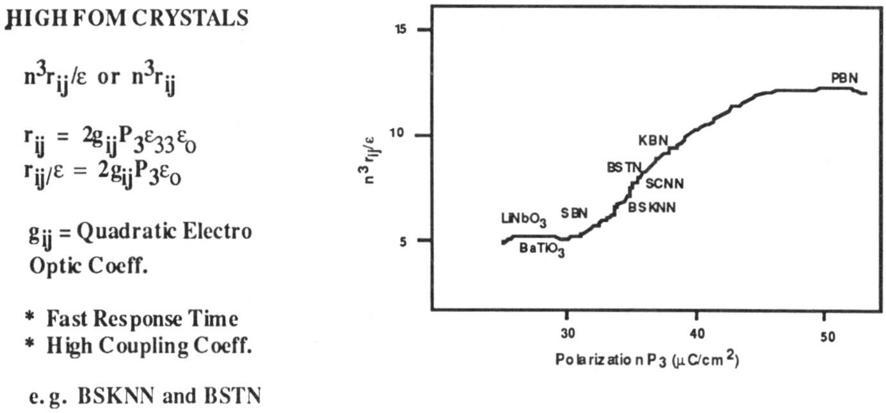

Fig. 3. Polarization P_3 vs $n^3 r_{ij}/\varepsilon$ for key photorefractive materials.

Table 1
Electro-Optic and Figure of Merit for New Tungsten Bronzes

Crystal	Electro-Optic Coeff. (r_{ij} x 10^{-12} m/V)	ε_{ij}	$n^3 r_{ij}/\varepsilon$	Crystal Size (cm.)
SBN:60	r_{33} = 240	ε_{33} = 1000	3.11	~ 6.0
SBN:40	r_{33} = 150	ε_{33} = 320	5.05	> 4.0
BSTN	r_{33} = 200	ε_{33} = 345	8.05	> 1.5
BSKNN	r_{51} = 400	ε_{11} =900	7.2	> 1.8
SCNN	r_{33} = 1200 r_{51} = 1150	ε_{33} = 1740 ε_{11} = 1700	8.63 8.32	~ 1.0

Depending on the processing conditions, SBN solid solution crystals have n-type dark conductivities in the 10^{-15} to 10^{-16} Ω^{-1}-cm^{-1} range, with highly reduced SBN crystals showing higher dark conductivities and fully oxidized crystals showing lower conductivities. Besides the control of valence states in this structure, the conductivity also depends on structural defects. SBN solid solution crystals have an unfilled bronze structure (15– and 12–fold sites are partially empty) while the other bronzes such as BSTN and BSKNN have a filled bronze structure. For this reason, both BSTN and BSKNN look promising in terms of long storage times.[22]

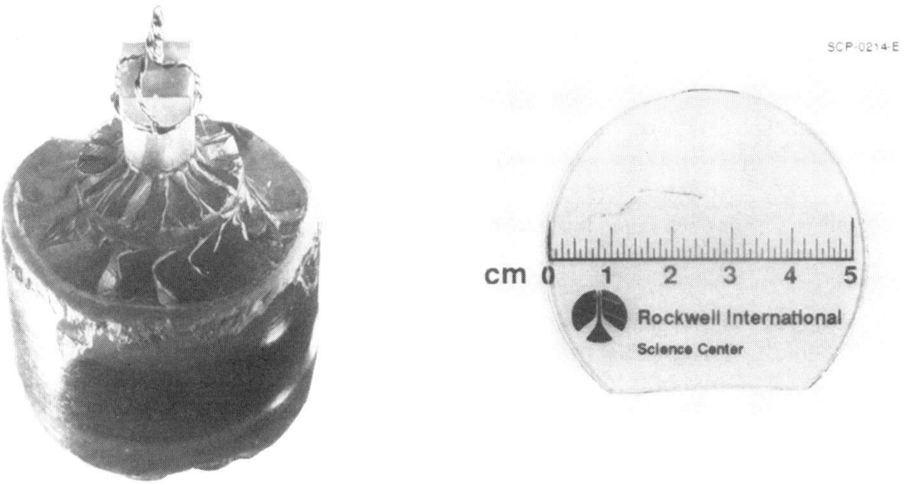

Fig. 4. Large-size T.B. SBN:60 crystal boule grown along [001] direction.

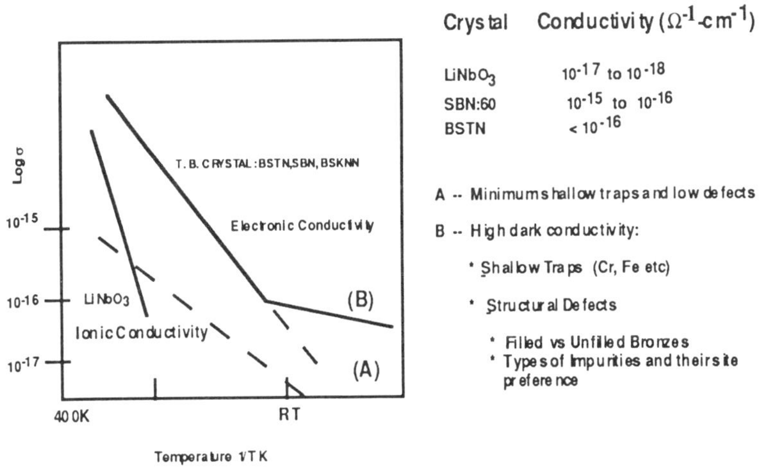

Fig. 5. Dark Conductivity vs reciprocal absolute temperature for photorefractive crystals.

At temperatures above 300°C, the dark conductivity of tungsten bronze crystals is determined by compensated deep donor levels in the band gap. In undoped crystals, the Fermi level is pinned near mid-gap (1.55–1.60 eV), whereas for doped crystals the Fermi level is typically shallower (e.g., 1.3 eV below the conduction band for Ce-doped crystals). Extrapolations of these high temperature conductivities (Fig. 5) down to room temperature on an Arrhenius plot generally predict far lower conductivities than are actually observed due to the presence of shallower defect and impurity levels. The elimination and/or compensation of these shallow levels can lead to significant reductions of the room temperature conductivities of the bronzes and thus a substantial improvement in image retention times. The filled lattice structure bronze crystals, such as BSKNN and BSTN, should be particularly effective in this regard because of their inherently low number of vacant A sites.[22]

Role of Dopants in T.B. Crystals

A proper balance of donor and acceptor levels is necessary to maximize the space charge densities $[N_{eff} = N_A (N_D - N_A)/N_D]$ and dark conductivities. Various dopants such as Ce^{3+}/Ce^{4+}, Fe^{2+}/Fe^{3+}, Cr^{3+}/Cr^{4+} and Rh^{3+}/Rh^{4+} have been studied in SBN and BSKNN,[19–21] and we found that these dopants have the desired valence states to achieve the appropriate photorefractive properties. However, using a single dopant system (e. g., Ce^{3+}/Ce^{4+} or Fe^{2+}/Fe^{3+}) complicates controling N_A and N_D. For instance, at growth temperature in an oxygen atmosphere, the presence of both Ce^{3+} and Ce^{4+} is difficult to control unless oxygen partial pressure is maintained precisely. To eliminate this problem, we are in the process of incorporating two independent species, one a donor and the other an acceptor, which have stable valence states under typical processing conditions.

Dopant combinations explored in our study have been as follows:

Donor	Acceptor	Site Preference
Ce^{3+}	Fe^{3+}	Ce^{3+} in 12- and Fe^{3+}- in 6 fold coordinated sites
Ce^{3+}	Mn^{3+}	Ce^{3+} in 12- and Mn^{3+} -in 6 fold coordinated sites
Cr^{3+}	Mo^{4+}	Cr^{3+} and Mo^{4+}-in 6 fold coordinated sites
Cr^{3+}	Rh^{3+}	Cr^{3+} and Rh^{3+} - in 6 fold coordinated sites

Cr^{3+} + Mo^{4+} – doped SBN:60 crystals exhibit fast response times and relatively high coupling coefficients (> 15 cm^{-1}). Singly Cr^{3+}–doped SBN:60 with similar doping levels exhibits similar response times but lower coupling (< 6 cm^{-1}). These properties are particularly suitable for sensor protection applications where fast response is mandatory. We expect that the speed of response can be improved further while maintaining high coupling by controlling the Cr^{3+}:Mo^{4+} ratio in SBN:60. Other dopant combinations listed above exhibit reasonable response times but much higher coupling coefficients (> 25 cm^{-1}) and dark conductivities <10^{-15} Ω^{-1} -cm^{-1}. These characteristics are promising for optical data storage applications.

As discussed earlier, both BSTN and BSKNN crystals have a filled bronze structure, and they have optical figures of merit superior to SBN, $LiNbO_3$ and $BaTiO_3$. Our recent work on these crystals shows that photorefractive response is much faster (> 10 times) than SBN:60 with comparable coupling (~ 25 cm^{-1}). At present, we are optimizing different dopant combinations to achieve suitable photorefractive properties as well as low dark conductivities (< 10^{-16} Ω^{-1} cm^{-1}) in these crystals. Figure 6 summarizes the roles of different dopants in tungsten bronze crystals.

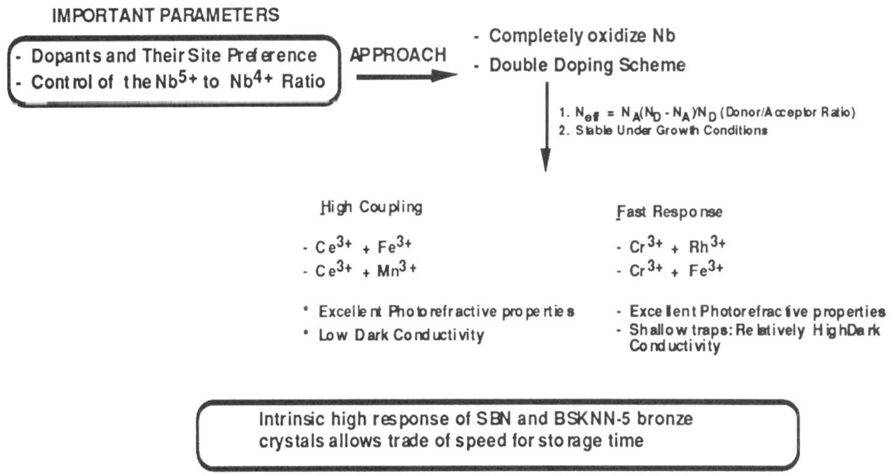

Fig. 6. The role of various dopants in T.B. crystals.

Effects of External Electric Field

The application of an external electric field along the the polar c-axis of tungsten bronze ferroelectric crystals substantially improves the photorefractive response, coupling coefficient, and diffraction efficiency.[10] Typical fields used are in the range of 3-10 kV/cm. The speed of response and coupling increase significantly under the external field, as shown below.[23-27]

Dopant	Speed of Response at 1 w/cm^2 Without Field	With Field
Ce^{3+}/Ce^{4+}	60–80 ms	10–20 ms
Rh^{3+}/Rh^{4+}	50–100 ms	10–25 ms
Cr^{3+}/Cr^{4+}	10–20 ms	1–5 ms

The two-beam coupling coefficients for Ce– and Cr–doped SBN:60 have been studied under an external field with significant improvements in coupling and diffraction efficiency. Figure 7 shows the two-wave coupling coefficient for 0.015 wt.% Ce-doped SBN:60 with and without an applied field.[10,23,24] The coupling is almost five times larger with an applied field, reaching 80 cm^{-1}. This result is in close agreement with theoretical predictions.[10] Recently we achieved high coupling (> 65 cm^{-1}) in SBN:60 by using higher Rh^{3+}/Rh^{4+} concentrations[24]; and here also we expect that with proper control of N_A and N_D, coupling can be raised above 100 cm^{-1} in SBN and other tungsten bronze crystals.

The work by Ford, et al.,[10] has demonstrated that under an external field, the erasing/writing time ratio improved by 25 times, which indicates that when multiple holograms are recorded, the diffraction efficiency of each hologram increases by 25^2.

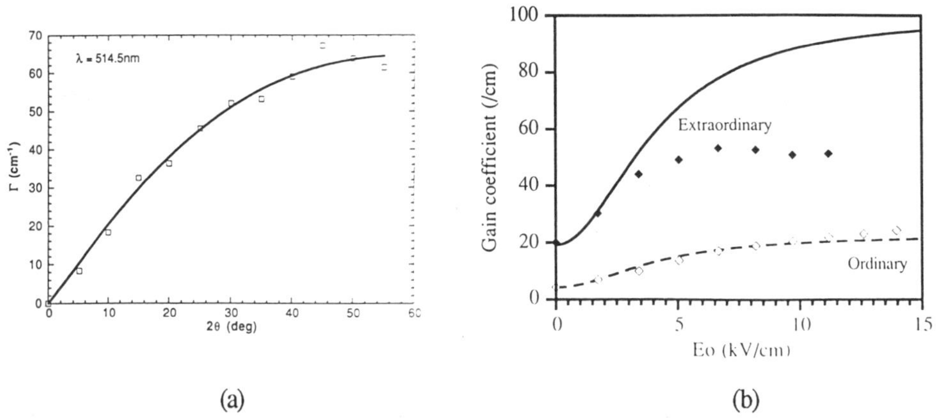

Fig. 7. Two beam coupling coefficients for T.B. SBN crystal: a) Gain coeff. as a function of 2-wave mixing angle 2θ. b) Gain coeff. as a function of external electric field for extraordinary and ordinary polarizations.

For photorefractive materials, the optical figure of merit is n^3r_{ij}/ε, where r_{ij} is proportional to the spontaneous polarization (P_3) and the static dielectric constant (ε_{33}). In a poled single domain crystal, P_3 and ε_{33} are not particularly sensitive to fields under 10 kV/cm. Consequently, we consider the improvement in speed of response could be due to enhanced charge transport with an applied dc field (5-10 kV/cm). Another important factor could be the prevention of microdomain reversal, since in the T.B. materials the coercive field is low, typically on the order of 2–5 kV/cm. Further investigations are under way to establish the mechanisms responsible for these improvements.

Holographic Fixation in T.B. Crystals

The low coercive field in tungsten bronze crystals can be used to advantage for electrical fixing of recorded holograms. The space charge field that is optically generated when the hologram is formed can reverse the direction of the microdomains in SBN:75, which has a low phase transition temperature (~ 50–54°C). This domain reversal is facilitated by the application of a negative external field. In the regions where the holographic space charge field is negative, the external applied field assists the space charge field in partially depoling the crystal partially. In the positive regions of the space charge field this depoling does not occur, and therefore the polarization grating that is formed is a given replica of the original space charge field. The polarization grating is not sensitive to light, and fixed holograms have been read out for several hours without any noticeable degradation. The fixed grating can be erased by applying a positive voltage pulse to the crystal, which then repoles it. The repoling eliminates the polarization grating, but it also reveals a strong electronic grating that had partially compensated the polarization grating.

This electronic grating can be erased optically. An example of a fixing experiment in SBN:75 that we performed[1] at Caltech using a crystal grown at Rockwell is shown in Figure 8, where the diffraction efficiency of the hologram is plotted as a function of time during the recording phase (A), negative pulse applied (B), end of negative pulse (C), optical erase begins (D), erasing beam blocked, and positive voltage pulse is applied (E), end of positive pulse (F), optical erasure begins (G).

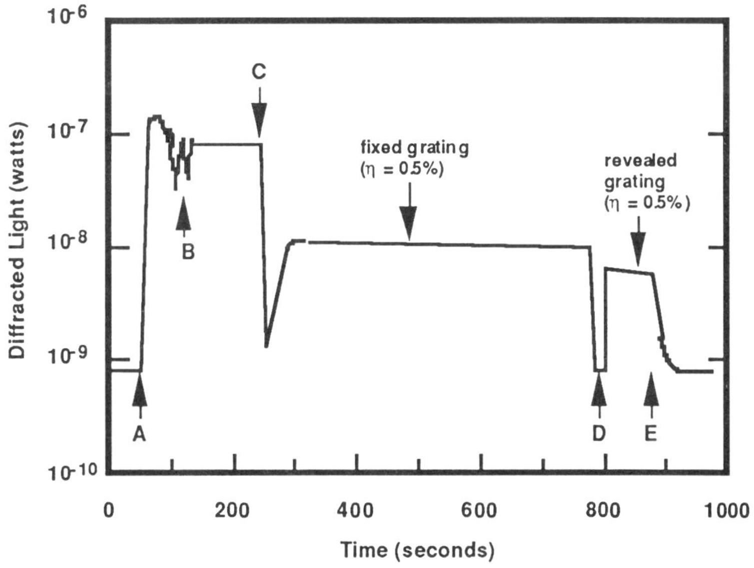

Fig. 8. Experimental demonstration of electrical fixing in SBN:75

3–D Holographic Disks

One of the more promising applications of tungsten bronze crystals is holographic storage. Of particular interest is the construction of 3–D disks[1] shown in Fig. 9. A 3-D disk is constructed from a holographic material shaped as a disk that has a thickness typically of several millimeters. Holograms are stored using angular or wavelength multiplexing in multiple locations on the disk. Each location typically has an area of several millimeters squared. A particular hologram stored on the disk is retrieved by mechanically rotating the disk and moving the head in the radial direction (as in conventional disks) and then selecting the angle or the wavelength of the readout beam. In excess of 100 bits per square micron can be stored on a 3–D disk of this type. SBN is a particularly interesting material for constructing 3–D disks because its crystal symmetry is such than an axis of rotation can be found which the electro-optic coefficients do not change as the disk rotates. This property allows the disks to be constructed with uniform efficiency over the entire disk surface, a property that is not achievable with $LiNbO_3$ crystals, for instance. Other advantages of tungsten bronze SBN crystals are: 1) they can be grown in large sizes in excess of 6 cm diameter, and 2) the speed of response (for recording) is much faster than $LiNbO_3$.

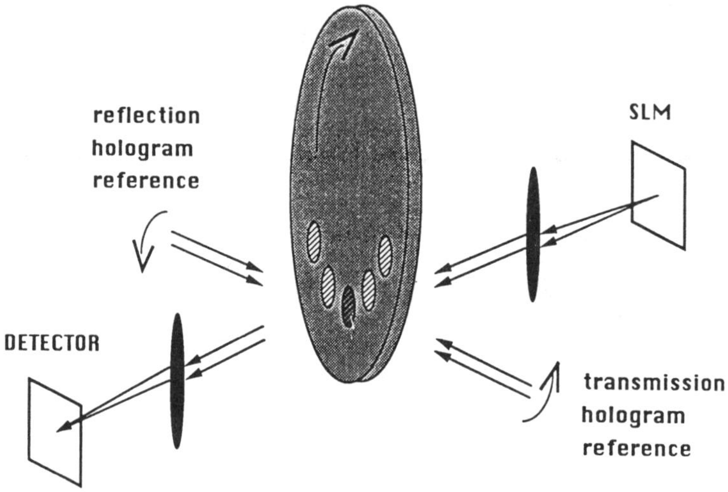

Fig. 9. 3–D Optical Disk Using Photorefractive Crystal.

SUMMARY

The excellent properties of doped tungsten bronze crystals for data storage applications have been established in this work. Specifically optimization of dopants as well as external factors to improve storage time has been successful, increasing hologram retention beyond 35 days. The tungsten bronze hosts BSTN and BSKNN appear to be the best choice for data storage. Because techniques have been developed to grow these crystals in large size, this family has a great future for data storage applications.

Acknowledgement

This work is supported under ARPA and Rockwell International IR&D programs.

REFERENCES

1. Y. Qiao, S. Orlov, D. Psaltis and R.R. Neurgaonkar, *Opt. Lett.* **18**, 1004 (1993).
2. F. Micheron and G. Bismuth, *Appl. Phys. Lett.* **20**, 79 (1972).
3. A. Kewitsch, M. Segev, A. Yariv and R.R. Neurgaonkar, *Opt. Lett.* **18**, 534 (1993).
4. S. Orlov, D. Psaltis and R.R. Neurgaonkar, "Dynamic electronic compensation of fixed gratings in photorefractive media", submitted to *J. Appl. Phys.* (1993).
5. A. Kewitsch, M. Segev, A. Yariv and R.R. Neurgaonkar, *Opt. Lett.* **18** **(15)**, 1262 (1993).
6. M.D. Ewbank, R.R. Neurgaonkar, W.K. Cory, J. Feinberg, *Appl. Phys. Lett.* **62**, 373 (1987).

7. E.J. Sharp, G.L. Wood, W.W. Clark III, G.J. Salamo and R.R. Neurgaonkar, *Opt. Lett.* **17**, 207 (1992).
8. G.J. Salamo, M.J. Miller, W.W. Clark III, G.L. Wood, E.J. Sharp and R.R. Neurgaonkar, *Appl. Opt.* **27**, 4356 (1988).
9. E.J. Sharp, W.W. Clark, M.J. Miller, G.L. Wood, B. Monson, G.J. Salamo, R.R. Neurgaonkar, *Appl. Opt.* **29**, 743 (1990).
10. J. Ford,Y. Taketomi, S. Lee, D. Bize, S. Feinman and R.R. Neurgaonkar, *Proc. SPIE,* **12** (1989).
11. G.J. Salamo, B. Monson, W.W. Clark III, G.L. Wood, E.J. Sharp and R.R. Neurgaonkar, *Appl. Opt.* **30**, 1847 (1991).
12. W.W. Clark III, G.L. Wood, E.J. Sharp, G.J. Salamo and R.R. Neurgaonkar, *Am.Ceram. Bull.* **74,** 311 (1991).
13. M.H. Francombe and B. Lewis, *Acta. Cryst.* **11**, 696 (1958).
14. L.G. Van Uitert, H.J. Levonstein, J.J. Ruben, C.D. Capio, W.A. Bonner, *Mat. Res. Bull.* **3**, 47 (1968).
15. P.B. Jamieson, S.C. Abrahams and J.J. Bernstein, *J. Chem. Phys.* **48**, 5048 (1968).
16. R.R. Neurgaonkar and W.K. Cory, *J. Opt. Soc. Am.* **3(B)**, 276 (1986).
17. R.R. Neurgaonkar, W.K. Cory, J.R. Oliver, M.D. Ewbank and W.F. Hall, *J. Opt. Eng.* **26**, 392 (1987).
18. R.R. Neurgaonkar, W.K. Cory, J.R. Oliver, E.J. Sharp, G.L. Wood and G.J. Salamo, *Ferroelectrics* **142**, 167 (1993).
19. R.R. Neurgaonkar, W.K. Cory, J.R. Oliver, E.J. Sharp, M.J. Miller, G.L. Wood and W.W. Clark III, *J. Cryst. Growth* **84**, 629 (1987).
20. R.R. Neurgaonkar, W.K. Cory, J.R. Oliver, M.J. Miller, E.J. Sharp, G.L. Wood and G.J. Salamo, *Mat. Res. Bull.* **24**, 589 (1989).
21. R.R. Neurgaonkar, W.K. Cory, W.W. Ho, W.F. Hall and L.E. Cross, *Ferroelectrics,* **51**, 185 (1984).
22. R.R. Neurgaonkar, W.K. Cory, J.R. Oliver, M. Khoshnevisan and E.J. Sharp, *Ferroelectrics* **102**, 3 (1990)
23. J.R. Oliver, R.R. Neurgaonkar, Y. Qiao and D. Psaltis, submitted to *Appl. Phys. Lett.* (1993).
24 R.A. Vazquez, F.R. Vachss, R.R. Neurgaonkar and M.D. Ewbank, *J. Opt. Soc. Am.* **B/8(9)**, 1932 (1991).
25. R.A. Vazquez, M.D. Ewbank and R.R. Neurgaonkar, *J. Opt. Soc. Am.* **B/52**, 898 (1992).
26. K. Sayano, A. Yariv and R.R. Neurgaonkar, *Opt. Lett.* **15(1)**, 9 (1990).
27. K. Sayano, A Yariv and R.R. Neurgaonkar, *Appl. Phys. Lett.* **67**, 1594, 1990.

Recent Advances in Holographic Data Storage Implemented With Photorefractive Fibers

Lambertus Hesselink
and
Matthew C. Bashaw
Department of Electrical Engineering
Stanford University
Stanford, CA 94305-4035

Phone: (415)-723-4850
Fax: (415)-725-3459

Abstract

In this invited paper we discuss recent advances in holographic data storage using ferroelectric SBN fibers as the recording medium. An optical architecture involving an array of SBN fibers is discussed that potentially allows 2-3 orders of magnitude faster access times than for conventional magnetic data storage devices of Gbyte-size . To realize this potential we are studying the interplay between fundamental materials issues related to fiber growth and photorefractive processes underlying storage and readout processes in SBN.

1 Introduction

In the early seventies, photorefractive media were investigated for use in holographic data storage devices because of the potentially high storage capacity [1, 2]. Theoretical investigations indicate that a potential storage density of 10^{13} bits/cm^3 could be achieved [3]. At that time, however, the necessary input and output devices were not available to take advantage of the high data bandwidths that page storage would afford. It was also difficult to fix information in the crystals, allowing repeated nondestructive readout of the same data and superposition of tens of pages inside the storage volume. The potential of holographic data storage was never realized.

Recently, substantial advances have been made in our understanding of photorefractive data storage, and the enabling I/O technologies are now capable of delivering data bandwidths commensurate with the space-time bandwidth product of ferroelectrics such as strontium barium niobate (SBN).

Therefore, holographic data storage in photorefractives deserves another look, emphasizing short data access time, high data transfer rates in addition to high storage capacity. These characteristics may overcome the current I/O bottleneck in digital computers. Presently, the processor speed far exceeds the ability of conventional rotating disk devices to import or to export data needed for manipulations. As a result, most computers are I/O bound. This I/O gap is widening at a steady rate as digital processor technology continually advances. Holographic data storage devices have potentially the ability to overcome this gap by offering much faster data access time and higher data bandwidth than for conventional rotating disk devices having similar capacity in the Gbyte range.

The approach we have taken involves superimposing 30-50 holograms in a small, rod-like volume of SBN. The diameter of the rods is approximately 1 mm and they are about 4-5 mm long. Hundreds of rods are arranged in a closed packed configuration, much like in a conventional microchannel plate using glass fibers. Holograms are superimposed using either angular multiplexing, wavelength multiplexing or phase encoding, as shown in Figure 1. Data are electronically input using a spatial light modulator which impresses an amplitude modulation onto the object beam. This pattern is Fourier transformed near the entrance face of the SBN rod by a positive lens. A reference beam, on which a binary phase code is impressed is incident on the medium from an off-axis position. Using a set of orthogonal phase functions, multiple holograms can be superimposed.

2 Performance Issues

One of the critical performance parameters for holographic data storage (HDS) is the signal-to-noise ratio or bit error rate for a desired storage capacity.

Following the work of Graf *et al.* [4] the storage capacity of a thin hologram of lateral size H using a detector array of lateral dimension D is calculated. Their estimate can be extended to 3-D volume recording in which N holograms are superimposed in a medium of thickness L to be:

$$\rho_{\mathrm{3DHmax}} = \frac{[D/4(H+D)]^2}{\lambda^2}\frac{N}{L} \qquad (\mathrm{bits/mm^3}) \qquad (1)$$

For $N = 10^4$, $L = 1.0$ mm, $H = D = 40$ mm, and $\lambda = 500$ nm, $\rho_{\mathrm{3DHmax}} = 6.25 \times 10^8$ bits/mm^3.

These theoretical capacities have not been realized, because noise tends to reduce the usable capacity long before geometric or wavelength limits are approached. The recording process is nonlinear and material noise tends to be additive. Bragg effects suppress nonlinear noise and crosstalk as a result

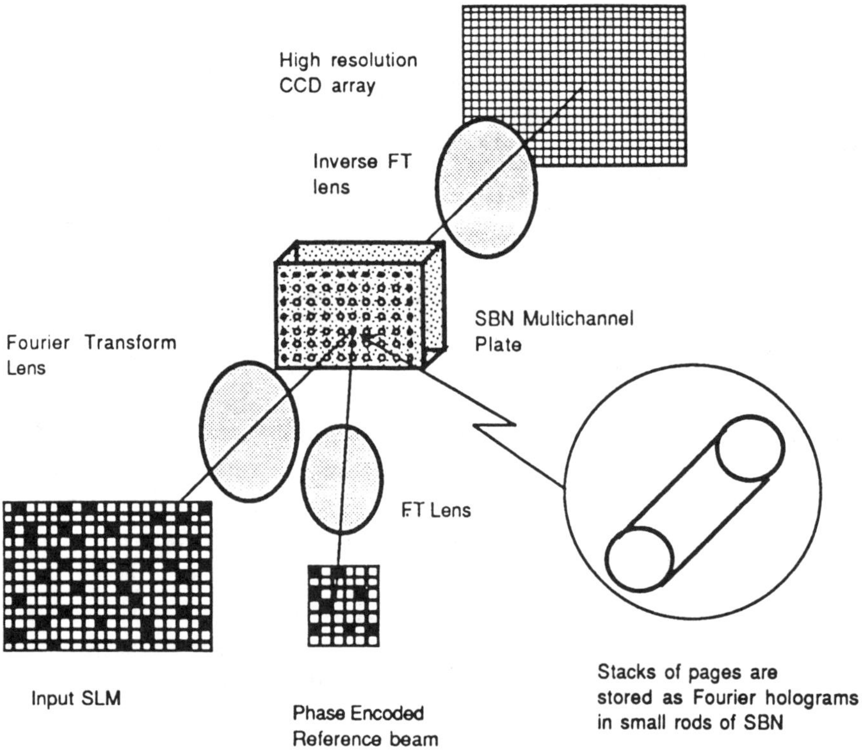

Figure 1: Optical architecture for holographic data storage in a photorefractive SBN microchannel plate

of hologram multiplexing. Material noise is usually the dominant source of noise, because the readout beam propagates throughout the volume of the storage medium. This implies that materials research related to growing low noise media is of the utmost importance.

Quantitatively we can express the useful storage capacity in terms of the SNR and the number of superimposed Fourier transform holograms, neglecting nonlinear noise and crosstalk as [5]:

$$\rho_{3\mathrm{DHmax}} \cdot \mathrm{SNR}_{3\mathrm{D}} = 2\eta(N)/\phi_n(0)\delta\nu[(1-\eta)\sigma_R + \eta\sigma_S] \qquad (2)$$

where $\mathrm{SNR}_{3\mathrm{D}}$ is peak signal power to rms media noise power ratio, $\eta(N)$ is the diffraction efficiency for N incoherent superpositions, $\phi_n(0)$ is the peak value of the normalized media noise power spectrum, $\delta\nu$ denotes the signal spatial frequency bandwidth, and σ_R and σ_S are media constants.

For small diffraction efficiency (large N), we find:

$$\rho_{3\mathrm{DHmax}} \cdot \mathrm{SNR}_{3\mathrm{D}} = \mathrm{constant}/N^2 \qquad (3)$$

The maximum number of holograms that can be superimposed for a prescribed N, and compensating for bias buildup is then [5]:

$$N = \sqrt[3]{[2\eta(N=1)(8\lambda)^2 L/\phi_n]/\mathrm{SNR}_{3D}(N)} \tag{4}$$

where $\phi_n = \phi_n(0)\delta\nu[(1-\eta)\sigma_R + \eta\sigma_S]$. For $\eta(N=1) = 0.1$, L = 1 mm, $\lambda = 500$ nm, $\phi_n = 10^{-12}$, and $\mathrm{SNR}_{3D}(N) = 10$ (10 dB), we find $N \approx 70$. The storage capacity under these conditions is approximately $\rho_{3DHmax} = 1/(8\lambda)^2(N/L) = 5 \times 10^6$ bits/mm$^3 \approx$ Gbyte/cm^3, four orders of magnitude smaller than the theoretical limit, showing that extremely low scatter levels in the recording materials are of paramount importance for achieving large storage capacity. Capacities in the tens of Gigabytes will require several cubic centimeter of material, which may be synthesized from rod arrays as shown in Figure 1. These materials must have extremely low scattering characteristics to make them useful in a practical system. In addition, the type of multiplexing technique used for HDS determines materials properties.

In general, no multiplexing scheme provides complete orthogonal recording (i.e., no crosstalk between superimposed holograms) and therefore the magnitude of the crosstalk noise terms must be compared with the intrinsic scattering noise when computing the signal-to-noise ratio (SNR). The SNR can be estimated in a variety of ways, and precise calculation of these cross talk terms for angular, phase-encoded, and wavelength multiplexing in thick Bragg media is a current topic of research [6]. As a general construct it is desirable to increase signal strength and reduce material and crosstalk noise, while maintaining a large storage capacity. For ROM applications, $LiNbO_3$ is probably still the recording medium of choice. For read-write applications, tungsten bronze crystals, such as SBN and related compounds appear attractive [7]. For example, photorefractive SBN:75 has a large electrooptic coefficient (r_{33}= 1245 pm/V) [8] and adequate time response (τ_{di} = 120 ms at 6W/cm^2) [7]. In addition, diffraction efficiency can be electrically enhanced [9] in SBN:75 by one to two orders of magnitude, and domain reversal allows electrical fixing, albeit at small diffraction efficiency (typically less than a few percent for large grating spacings, which reduces storage capacity).

Since large size tungsten bronze crystals are difficult to grow by the Czochralski technique, a fiber array offers an attractive alternative because small rods or fibers of good optical quality are relatively easily obtained using by the laser heated pedestal growth technique. The particular crystallographic orientation of the fibers is largely determined by architecture issues.

For wavelength multiplexing, c-axis fibers are preferred, because they afford the least crosstalk between superimposed reflection holograms. These fibers are of excellent optical quality, but the limiting space charge field is relatively small. In addition, electrical effects are not effective in this arrangement, thereby making electrically enhanced diffraction difficult to realize. On the other hand, a-axis fibers are attractive for phase and angular encoding. Since the c-axis is perpendicular to the fiber axis, transmission holograms are optimal, and only small voltages (on the order of tens of volts) are needed to achieve large field strengths inside the medium. The difficulty is that a-axis fibers of good optical quality are not easily grown. In Ce doped fibers, there appears a core of slightly different composition and refractive index which scatters light, thereby significantly decreasing the SNR. Despite an intense international research effort, this problem has not been resolved to date [10].

These tradeoff considerations point in the direction of reflection holography using SBN rods or fibers; the final choice, however, is probably going to be determined by other system considerations, such as the availability of a tunable source and compatibility between material sensitivity and laser wavelength. Most tunable sources now operate in the near infrared regime of the spectrum where materials are usually not very sensitive. For example, recently, appropriately doped $LiNbO_3$ has been found to have enhanced sensitivity in the 700 nm and above wavelength range, making operation of tunable diode laser sources feasible. Although $LiNbO_3$ is an attractive materials for ROM applications, much materials related work is still needed, to make $LiNbO_3$ faster responding. In addition, temperature fixing, which is an attractive property of green sensitive $LiNbO_3$ appears to be much more difficult to achieve in infrared sensitive materials. Similar studies are also needed for SBN.

The kind and level of dopants in the material profoundly influence the photorefractive response of the medium. Since the photorefractive response is dispersive with respect to grating vector, different dopant levels are required for transmission and reflection applications. For most applications using SBN, the index perturbation arises from the Pockels effect and is proportional to the space-charge field. The dispersive nature of the photorefractive response is therefore best illustrated by considering the steady-state amplitude of the space-charge field, given by

$$E_{SC} = -m\frac{E_0 + iE_D}{1 + E_D/E_q - iE_0/E_q}, \tag{5}$$

where the modulation depth of a periodic intensity pattern is m, the applied field is E_0, the diffusion field is $E_D = Kk_BT/e$ for a grating vector of magnitude K, and the limiting space-charge field is $E_q = eN_R/\epsilon\epsilon_0 K$ for $N_R = N^\circ N^\bullet/N$, in which the total density of states (N) consists of filled traps ($N^\bullet$) and empty traps (N°). Here k_B is the Boltzmann constant, T is temperature, e is the unit of electron charge, ϵ_0 is the permittivity of free space, and e is the relative dielectric constant. In media with a strong photovoltaic effect, asymmetric photoionization results in a constant photovoltaic

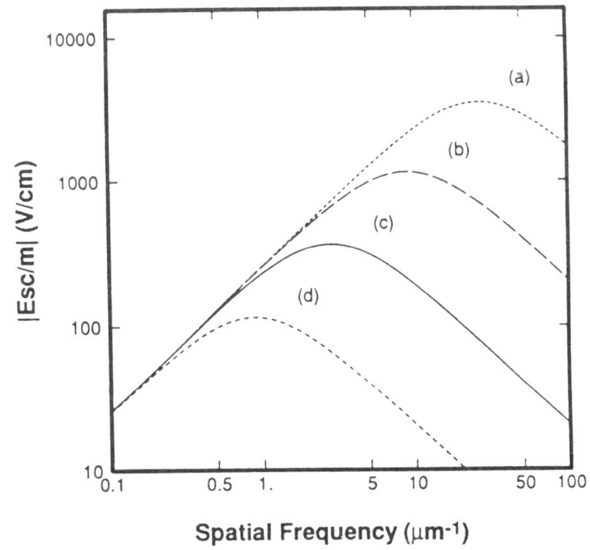

Figure 2: The dispersion of the space charge field as a function of grating spatial frequency , K, in SBN:61 for different oxidation states. The field is oriented in the direction of the c-axis, and the relative dielectric constant is taken to be ϵ_{33}= 880. The number of photorefractive centers, N, is 10^{19} cm^{-3}, and the number of empty centers serving as free carrier traps, $N°$, is (a) 10^{18} cm^{-3}, (b) 10^{17} cm^{-3}, (c) 10^{16} cm^{-3}, and (d) 10^{15} cm^{-3}.

field, which acts qualitatively as an applied field but leads to a space-charge field with a significantly more complex functional form [11, 12].

The space-charge field in Eq. (5) may also be expressed as

$$E_{SC} = -m\zeta(K)(E_0 + iE_D), \tag{6}$$

where

$$\zeta(K) = (1 + l_D^2 K^2 - il_E K)^{-1}. \tag{7}$$

and l_D is the Debye screening length, and l_E denotes the applied field screening length.

The steady-state field is bounded for small spatial frequencies by diffusion, through E_D, and for large spatial frequencies by the charge available for redistribution, through the screening factor $\zeta(K)$. For the diffusion case, the peak space-charge field occurs for $K = l_D{}^{-1}$. These features are illustrated in Figure 2 for SBN:61 with different concentrations of active species with reduction states. As the availability of empty traps ($N°$) increases, the space-charge field at high grating spatial frequencies increases, but does

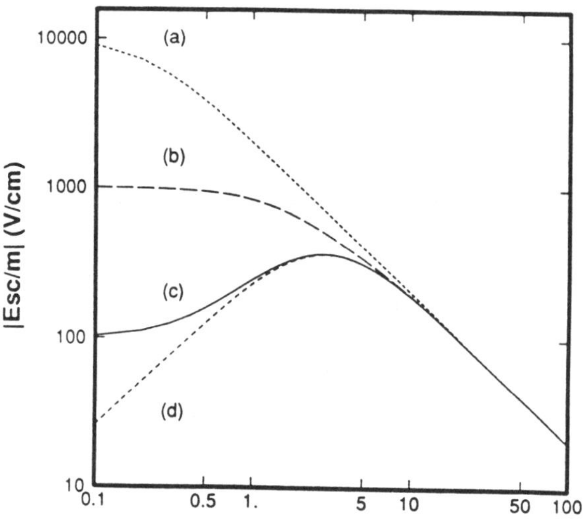

Figure 3: The dispersion of the space charge field as a function of the grating spatial frequency, K, in SBN:61 for different applied fields. The field is oriented in the direction of the c-axis, and the relative dielectric constant is taken to be $\epsilon_{33}= 880$. The number of photorefractive centers, N, is 10^{19} cm^{-3}, and the number of empty centers serving as free carrier traps, $N°$, is 10^{16} cm^{-3}. The applied field, E_0 is (a) 10,000 V cm^{-1}, (b) 1000 V cm^{-1}, (c) 100 V cm^{-1}, and (d) 0 V cm^{-1}.

not exceed mE_D. For low spatial frequencies, however, diffusion dominates, and the space-charge field is essentially independent of the concentration of empty traps. For grating spatial frequencies less than the inverse Debye screening length, application of a high electric field significantly alters the behavior of photorefractive media. For $E_D \ll |E_0| \ll E_q$, the magnitude of the space-charge field is proportional to the magnitude of the applied field, as shown in Figure 3, and consequently the diffraction significantly increases. Furthermore, the spatial frequency response is essentially flat for this regime. As $|E_0| \sim E_q$, however, the magnitude of the space-charge field saturates at mE_q. Similarly, for $E_D \sim E_q$, the magnitude of the space-charge field saturates at mE_q, and an applied field makes little further contribution.

From this discussion we note that the materials properties should be tailored to the application. As an example, the transmission geometry is attractive because an applied field increases the photorefractive response for large grating spacings. It is, however, difficult to achieve good optical a-axis SBN fibers needed for this geometry. For wavelength encoding, reflection gratings are optimal for reducing crosstalk, but the photorefractive response is limited by the diffusion field. By oxidation and reduction treatment the Debye screening length can be reduced, thereby increasing the space charge field for grating periods near this length. A more in-depth treatment of these issues is found in Hesselink and Bashaw [6], where the close relationship between materials properties and device architecture are further analyzed and discussed. Since useful c-axis SBN fibers are currently available, we finish this discussion with a brief description of capacity considerations for reflection holograms using wavelength multiplexing.

For reflection holograms with a normally incident reference wave, the wavelength sensitivity is given by [13]:

$$(\delta\lambda)^2 = \frac{\lambda^2}{4n^2}[\frac{\lambda^2}{L^2} + \delta n^2] \tag{8}$$

Assuming a crystal thickness of approximately 4 mm, $n \approx 2.5$, and $\delta n < 10^{-4}$, we find that $\delta\lambda \approx 0.01$–$0.1$ nm for light from visible to near infrared wavelengths. Crosstalk in this case is again found by summing all components of superimposed holograms that overlap with the reconstruction wavelength. As an example, for a superposition of 200 holograms the SNR is better than 800 [14]. The SNR remains quite constant for up to a few thousand holograms, suggesting longer material length, and optimized dopant levels to maximize diffraction efficiency. This leads to approaches involving either fibers or long slabs of photorefractive material.

Diode lasers are now commercially available with a tuning range of 20 nm, allowing superposition of 200 holograms with a theoretical SNR of better than 800, neglecting material scatter noise. For a page size of 220,000 pixels this corresponds to 40 Mbytes for grayscale recording. To achieve a 10 Gbyte storage capacity requires 250 – 2000 rods in the array, or larger page sizes and a smaller number of superimposed holograms. This leads to a tradeoff between page size, the number of superimposed holograms, total capacity and the need for spatial multiplexing. An experimental arrangement to measure the actual SNR is currently being implemented at Stanford University.

3 Summary

Materials and architecture issues related to holographic data storage are closely related. The photorefractive response is strongly dependent on the orientation of the grating vector in relation to the crystallographic axes. For angular and phase encoding multiplexing, the 90 degree configuration, by which the object and reference beams make an angle of 90 degrees with respect to each other, is optimal from a Bragg selectivity viewpoint. This arrangement can be approached using reference waves incident at large angles and an object waves propagating along the fiber axis, requiring an a-axis SBN fiber for optimal response. These fibers are difficult to grow, necessitating a substantial fiber growth research effort. On the other hand, wavelength multiplexing is optimally achieved in a reflection configuration, suggesting c-axis SBN fibers. These are relatively easily grown, making them quite suitable for this purpose. Large storage capacity can then be obtained by making fiber arrays of heavily doped material for maximal photorefractive response. Unfortunately, electrical enhancement is not effective in this configuration, favoring a-axis fibers. A detailed systems study is therefore needed to determine the tradeoffs between materials and systems issues for HDS.

4 Acknowledgements

Helpful discussions with Dr. Jeff Wilde and John Heanue are gratefully acknowledged. The SBN fibers are grown at Stanford University by Ludwig Galambos. The Stanford research is funded by the Advanced Research Project Administration under grant N00014-92-J-1903.

References

[1] J.B. Thaxter and M. Kestigian, "Unique properties of SBN and their use in a layered optical memory," *Appl. Opt.* **13**, 913 (1974).

[2] L. d'Auria, J.P. Huignard, C. Slezak, and E. Spitz, "Experimental holographic read-write memory using 3-D storage," *Appl. Opt.* **13**, 808 (1974).

[3] P.J. van Heerden, "Theory of optical information storage in solids," *Appl. Opt.* **2**, 393 (1963).

[4] P. Graf and M. Lang, "Geometrical aspects of consistent holographic memory design," *Appl. Opt.* **11**, 1382 (1972).

[5] R.G. Zech, "Volume Hologram Optical Memories," *Opt. Photon. News* **3**(8), 16 (August, 1993).

[6] L. Hesselink and M.C. Bashaw, "Optical memories implemented with photorefractive media," *Opt. Quant. Electron.* **25**, S611 (1993).

[7] W.K. Neurgaonkar, W.K. Cory, J.R. Oliver, M.D. Ewbank, and W.F. Hall, "Development and modification of photorefractive properties in the tungsten bronze family crystals," *Opt. Eng* **26**, 392 (1987).

[8] P.V. Lenzo, E.G. Spencer, and A.A. Ballman, "Elcctro-optic coefficients of ferroelectric strontium barium niobate," *Appl. Phys. Lett.*. **11**, 23 1967.

[9] J. P. Wilde and L. Hesselink, "Electric-field-controlled diffraction in photorefractive strontium barium niobate", *Opt. Lett.* **17**, 853 (1992).

[10] L. Hesselink, "New Applications for Photorefractive Fibers," *Opt. Photon. News* **4**(4), 9 (April, 1993).

[11] N.V. Kukhtarev, V.B. Markov, S.G. Odulov, M.S. Soskin, and V.L.Vinetskii, "Holographic storage in electrooptic crystals. I. Steady state," *Ferroelectrics* **22**, 949 (1979).

[12] C. Gu, J. Hong, H.-Y. Li, D. Psaltis, and P. Yeh, "Dynamics of grating formation in photovoltaic media," *J. Appl. Phys.* **69**, 1167 (1991).

[13] H. W. Kogelnik, "Coupled wave theory for thick hologram gratings," *Bell Syst. Tech. J.* **48**, 2909 (1969).

[14] K Curtis, C Gu, and D Psaltis, "Crosstalk in wavelength-multiplexed holographic memories," *Opt. Lett.* **18**, 1001 (1993).

Fixing and IR response of doped barium titanate

T.W. McNamara*, S.G. Conahan*, I. Mnushkina**, M.H. Garrett**, H.P. Jenssen** and C. Warde*

* Department of Electrical Engineering and Computer Science, Massachusetts Institute of Technology, Cambridge, MA, 02139 ** Center for Materials Science and Engineering, MIT.

ABSTRACT

Two key obstacles to the development of high-capacity three-dimensional information mass storage systems using photorefractive crystals are: (1) readout of holograms stored in the conventional manner is destructive, and (2) undoped photorefractive crystals are typically quite insensitive to red and infrared write light from diode lasers (the preferred source). As a consequence, bulky, expensive ion lasers have been necessary for hologram writing. Thus hologram fixing and high infrared sensitivity in these materials are both important if practical systems are to be built. We present our own high-temperature (80–110°C) photorefractive grating writing results for a variety of pure and doped barium titanate ($BaTiO_3$) crystals. We found strong fixable secondary gratings that correlated with increasing levels of Fe and Ni in doped crystals. Fixing was not observed in our pure crystals or those doped with V, Rh, or Co. Fixing was enhanced in an iron-doped crystal reduced by high-temperature annealing at low oxygen partial pressures. We also found that the readout diffraction efficiency of a fixed grating in $BaTiO_3$ is a rapidly increasing function of readout temperature above a threshold temperature, which depends upon the crystal orientation. Lower temperatures lead to longer storage times, as would be expected if the carriers forming the fixed grating have a thermally activated mobility. We also report preliminary experimental results on blue Rh-doped $BaTiO_3$ crystals with fast and high-gain infrared (840 nm) response.

1. INTRODUCTION

1.1 Fixing

Photorefractive phenomena can be observed in any material which is photoconductive, transparent to the wavelengths of interest, and electrooptic. When a transparent photoconductor is exposed throughout its volume to a three-dimensional interference pattern, charge carriers are generated in the bright regions and recombine in the dark. The result is a faithful three-dimensional recording of the original light intensity, stored as a space-charge pattern. The electric field due to this charge separation modulates the refractive index through the electrooptic effect, so that a phase-only volume hologram develops in real time. These photoconductive holograms persist in the dark but can be erased under uniform illumination, so that photorefractive media are fully reusable. Historically, the photorefractive effect was discovered as "optical damage" during second harmonic generation experiments in

$LiNbO_3$; its first application to holographic storage was reported in 1968 by Chen et al. [1]. Details on the physical mechanism behind the photorefractive effect are described in the review chapters by Günter [2] and Staebler [3].

Multiplex holographic storage in photorefractive crystals has been under study for over 25 years; some of the most recent results in lithium niobate include storage of 5000 images of 200 × 300 pixels each in a single 1 cm^3 crystal [4], an information density of 3×10^8 bits/cm^3. It is believed that the ultimate theoretical limit on the information density is on the order of one bit per cubic wavelength: 10^{12} bits/cm^3 at a wavelength of 1 μm [5]. At the same wavelength, diffraction-limited optical disks are limited to 10^8 bits/cm^2, so there are in principle storage capacity advantages to 3-D holographic memories over non-holographic 2-D memories. The high storage density of volume holographic media has been achieved by angular multiplexing: each image is recorded as a volume hologram with a unique reference wave. Each hologram can then be retrieved by re-illuminating the crystal with the appropriate reference wave. Since each hologram is Bragg matched to a unique reference wave, selective retrieval is possible with very low crosstalk [4, 6]. The requirement for addressing a particular hologram is merely angular scanning of a collimated beam (the storage medium itself need not move). The seek time for an entire page of data can be short by the standards of conventional mass storage devices (less than 1 ms with commercially available beam scanners).

Unfortunately, under normal circumstances this readout is destructive: in a crystal with high photoconductivity, the stored gratings are all progressively erased during readout. This problem can be partially solved by using a "safe" readout wavelength to which the crystal is not very sensitive. The disadvantages of this procedure are that: (1) readout is still destructive over many readout cycles, particularly at high readout intensity levels, (2) two lasers are necessary, and (3) the stored image is distorted on readout. The life of the hologram may be extended under destructive readout by optimizing the applied electric field [7]. Alternatively, storage of an array of binary pixels can be maintained by optical "refresh" or positive-feedback bistable imaging systems employing one or more photorefractive crystals [8, 9, 10]. However, the storage densities demonstrated so far in these systems have been very low, and the systems are relatively complex.

All these restrictions can be eliminated by using a fixing procedure to render recorded holograms insensitive to photoconductive erasure. We adopt the definition here that a fixing process is one whose storage time (τ_s) is much longer than its writing time (τ_w) at constant light intensity. For convenience of use this asymmetry should be large ($\tau_s/\tau_w = 10^6$ or more). For instance, if τ_s =10 years and $\tau_w = 1$ minute, $\tau_s/\tau_w \simeq 5\times10^6$. Fixing is usually accomplished by techniques which somehow free light-insensitive secondary charge species to move in the space-charge electric field of the primary photorefractive grating.

Microscopic physical mechanisms which have been found to explain the properties of secondary gratings include thermally activated ionic drift and formation of alternating microdomains. Thus most techniques for fixing photorefractive gratings belong to one of two broad categories: (1) those involving high-temperature recording and low-temperature readout (fixing by thermal cycling), and (2) those involving applied electric fields exceeding the coercive field of ferroelectric photorefractive media (electrical fixing) leading to microdomains. In thermal fixing, at sufficiently high temperatures, in some materials, secondary carriers are attracted to primary charges of the opposite sign, thus locally neutralizing some or all of the space charge which formed the original photorefractive hologram. When optical hologram writing ceases and the secondary carriers are once more fixed in place, uniform illumination is used to distribute the primary carriers evenly throughout the crystal, revealing the secondary grating. This fixed grating is not optically erasable. It may persist for years or longer. Long-range ionic drift can often be distinguished from microdomain formation in photorefractive ferroelectrics by observing whether the process is sensitive to the polarity of fields applied along the crystal's axis of polarization. Repoling with fields well above the coercive field in the material will erase gratings consisting of microdomains [11, 12].

Historically, thermal fixing methods were reported first in $LiNbO_3$ crystals, by Amodei and Staebler [13]. Writing took 30 minutes at 100°C or above and fixing was accomplished by cooling the crystal to room temperature. This froze the secondary grating in place, and illumination with uniform light to redistribute the photorefractive carriers uniformly revealed a phase grating which was not optically erasable. It has since been shown that the fixable gratings in these crystals were formed by charge compensation of photorefractive gratings written at high temperature by hydrogen ions with a temperature-activated mobility [14]. These bare protons formed OH^- centers which were detected by their infrared absorption near $3\mu m$. The concentration of these ions could be increased by annealing the crystals in a water vapor atmosphere, or decreased by heating in a dry atmosphere [15]. The fixing properties were found to improve or disappear accordingly.

Holograms have been thermally fixed in $Bi_{12}SiO_{20}$ (BSO) and potassium niobate ($KNbO_3$). Arizmendi [17] reported fixing in BSO by thermal cycling between room temperature and 260°C. Since BSO is not ferroelectric, ionic drift was suspected as the cause of the observed secondary gratings, but the nature of the ions was uncertain. Fixing by thermal cycling was accomplished in $KNbO_3$ by Montemezzani and Günter [18]. Montemezzani et al. based theoretical work including a set of transport equations for the coupled photoconductive-ionic system on their observations in that material [19].

$BaTiO_3$ has very large electrooptic coefficients and low absorption at red and infrared wavelengths, making it one of the most promising materials for nonvolatile storage of high-diffraction-efficiency volume phase holograms. To this end, several workers have demonstrated techniques for fixing holograms in $BaTiO_3$ [11, 20]. Electrical fixing in $BaTiO_3$ was first demonstrated by Micheron and Bismuth [11] as well as in strontium barium niobate (SBN) [12]. They used a four-step technique: (1) a photorefractive hologram was written to saturation in a poled crystal; (2) a negative (opposite to the sense of the poling field) voltage pulse exceeding in magnitude the coercive field for ferroelectric switching ($\sim$ -1 kV/cm) was applied to form a fixed compensating secondary grating, resulting in a lowered diffraction efficiency as primary and secondary gratings cancelled out; (3) the crystal was illuminated uniformly to redistribute the carriers of the primary grating, revealing a fixed grating with a high diffraction efficiency; (4) erasure was accomplished by applying a positive repoling field which was several times the coercive field of the crystal. Micheron and Bismuth reported much higher diffraction efficiencies in SBN than in $BaTiO_3$, but recent workers have had difficulty achieving these values [16].

In $BaTiO_3$, fixing has also been accomplished by heating crystals to about 100°C without any applied field by Kirillov and Feinberg [20]. Kirillov and Feinberg have argued that the secondary grating in the thermal cycling experiments were ions [20], but did not report any attempts to erase their fixed gratings by applying an electric field. Practical problems encountered in their pioneering work led to low fixed-hologram diffraction efficiencies, and the most effective crystals used were of unknown composition [20].

Preliminary results of our research indicate that thermal fixing is effective in iron- and nickel-doped $BaTiO_3$, but not to any significant degree in undoped crystals or those doped with cobalt or vanadium. One sample, doped with 1000 parts per million atomic (ppma) of iron and annealed at 800°C in a reducing atmosphere with an oxygen partial pressure of 10^{-4} atm, seems to be the most promising. In this crystal, fixed-grating diffraction efficiencies of several percent have been stored over times far in excess of 24 hours by exploiting crystal orientation and readout temperature. The orientations explored were limited by geometrical considerations in our original setup; we expect to see significant improvement in diffraction efficiency when the optimal orientation is found. Further, we have noticed that the writing intensity can be varied to enhance the diffraction efficiency of the fixable holograms. By optimizing crystal composition, reduction state, orientation, writing temperature, writing intensity, readout/storage temperature and applied electric fields, we hope to obtain diffraction efficiencies in excess of 30% and storage times of months to years.

It is worth noting that thermal fixing processes generally have the following disadvantage: during high-temperature writing to update a fixed hologram, all other holograms stored in the same crystal begin to decay. Therefore, thermal fixing processes developed to date are most suitable for read-only applications. Selective erasure of electrically fixed gratings has been reported recently in SBN crystals [21], and opens up the possibility of fully erasable/rewriteable nonvolatile mass storage in photorefractive crystals. Electrical fixing also has the advantage of room-temperature operation; but application of electric fields can cause cracking near the electrodes in many ferroelectrics, limiting crystal lifetime.

Another practical consideration is that the most useful known photorefractive crystals, semi-insulators with high electrooptic coefficients, have wide bandgaps (typically around 3 eV) and thus in the undoped form are most sensitive to green or blue light. Unfortunately the most compact and efficient laser sources available, diode lasers, are not yet available for these wavelengths. Many workers are currently carrying out research on doping photorefractive crystals such as $BaTiO_3$ to improve their infrared photoconductivity. This opens the door to replacing ion lasers with inexpensive, compact, highly efficient diode laser sources in optical information storage systems with photorefractive components.

1.2 Infrared Sensitization

A variety of applications for high-infrared-sensitivity photorefractive crystals have been demonstrated. Passive phase conjugation (i.e., without any externally supplied pump beams) has been achieved in undoped $BaTiO_3$ by many workers, both with an external cavity [22, 23] and by total internal reflection (TIR) [24]. The latter is commonly called self-pumped phase conjugation (SPPC) [25]. The TIR geometry eliminates Fresnel reflection losses between the active region and the mirrors, and generally yields higher phase-conjugate reflectivities. Either variety of passive phase conjugation can be used in applications such as compensation for inhomogeneities in laser gain media (and other real-time optical signal processing operations), particularly where efficient use of optical power is critical. Alignment-sensitive external cavity arrangements of diode lasers, which make mode locking and linewidth reduction possible can be made self-aligning by using passive phase-conjugate mirors [22]. Thus far, these applications have been limited by the long response times and relatively low infrared gains of undoped $BaTiO_3$. Some progress has been made by doping with Co to improve gain and to some extent speed [26]. Since Rh-doped $BaTiO_3$ can have high gain and an order of magnitude faster response time than the Co-doped material, presumably further advances will occur in this area [28].

An extension of the external cavity application of photorefractive phase conjugate mirrors is injection locking of multiple diode lasers by phase conjugation. This has been reported by several groups [27, 29], using different techniques. Cronin-Golomb et al. coupled two diode lasers using an undoped $BaTiO_3$ crystal. Later, using a Cr-doped $BaTiO_3$ crystal grown in our laboratory, MacCormack and Feinberg [29] have reported self-starting injection locking of a linear array of infrared semiconductor lasers operating at 795 nm. Their scheme uses self-pumped phase conjugation in a $BaTiO_3$:Cr crystal. Mode combining of laser diodes injection-locked by some other means can also be accomplished efficiently by a high-gain photorefractive crystal [30]. The inevitable phase variations between the two lasers can be compensated, but only when the variations are slower than the photorefractive response time.

In yet another application, Lambelet et al. [31] have proposed a beam-cleanup device which can work in isolation from the driving laser. It is based on a novel setup in which a mirror in physical contact with the $BaTiO_3$ crystal reflects part of the input beam to make it a signal beam, while the remainder of the input beam passes directly into the crystal and acts as pump.

It has been suggested that beam coupling in a high-gain photorefractive crystal with an intensity-dependent gain characteristic be used for predetection dynamic range compression [32]. This might be used to prevent saturation and blooming on detector arrays. Since operation at television rates ($\sim$ 30 Hz) or above is critical for this application, doping the photorefractive crystal to obtain maximum response speed is clearly valuable. The high gains obtainable in $BaTiO_3$ would allow thin crystals to be used, thus improving the spatial resolution of the compression operation.

Clearly in many of these applications, the high gain and high phase-conjugate reflectivity of $BaTiO_3$ are useful or essential. Constant progress has been made over the last ten years in doping $BaTiO_3$ for improved infrared sensitivity. In 1989, Beckwith and Christian [33] reported two-beam coupling measurements on a nominally undoped $BaTiO_3$ crystal using a GaAlAs laser diode (λ = 830 nm). They achieved a beam-coupling gain of $\sim$ 18 cm^{-1} and a response time of about 50 s at an intensity of 680 mW/cm^2 in the extraordinary polarization. This high gain was encouraging, but the response time was about two orders of magnitude longer than the response for 488 nm light of similar intensity. Beckwith and Christian attributed the increase of the response time at long wavelengths to a decrease of the photoionization cross-section in the near infrared region. This was supported by the very low measured absorption coefficient of 0.08 cm^{-1} at 830 nm.

Rytz et al. used HeNe, Ti:sapphire, and GaAlAs lasers to measure the red and infrared self-pumped phase conjugate reflectivity for selected wavelengths between 633 and 933 nm in $BaTiO_3$:Co [26]. For crystals doped with Co to levels above 25 ppma, they were able to find SPPC reflectivities above 50% at all wavelengths measured. Response times in the heavily-doped crystals varied linearly with input power, but because the beam was focused to a small spot in the crystal, direct comparison with infrared response-time measurements made in other crystals is difficult.

We have recently made progress toward improving the infrared speed and gain of photorefractive $BaTiO_3$ crystals in our own laboratory, primarily through doping with rhodium. Section 3 below contains details on this work. The next step in our research program will be to prepare infrared-sensitive crystals which also have good fixing properties. Unfortunately the rhodium-doped blue crystals grown thus far do not seem to be fixable, at least in the as-grown state.

2. FIXING STUDIES

2.1 Fixing technique

In an effort to discover the influence of doping on the optical properties of fixed gratings in $BaTiO_3$, we conducted high-temperature (80–110°C) grating writing experiments on a variety of pure and doped $BaTiO_3$ crystals. These crystals were grown in our own laboratory by the top-seeded solution growth (TSSG) method [34] over the past ten years (with the exception of the Ni-doped samples, which were also grown by TSSG, at Sanders Inc., Nashua, NH).

In the hologram writing system shown schematically in Fig. 1, writing beams W_1 and W_2 (514 nm wavelength) from a single argon-ion laser interfered to produce a sinusoidal intensity pattern in a photorefractive $BaTiO_3$ crystal X. The intensities of both writing beams were approximately 600 mW/cm^2, so that the fringe modulation, $m = 2(I_1 I_2)^{1/2}/(I_1 + I_2)$, was approximately unity. Shutters were arranged so that the beams could be turned on and off at will. The resulting refractive index grating was read out at the Bragg angle by a 633 nm helium-neon laser beam R,

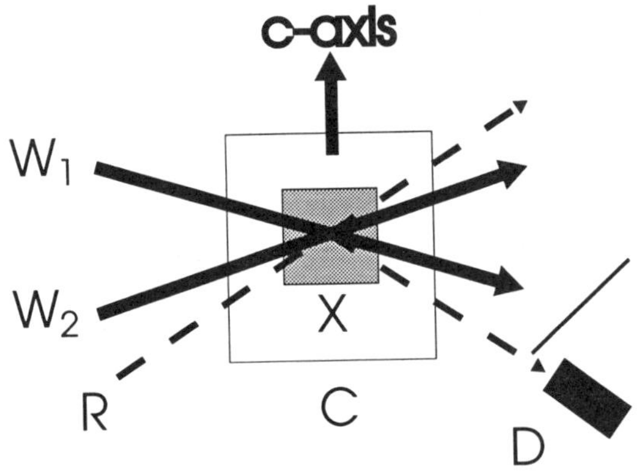

Figure 1: Experimental geometry for writing semipermanent holograms in $BaTiO_3$ at elevated temperatures. Legend: W_1 and W_2 are 514 nm writing beams, R is a 633 nm readout beam, S is the 633 nm beam diffracted by gratings in the crystal, D is a detector, X is the $BaTiO_3$ crystal, and C is the temperature-controlled oil cell.

which was Bragg matched to the index grating. The diffracted beam S was monitored by detector D. During writing of the fixable grating, the temperature of the $BaTiO_3$ crystal was set and maintained between 80°C and 110°C by a Peltier-effect heating/cooling module thermally coupled to miniature oil cell C (1 cc in volume). Once the cell settled (about 2 min after changing temperatures), the selected oil temperature was maintained by a linear proportional controller to $\pm$ 0.2 C°. The oil cell was surrounded by a small insulating box with glass windows to let the beams pass in and out. We did not apply any voltage to the crystal during these experiments.

At low temperatures ($T < 80$°C), with W_2 and R left on at all times, the diffraction efficiency rose monotonically in the familiar fashion for a single-carrier photorefractive grating (see Fig. 2) when writing beams W_1 was turned on. After writing beam W_1 was turned off, we observed a rapid monotonic decay of the diffraction efficiency.

In several of the crystals investigated, heating above about 80°C lead to a dramatic change in the time evolution of the diffraction efficiency. Experimental results are shown in Fig. 3 for a reduced 1000 ppm Fe-doped crystal heated to 103.3°C. The crystal temperature was maintained at this value and R and W_2 were left on at all times. W_1, initially off, was turned on at $t = 0$. The resulting diffraction of R into S was monitored with a photodiode. The diffraction efficiency initially rose but then slowly dropped off. After 80 sec., W_1 was turned off and the uniform intensity of W_1 quickly erased any photoconductive gratings. Nevertheless, the diffraction efficiency rebounded quickly to a peak before decaying slowly to zero (see Fig. 3).

The differences between low- and high-temperature responses can be explained by the appearance of a second mobile charge-carrying species at high temperatures

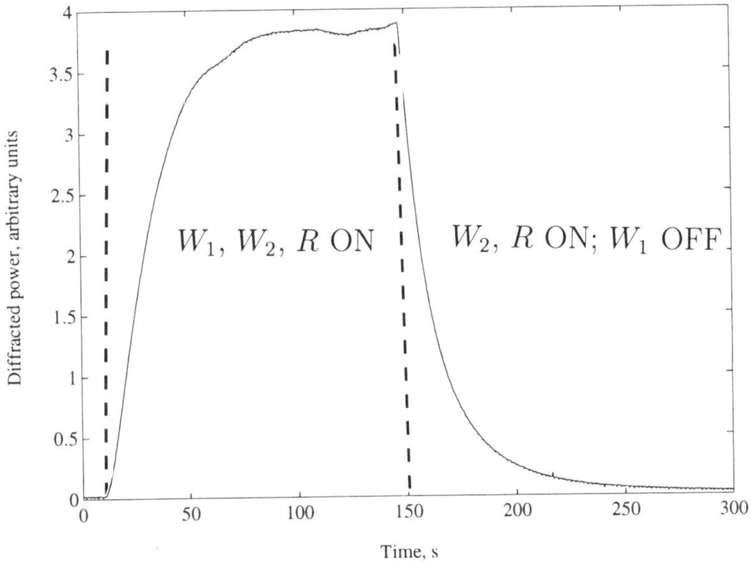

Figure 2: Measured diffracted power versus time at low temperature ($T = 60.1°C$), where no complementary gratings form. When writing beam W_1 was turned on the diffraction efficiency rose monotonically. When it was turned off, the diffraction efficiency decayed monotonically.

[13, 19]. These charges drift in the space-charge field formed by motion of the primary photorefractive carriers. Since the secondary carriers are attracted toward primary carriers of the opposite sign, the secondary carriers form a complementary grating which partially cancels the electric field of the primary grating, explaining the first slow decay of the diffracted power. Erasing the primary grating at high temperature revealed slowly-decaying complementary gratings formed by carriers insensitive to light (perhaps mobile ionized defects), thus explaining the second rise in diffracted power seen in Fig. 3 [19]. We found no evidence of a coercive field threshold: complementary gratings showed similar response shapes at small and large amplitudes. Secondary charge transport is apparently due to mobile ionized defects rather than local depoling of the crystal.

Not all crystals showed this departure from a single-carrier time response at high temperature (see Table 1). Our Fe- and Ni- doped samples showed significant fixable secondary gratings at the highest doping levels (1000 ppma and 50 ppma, respectively), while undoped or lightly doped crystals from each of these series showed little or no fixing. The reduced 1000 ppma Fe-doped sample showed much stronger fixing than the as-grown 1000 ppma Fe-doped sample. Pure and Co-, Rh-, and V-doped samples likewise showed little or no secondary grating formation or fixing. The reasons for these apparent composition and processing dependences are not yet understood.

The complementary gratings described above were fixed by rapidly cooling the

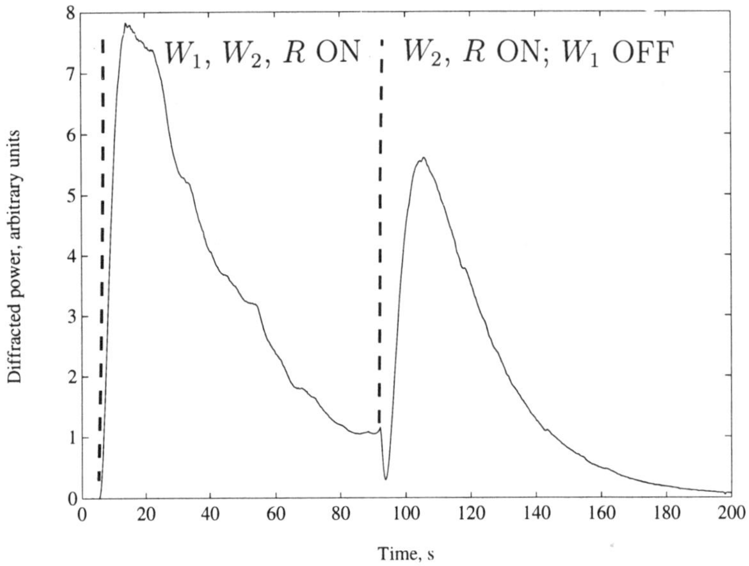

Figure 3: Diffraction efficiency of a reduced $BaTiO_3$:Fe (1000 ppm) crystal over time, showing effects of large complementary gratings at 103°C. The trend is a rapid increase followed by a slow decay when W_1 is turned on, and a rapid drop to a small value, followed by a rapid rise and a slow decay when W_1 is turned off.

crystal to temperatures between 10° and 40°C with the writing beams W_1 and W_2 either both on or both off. In our experiments so far, the usual write-fix-readout-erase cycle for ionic gratings has been as follows:

1. **Write:** With the crystal maintained at a constant temperature (typically between 80 and 110°C), writing beams W_1 and W_2 are unblocked, and the diffracted beam S is monitored until steady-state is reached.

2. **Fix:** With the writing beams still on, the crystal is quickly (within 30 s) cooled to below the writing threshold temperature Θ_W (about 80°C for 1000 ppm reduced $BaTiO_3$:Fe) so the ionic grating will no longer decay rapidly. Then W_1 and W_2 are blocked and cooling is continued until the selected readout temperature is reached.

3. **Store:** With crystal temperature maintained as low as possible within the tetragonal phase (i.e. greater than about 9°C), the fixed gratings can be stored for very long periods. However, we have observed that readout will be inefficient unless the storage temperature is above a readout temperature threshold Θ_R (about 40°C for the reduced 1000 ppm Fe-doped crystal with its c-axis oriented 14° from the grating vector).

4. **Readout:** With crystal temperature maintained at the desired readout value above Θ_R, the diffracted signal beam S can be read out nondestructively by the reference beam R. The readout Bragg angle may be slightly different

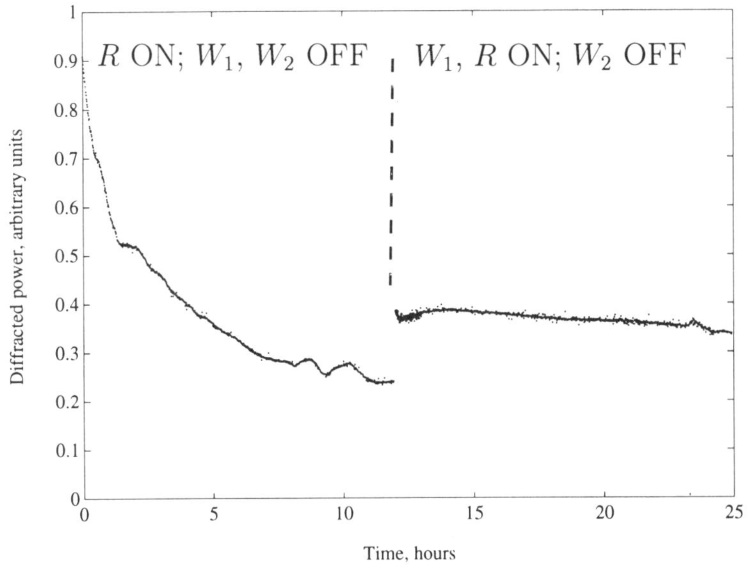

Figure 4: Diffraction efficiency versus time after fixing by cooling to a readout temperature of 39.7°C, for an ionic grating formed oblique to the c-axis ($\theta(\mathbf{k}_g, \hat{\mathbf{c}}) = 14°$) of a reduced $BaTiO_3$:Fe (1000 ppm) crystal. The single erasing beam W_1 was off for the first 12 hours and on for the last 12 hours. The peak diffraction efficiency shown was measured to be roughly 2% .

from its former value at the writing temperature, due to thermal expansion and other effects. Storage time is longer at lower temperatures.

5. **Erase:** The crystal is reheated to the writing temperature with the writing beams still blocked, and the diffracted signal beam is monitored until it effectively drops to zero. At this point the ionic grating recorded in steps 1–3 above has been erased and the cycle can be repeated.

2.2 Optical characterization of fixed gratings in doped $BaTiO_3$

In the Fe- and Ni-doped crystals with substantial complementary-grating response, gratings could be fixed by rapid (about 30 s) cooling to temperatures below 80°C (see Fig. 4). Good fixing results were obtained with the writing beams either on or off during cooling. As shown in the Fig. 4, 514 nm light cannot erase the fixed grating, at least at intensities near 1 W/cm^2. In fact, for the crystal orientation shown the diffraction efficiency actually increased on exposure to uniform writing beam W_1. We believe that this might be explained by photorefractive two-beam coupling between the writing beam and the beam diffracted from the writing beam by the fixed grating [13, 35]. In our model: (1) a single writing beam is diffracted by the fixed grating, reconstructing the other writing beam within the crystal, (2) beam coupling takes place between the original and the reconstructed writing beams, (3) a photorefractive grating is written: if the diffracted beam propagates

Crystal	Max. Percent fixed diffraction efficiency (with respect to transient hole grating)	Temp.,°C	Λ (Grating Period)	Measurement Technique
1000 ppm Fe, reduced (1462c)	29%	100	3.9 μm	Beam coupling
	80%	108	1.4 μm	Diffraction eff. at 633 nm
1000 ppm Fe, as grown (1462)	NONE	110	3.9 μm	Beam coupling
	small <1%	103	1.4 μm	Diffraction eff. at 633 nm
50 ppm Fe, reduced (1446B)	<0.1%	110	3.9 μm	Beam coupling
50 ppm Ni (146A)	15%	108	0.4 μm	Diffraction eff. at 633 nm
	48%	100	0.6 μm	Beam coupling
	20%	103	1.4 μm	Diffraction eff. at 633 nm
	NONE	100	3.9 μm	Beam coupling
25 ppm Ni (144A)	4.5%	106	1.4 μm	Diffraction eff. at 633 nm
	NONE	101	3.9 μm	Beam coupling
25 ppm Ni (144B)	small (<1%)	84	0.6 μm	Beam coupling
10 ppm Ni (141C)	NONE	86	0.6 μm	Beam coupling
Undoped,reduced (1434A)	NONE	85	0.6 μm	Beam coupling
Purified,reduced (2oF#16-2)	NONE	86	0.6 μm	Beam coupling
50 ppm V (1o#25A)	NONE	110	1.4 μm	Diffraction eff. at 633 nm
		100	3.9 μm	(low-gain crystal)
'Blue' $BaTiO_3$ (2oF#18)	NONE	108	3.9 μm	Beam coupling

Table 1: Fixing properties of several $BaTiO_3$ crystals as a function of doping and experimental conditions

toward the +c-axis of the crystal, the photorefractive grating written as W_1 and the diffracted beam interfere to selectively reinforce the fixed grating; (4) the 633 nm readout beam diffracts from the combined fixed and photorefractive gratings more efficiently than it did from the fixed grating alone.

We also found that when a crystal supporting fixed gratings is cooled to room temperature, the diffraction efficiency is very low ($\ll 1\%$). But when the crystal is reheated to the writing temperature, the diffraction efficiency of the fixed grating recovers to values comparable to those of the complementary gratings orig-

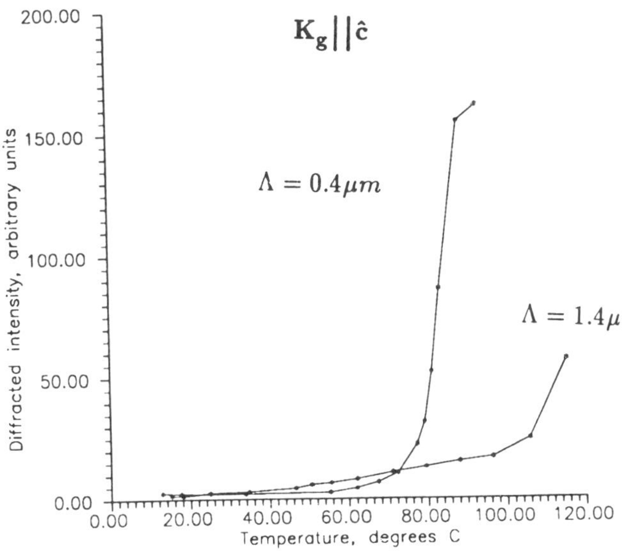

Figure 5: Diffraction efficiency of a fixed grating in an as-grown $BaTiO_3$:Ni (50 ppm) crystal as a function of reheating temperature for several grating periods, at constant intensity. Because these curves are reversible without rewriting the grating, charge compensation at low temperatures is believed to be responsible for the drop-off in diffraction efficiency.

inally observed before cooling (see Fig. 5). In this experiment, the Bragg readout angle was maintained by re-maximizing the diffracted beam at several different temperatures (to compensate for thermal expansion). In particular, no amount of care in the Bragg angle adjustment gave any significant improvement in the room-temperature diffraction efficiency (which was very low). The near disappearance of the diffracted beam at low temperatures and its return at high temperatures could be observed repeatedly without rewriting the grating. This suggests that our initial cooling was rapid enough to fix most of the charge of the complementary grating in place before it could decay, but that a temperature-dependent compensation mechanism cancels the fixed component of the electric field at room temperature. Temperature-dependent reinforcing gratings due to beam coupling between the readout and diffracted beams may also be a factor.

The curve of diffraction efficiency as a function of temperature, shown in Fig. 5, was found to depend upon the period of the fixed grating. Impractically small grating periods would have been required to obtain high diffraction efficiency at or near room temperature for this crystal orientation.

Since high diffraction efficiency simultaneous with long storage time (which demands low-temperature operation) was not possible in the standard orientation, fixing experiments were carried out in several oblique orientations. In these experiments, the angle θ between the grating vector and the c-axis, zero in the standard orientation,

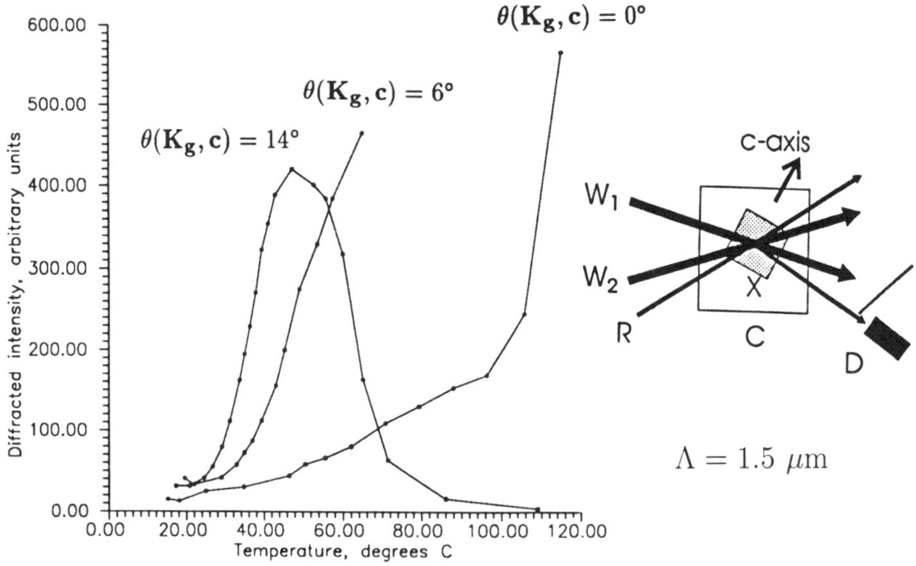

Figure 6: Power diffracted by a fixed grating as a function of reheating temperature for several crystal orientations, at constant intensity. Reduced $BaTiO_3$:Fe (1000 ppm).

was increased by rotating the crystal about its vertical a-axis (see Fig. 6). The goal here was to increase the effective dielectric constant $\epsilon_{\text{eff}} = \epsilon_c \cos^2\theta + \epsilon_a \sin^2\theta$, where $\epsilon_c \simeq 150$ and $\epsilon_a \simeq 3600$ at room temperature. The results were very rewarding (see Fig. 6). We found that the temperature threshold for efficient readout could be lowered from ~80°C at least to ~40°C by rotating the crystal to make $\theta = 14°$. (Further rotation was not practical in the setup used.) This made the long storage time shown in Fig. 4 possible in conjunction with reasonably high diffraction efficiency. Another advantage of oblique orientations is that the large electrooptic coefficient ($r_{42} \simeq 1640$ pm/V at room temperature) can be accessed for efficient readout of the charge gratings.

Finally, we observed a different secondary grating response in another crystal, $BaTiO_3$:Fe (1000 ppm), as-grown (crystal 1462). In this crystal, (1) significant secondary gratings became apparent at 55°C (see Fig. 7), while in other crystals temperatures of 80°C or more were necessary, and (2) the complementary grating response clearly saturated short of completely cancelling the photorefractive grating. This same crystal also showed weak complementary gratings of the more familiar type seen in the other Fe- and Ni-doped $BaTiO_3$ crystals in our study. This was confirmed by the fact that complementary gratings recorded above 100°C in this crystal persisted for hours at 55°C, while those recorded below 80°C decayed in minutes at 55°C. In view of the above observations, it is reasonable to assume that two distinct light-insensitive charge carriers can be activated in $BaTiO_3$:Fe.

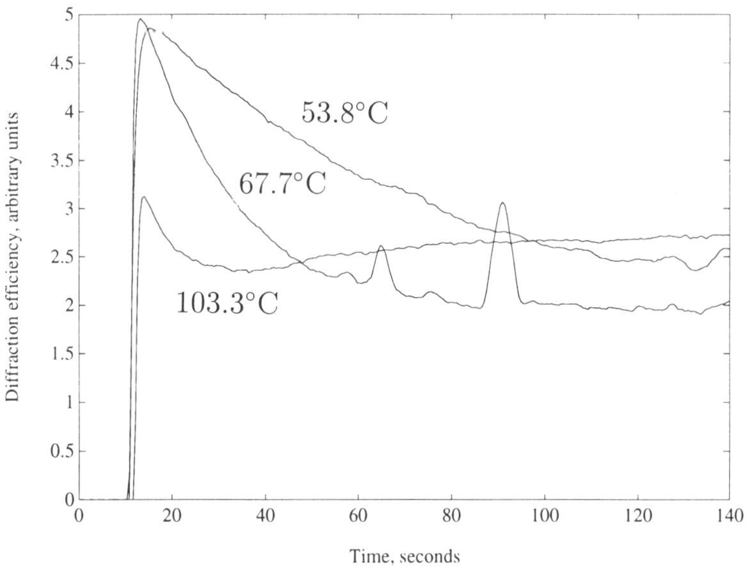

Figure 7: Time response of diffraction efficiency in the as-grown $BaTiO_3$:Fe (1000 ppm) crystal #1462 as a function of temperature at a grating period of 1.4μm, showing saturated complementary gratings, apparently due to a second type of ionic carrier, highly mobile even at 54°C. The complementary gratings can be seen to respond more rapidly at higher temperatures.

3. INFRARED-SENSITIVE $BaTiO_3$

3.1 Rhodium doping for infrared sensitization

Recently, we have grown $BaTiO_3$:Rh crystals which have high gains and fast writing response at near-infrared wavelengths. These blue crystals, originally produced by accident when Pt-Rh wire attaching the seed crystal to the pulling rod fell into a growth melt in a Pt crucible, have been reproduced by adding measured amounts of Rh metal to an otherwise undoped $BaTiO_3$ growth melt. The absorption spectrum of one of these blue crystals is shown in Fig. 8. The absorption at 850 nm is several inverse centimeters, which could account for the greatly enhanced photoconductivity at 850 nm as compared with crystals grown from the same original source material before Rh doping. Face reflections were accounted for by assuming a single reflection at each interface and using the Sellmeier formula and coefficients given by Wemple et al. [38].

The dominant impurity concentrations detected in a spark-source mass spectroscopy analysis of this material are listed in Table 2. The boules were aligned by x-ray diffraction, then sawed into rectilinear crack-free pieces. These pieces were poled by the etching method [39] and polished on opposite faces for optical characterization.

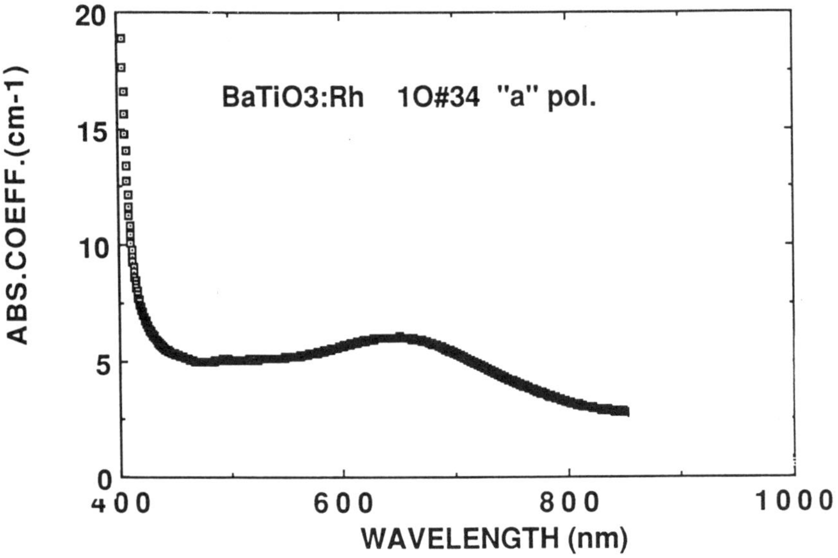

Figure 8: Absorption spectrum of as-grown blue $BaTiO_3$:Rh crystal #1o34, showing peak at 650 nm.

3.2 Photorefractive characterization of $BaTiO_3$:Rh

Because no convenient safe readout wavelength could be found for diffraction-efficiency measurements in these infrared-sensitive crystals, photorefractive characterization was carried out by beam coupling. We measured the extraordinary-wave beam-coupling gain of the blue $BaTiO_3$:Rh crystal #1o34 at 840 nm, for several grating vectors parallel to the c-axis, using the setup shown in Fig. 9. The intensity of the pump beam W_1 was 550 mW/cm^2 and the intensity of the probe beam W_2 was 71 mW/cm^2. Both beams were controlled by shutters. The crystal under test was placed in an oil-filled cuvette to minimize heating by absorbed laser light. The K-vector of the grating was kept parallel to the c-axis of the crystal at all times, and the grating period was varied by moving a symmetrical pair of mirrors. To ensure that the two infrared writing beams were mutually aligned, they were first both adjusted to shine through a pinhole adjacent to the crystal. After this was done, the pinhole, cuvette and crystal were moved sideways together on a micrometer-driven translation stage so that the pinhole was replaced by the crystal, and the writing beams intersected accurately within the crystal. In each beam-coupling run, the detected intensity was first allowed to stabilize with beam W_2 on and W_1 off. After this occurred, W_1 (the pump beam) was turned on and the time evolution of the detected intensity was monitored using a silicon photodiode.

Fig. 10 shows the time evolution of the transmitted probe beam in a representative beam coupling measurement using an extraordinarily polarized 830 nm wavelength beam from a laser diode. The dashed curve in Fig. 10 is a theoretical fit to the

Element	PPM by weight, undoped $BaTiO_3$ crystal 2oF#4	PPM by weight, blue $BaTiO_3$:Rh crystal 2oF#18	Ratio: ppmw after Rh doping/ ppmw before Rh doping
Na	≤ 25	≤ 25	~1
Mg	≤ 25	≤ 25	~1
Al	0.3	10	30
Si	≤ 3	≤ 3	~ 1
S	1	1	1
Cl	0.8	0.8	1
Ca	30	6	0.2
V	0.2	<0.2	~ 1
Fe	0.6	1	1.7
Ni	$\leq$0.7	0.7	~ 1
Cu	0.3	0.3	1
Sr	>2000	>2000	uncertain
Rh	0.5	2	4
La	$\leq$0.9	$\leq$0.9	~ 1
Pb	0.5	<0.5	~ 1

Table 2: Measured impurity concentrations by weight in undoped $BaTiO_3$ crystal 2oF#4 and blue $BaTiO_3$ crystal 2oF#18 (impurities present at levels ≥ 0.2 ppmw, as measured by spark-source mass spectroscopy analysis). The third column shows ratios of detected impurity concentrations before and after doping. The high level of Al is due to the alumina seed rod making contact with the melt. Infrared sensitive blue crystals have been reproduced by adding Pt-Rh wire to the melt without allowing this Al contamination to occur.

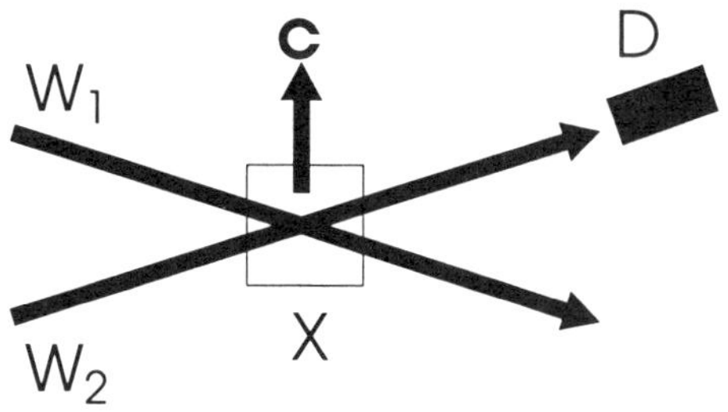

Figure 9: Setup for measuring the infrared beam coupling response of $BaTiO_3$:Rh at 840 nm. Legend: W_1 and W_2 are 840 nm writing beams, D is a detector, X is the $BaTiO_3$ crystal, and C is an oil-filled cuvette at room temperature.

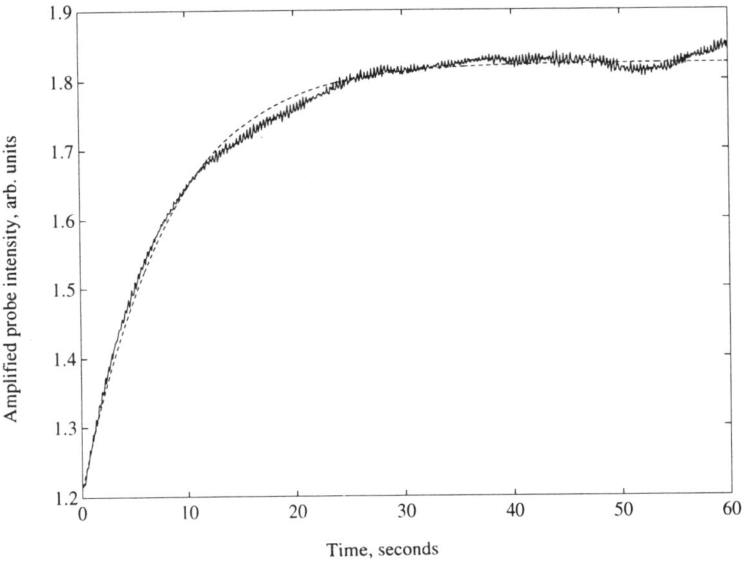

Figure 10: Extraordinary-wave beam coupling time response of as-grown blue $BaTiO_3$:Rh crystal #1o34 at a wavelength of 830 nm, a spatially-averaged intensity of 620 mW/cm^2 and a grating period of 1.2 μm (solid curve). Dashed curve: least-squares theoretical fit yielding time constant $\tau = 7.5$ s and gain $\gamma = 3.2$ cm^{-1}. The crystal thickness is 1.2 mm.

derived form of the beam-coupling response [40, 41]

$$I(t) = I_0 \exp\left(\gamma d[1 - \exp(-t/\tau)]\right), \tag{1}$$

where I_0 is the detected intensity before the pump is turned on, the interaction length $d = L/\cos\theta_i$, L is the crystal thickness, θ_i is the half-angle between the beams inside the crystal, and the gain γ and rise time τ are parameters determined by the least-squares fit between the measured response and Eq. 1. The beam coupling gain of our blue crystal as a function of grating vector was measured by repeating this experiment, varying the angle between the writing beams. Shown in Fig. 11 is the dependence of the beam-coupling gain on the grating wavevector. At a spatially-averaged intensity of 620 mW/cm^2, extraordinary-polarization beam coupling gain (i.e., the second-smallest electrooptic coefficient, r_{33}, was used) was $\sim$ 3.2 cm^{-1}. The response time was $\sim$ 7.5 s, which is much faster than had been previously reported for beam coupling in doped or undoped $BaTiO_3$ in the infrared at similar intensities. Optimizing the Rh doping level and reduction treatment should allow further improvements in speed beyond that measured to date.

4. SUMMARY

To summarize our work on fixing to date, the fixing process used and the time-responses we obtained were similar to those reported by earlier workers on fixing

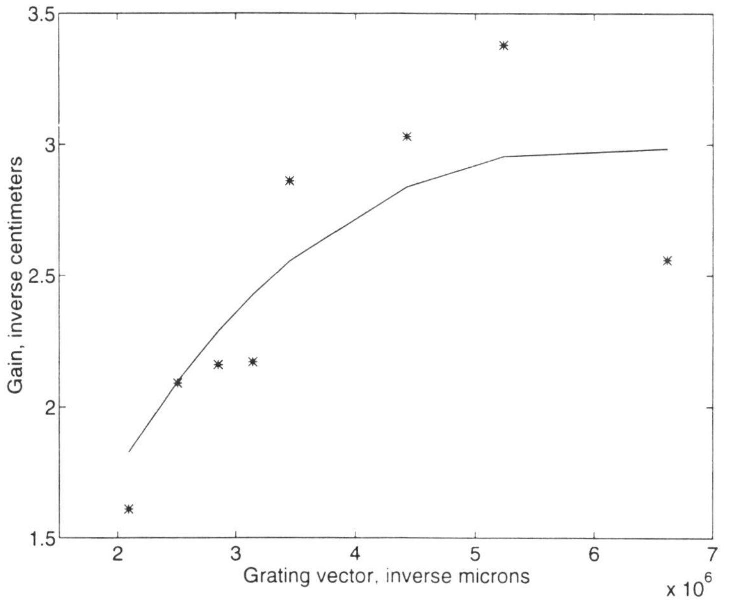

Figure 11: Preliminary beam coupling gain data for blue crystal at 850 nm, as a function of grating period. Also shown is a least squares curve fit to the function [2] $\gamma(k) = 2\gamma_0 k_0 k/(k^2 + k_0^2)$, where $\gamma_0 = 3.0\ \text{cm}^{-1}$ is the peak gain on the theoretical fit curve, $k_0 = 6.15\ \text{cm}^{-1}$, and k is the grating vector.

by thermal cycling of $BaTiO_3$ [11, 20] and other materials [13, 36, 37, 17, 18, 19]. We have extended the results of Kirillov and Feinberg [20] by (1) finding a correlation between specific dopants (Fe and Ni) and strong fixable complementary grating response, (2) showing that the diffraction efficiency is strongly temperature dependent, and (3) finding that proper crystal orientation can greatly increase the diffraction efficiency of the fixed gratings at the low temperatures necessary for long-term fixing.

Results of our infrared beam-coupling measurements show that Rh can be considered the most effective dopant discovered to date for enhancing the infrared sensitivity of $BaTiO_3$. However, it must be emphasized that the mechanism for the increased photoconductivity observed in Rh-doped crystals is not yet understood on the microscopic level and interactions with other defects or impurities might play an important role in explaining the success of this doping procedure.

Practical miniature photorefractive high-density optical data storage systems will require crystals with a combination of good infrared sensitivity and strong fixable secondary gratings. Our objective is therefore demonstration of fixing in blue $BaTiO_3$ crystals. Unfortunately our as-grown blue crystals do not show any fixing. Reducing blue $BaTiO_3$:Rh crystals or annealing them in a water-vapor atmosphere to incorporate mobile hydrogen ions are two possibilities which we are pursuing.

5. ACKNOWLEDGEMENTS

The authors would like to thank Dr. M.H. Garrett for relating to them his initial observation of blue $BaTiO_3$ crystals. This research was supported in part by the Defense Advanced Research Projects Agency through the National Center for Integrated Photonic Technology under contract MDA 972-90-C-0037.

6. REFERENCES

References

[1] F.S. Chen, J.T. LaMacchia, and D.B. Fraser, "Holographic storage in lithium niobate," Appl. Phys. Lett. **13**, 223 (1968).

[2] P. Günter, "Holography, coherent light amplification, and optical phase conjugation in photorefractive materials," Phys. Reports **93**, 199 (1982).

[3] D.L. Staebler, "Ferroelectric Crystals," in *Holographic Recording Materials*, H. Smith, ed., Springer-Verlag, Berlin, 1977.

[4] F.H. Mok, "Angle-multiplexed storage of 5000 holograms in lithium niobate," Opt. Lett. **18**, 915 (1993).

[5] R.G. Zech, "Volume hologram optical memories: mass storage future perfect?" Optics & Photonics News, August, 1992, p. 16.

[6] H. Kogelnik, "Coupled wave theory for thick hologram gratings," Bell Syst. Tech. J. **48**, 2909 (1969).

[7] S. Redfield and L. Hesselink, "Enhanced nondestructive holographic readout in strontium barium niobate," Opt. Lett. **13**, 880 (1988).

[8] D. Brady, K. Hsu, D. Psaltis, "Periodically refreshed multiply exposed photorefractive holograms," Opt. Lett. **15**, 817 (1990).

[9] H. Sasaki, Y. Fainman, J.E. Ford, Y. Taketomi, and S.H. Lee, "Dynamic photorefractive optical memory," Opt. Lett. **16**, 1874 (1991).

[10] S. Boj, G. Pauliat, and G. Roosen, "Dynamic holographic memory showing readout, refreshing, and updating capabilities," Opt. Lett. **17**, 438 (1992).

[11] F. Micheron and G. Bismuth, "Electrical control of fixation and erasure of holographic patterns in ferroelectric materials," Appl. Phys. Lett. bf 20, 79 (1972).

[12] F. Micheron and G. Bismuth, "Field and time thresholds for the electrical fixation of holograms recorded in $(Sr_{0.75}Ba_{0.25})Nb_2O_6$ crystals," Appl. Phys. Lett. **23**, 71 (1973).

[13] J.J. Amodei and D.L. Staebler, "Holographic pattern fixing in electro-optic crystals," Appl. Phys. Lett. **18**, 540 (1971).

[14] H. Vormannn, G. Weber, S. Kapphan and E. Kratzig, "Hydrogen as origin of thermal fixing in $LiNbO_3$:Fe," Solid State Commun. **40**, 543 (1981).

[15] W. Bollmann and H.-J. Stöhr, "Incorporation and mobility of OH^- ions in $LiNbO_3$ crystals," Phys. Stat. Sol. (a) **39**, 477 (1977).

[16] Y. Qiao, S. Orlov, D. Psaltis, and R. Neurgaonkar, "Electrical fixing of photorefractive holograms in $Sr_{0.75}Ba_{0.25}Nb_2O_6$," Opt. Lett. **18**, 1004 (1993).

[17] L. Arizmendi, "Thermal fixing of holographic gratings in $Bi_{12}SiO_{20}$," J. Appl. Phys. **65**, 423 (1989).

[18] G. Montemezzani, G. and P. Günter, "Thermal hologram fixing in pure and doped $KNbO_3$," J. Opt. Soc. Am. B **7**, 2323 (1990).

[19] G. Montemezzani, M. Zgonik and P. Günter, "Photorefractive charge compensation at elevated temperatures and application to $KNbO_3$," J. Opt. Soc. Am. B **10**, 171 (1993).

[20] Kirillov, D. and J. Feinberg, "Fixable complementary gratings in photorefractive $BaTiO_3$," Opt. Lett. **16**, 1520 (1991).

[21] A. Kewitsch, M. Segev, A. Yariv and R. Neurgaonkar, "Selective page-addressable fixing of volume holograms in $Sr_{0.75}Ba_{0.25}Nb_2O_6$ crystals," Opt. Lett. **18**, 1262 (1993).

[22] M. Cronin-Golomb and K.Y. Lau, "Infrared photorefractive passive phase conjugation with $BaTiO_3$: demonstrations with GaAlAs and 1.09-μm and Ar^+ lasers," Appl. Phys. Lett. **47**, 567 (1985).

[23] B.T. Anderson, P.R. Forman, and F.C. Jahoda, "Self-pumped phase conjugation in $BaTiO_3$ at 1.06 μm," Opt. Lett. **10**, 627 (1985).

[24] H.Y. Zhang, X.H. He and Y.H. Shih, "Two-wave coupling and self-pumped phase conjugation at near infrared wavelengths in $BaTiO_3$," Opt. Commun. **88**, 424 (1992).

[25] J. Feinberg, "Continuous-wave self-pumped phase conjugator with wide field of view," Opt. Lett. **8**, 480 (1983).

[26] D. Rytz, R.R. Stephens, B.A. Wechsler, M.S. Keirstead and T.M. Baer, "Efficient self-pumped phase conjugation at near-infrared wavelengths using cobalt-doped $BaTiO_3$," Opt. Lett. **15**, 1279 (1990).

[27] M. Cronin-Golomb, A. Yariv, and I. Ury, "Coherent coupling of diode lasers by phase conjugation," Appl. Phys. Lett. **48**, 1240 (1986).

[28] G.W. Ross, Petr Hribek, R.W. , M.H. Garrett, and D. Rytz, "Impurity-enhanced self-pumped phase conjugation at near infrared wavelengths using "blue" $BaTiO_3$," CLEO 1993, Paper CFC4.

[29] S. MacCormack and J. Feinberg, "High-brightness output from a laser-diode array coupled to a phase-conjugating mirror," Opt. Lett. **18**, 211 (1993).

[30] W.R. Christian, P.H. Beckwith, and I. McMichael, "Energy transfer between injection-locked single-mode diode lasers by two-beam coupling in $BaTiO_3$," Opt. Lett. **14**, 81 (1989).

[31] P. Lambelet, M. Glick, R.P. Salathé, and D. Rytz, "Novel configuration of two-wave mixing of laser-diode radiation in cobalt-doped $BaTiO_3$," IEEE Phot. Tech. Lett. **4**, 1041 (1992).

[32] H.-K. Liu and L.-J. Cheng, "Infrared predection dynamic range compression via photorefractive crystals," Appl. Opt. **27**, 1006 (1988).

[33] P. H. Beckwith and W. R. Christian, "Two-wave and phase conjugation at 830 nm in $BaTiO_3$." Opt. Lett. 14, 642 (1989).

[34] V. Belruss, J. Kalnajs, A. Linz, and R.C. Folweiler, "Top-seeded solution growth of oxide crystals from non-stoichiometric melts," Mat. Res. Bull. **6**, 899 (1971).

[35] C. Gu and P. Yeh, "Diffraction properties of fixed gratings in photorefractive media," J. Opt. Soc. B **7**, 2339 (1990).

[36] R. Sommerfeldt, R.A. Rupp, H. Vormann, E. Krätzig, "Thermal fixing of volume phase holograms in $LiNbO_3$:Cu," Phys. Stat. Sol. (a) **99**, K15 (1987).

[37] P. Hertel, K.H. Ringhofer, and R. Sommerfeldt, "Theory of thermal hologram fixing and application to $LiNbO_3$:Cu," Phys. Stat. Sol. (a) **104**, 855 (1987).

[38] S.H. Wemple, M. Didomenico, Jr., and I. Camlibel, "Dielectric and optical properties of melt-grown $BaTiO_3$," J. Phys. Chem. Solids **29**, 1797 (1968).

[39] M.H. Garrett, J.Y. Chang, H.P. Jenssen, and C. Warde, "A method for poling barium titanate, $BaTiO_3$," Ferroelectrics (1991).

[40] R.S. Hathcock, "Optical and photorefractive properties of iron-doped barium titanate," Ph.D. Dissertation, Massachusetts Institute of Technology, 1989.

[41] D.W. Vahey, "A nonlinear coupled-wave theory of holographic storage in ferroelectric materials," J. Appl. Phys. **46**, 3510 (1975).

Applications I

Holographic storage and interconnection using a (Ce:Fe: doped $LiNbO_3$) photorefractive crystal fiber with a tunable visible-light diode laser

Francis T. S. Yu and Shizhuo Yin

Department of Electrical and Computer Engineering
The Pennsylvania State University
University Park, PA 16802

ABSTRACT

In this paper, we shall briefly review some recent works done at the Pennsylvania State University on the photorefractive (PR) fibers and their applications. Firstly, a laser heated pedestal growth (LHPG) technique to grow single crystal PR fiber is briefly described. A technique of constructing phase conjugate PR fiber hologram is introduced. Angular and wavelength selectivities for the fiber holograms are evaluated, in which we have shown that the reflection-type wavelength-multiplexed fiber hologram offers a uniform and higher selectivity, as in contrast with the transmission-type. Cross-talk noise for reflection-type wavelength-multiplexed fiber hologram is also calculated, in which we have shown that the noise level would be lower by using a narrower linewidth of the light source. An application of the PR fiber hologram to optical interconnection is advocated, for which a 5-dimensional holographic memory is addressed. Finally, a simple experimental demonstration using a Ce:Fe: doped $LiNbO_3$ PR fiber with a tunable visible-light diode laser is given, in which we have expressed that the potential application to large capacity holographic memory, reconfigurable interconnects, and many others are possible.

1. INTRODUCTION

The use of photorefractive (PR) materials for data storage and interconnection has attracted a myriad of applications to real-time optical signal processing and computing.[1-8] The salient features of PR material must be the large storage capacity, high diffraction efficiency, and the high angular and wavelength selectivities. The growth of PR fibers[4,9-11] in recent years has also provided an alternative to a variety of applications. Although there are many good works in this area, due to page limitation, we shall concentrate our discussion on the recent works done at the Pennsylvania State University. Firstly, we shall briefly introduce a LHPG technique for single crystal fiber fabrication. Secondly, a phase conjugate fiber holographic construction architecture is presented, which has the advantage of minimizing the intra-modal distortion noise.[12] Angular and wavelength selectivities of PR fiber hologram, using coupled wave theory of Kogelnik,[13] are provided, in which the calculated results are consistent with the experimental data as reported by Hesselink et al.[7] Since the reflection-type wavelength-multiplexed hologram offers a higher and uniform selectivity,[1] we shall use this type of fiber hologram for our experimental demonstration. Cross-talk noise of the fiber hologram among the fiber channels is also evaluated, in which we have seen that the narrower the laser linewidth the lower the cross-talk would be. Reconfigurable interconnection network using fiber bundle hologram is proposed, for which a 5-dimensional addressable holographic memory can be, in principle, constructed . Finally, an experimental demonstration of the wavelength-multiplexed reflection-type PR fiber hologram using a visible-light tunable diode laser is provided, in which we have shown the potential applications to large capacity holographic memory, switching and reconfigurable interconnects.

2. SINGLE CRYSTAL PR FIBER GROWTH

The Laser Heated Pedestal Growth (LHPG) technique is a relatively new crystal fiber growth technology that can fabricate 50 - 1000 μm diameter and up to 5-10 cm long fibers.[14,15] The groups at the Stanford University and the Pennsylvania State University have independently grown PR fibers by using this technique.[7,9] The major advantages of growing single crystal fiber using LHPG system are:

[1] It does not require any holding device (e.g., a crucible), by which the contamination from the holder can be avoided.

[2] Some growing limitations, such as the melting point, shape, etc. can be alleviated.

[3] Since the LHPG system uses CO_2 laser as the heating source, the growing process is much easier to control. For example, the laser beam has a critical taper form, with a annular section so that the crystal fiber will only be grown in the vertical direction, which is crucial for fiber growing.

[4] The LHPG technique can also be used in a variety of applications, for example, as applied to high temperature superconductor wire-rod and other novel materials.

A LHPG system, constructed in the Material Research Laboratory (MRL) at the Pennsylvania State University, is depicted in Fig. 1 with a schematic diagram described in Fig. 2. The growth chamber is shown in Fig. 3, in which a 0.9x0.9 mm (Ce: Fe: $LiNbO_3$) preform is placed in a holder that maintains a precise vertical orientation. A 50 watt CO_2 laser is brought to focus by first forming an annular beam with a reflaxicon. Then by steering the annular beam with a 45° flat mirror, a highly intensive focusing beam can be formed using the concave parabolic mirror on the PR rod (preform) as shown in Fig. 3. A seed rod is brought in contact with the molten tip of the preform, then fiber growth can be initiated by slowly pulling the seed rod upward with a computer control driver, as shown in Fig. 4

3. PHASE CONJUGATE FIBER HOLOGRAM

We now describe a method for recording an axial hologram in a photorefractive fiber.[12] A linearly polarized monochromatic plane wavelength is coupled into a photorefractive fiber using a condensing lens. A cone of light beams having different polarizations[16], caused by the anisotropism and the Fresnel reflection inside the fiber, exits from the other end of the fiber. Thus, as the light beam propagates along the fiber, the spatial information of the incident light beam is scrambled on the exiting end of the fiber. If the incident light beam, called an object beam, interferes with a reference beam which is coupled into the opposite end of the fiber, a hologram (similar to that of a reflection hologram) can be recorded in the nonlinear photorefractive material of the fiber. However, the object beam cannot be read out by merely using the reference beam. If the recording is made using a conjugate reference beam, which can be derived with a polarization preserving conjugating mirror, as shown in Fig. 5, a transmission type hologram can be recorded in the photorefractive fiber. Notice that the writing reference beam R can be made to exceed the coherence length of the object beam O, in which R* is coherent with O. In this manner, it is apparent the spatial information of the object beam can be read out using the reference beam R.

Figure 5 shows a schematic diagram of the hologram recording system. The object transparency is illuminated by a monochromatic plane wave that is coupled into the PR fiber by an imaging lens and forms the object beam O that propagates along the fiber. The reference beam R which is incoherent with the object beam O is coupled into the other end of the fiber. A polarization preserving phase conjugate mirror is used to generate the phase conjugate reference beam R* that is within the coherence length of the object beam, Then a transmission type hologram formed by the object beam and the phase conjugate reference beam inside the fiber can be obtained, which is expressed as

$$H = |O|^2 + |R|^2 + O^*R^* + OR. \qquad (1)$$

Note that if the c-axis of the PR crystal is along the axis of the fiber, either the reference beam R or the object beam O should be off-axis.

Now, if the recorded photorefractive fiber is illuminated by the reference beam R, the phase conjugate object O* will be reconstructed as shown by the dashed lines in Fig. 5. Notice that in the readout process, no phase conjugate mirror is needed.

4. ANGULAR AND WAVELENGTH SELECTIVITIES

By referring to the coupled wave theory,[13] the normalized diffraction efficiency η for an unslanted (transmission-type and reflection-type) holograms can be expressed as

$$\eta_t = \frac{\mathrm{Sin}^2 (v_t^2 + \xi_t^2)^{1/2}}{1 + \frac{\xi_t^2}{v_t^2}}, \tag{2}$$

$$\eta_r = \frac{1}{1 + \frac{1 - \xi_r^2 / v_r^2}{\mathrm{Sh}^2 (v_r^2 - \xi_r^2)^{1/2}}}, \tag{3}$$

where $v_t = \frac{\pi \Delta n d}{\lambda \mathrm{Cos}\theta}$, $\xi_t = \frac{2\pi n d \mathrm{Sin}\theta}{\lambda} \Delta\theta$,

$v_r = \frac{\pi \Delta n d}{\lambda \mathrm{Sin}\theta}$, $\xi_r = \frac{2\pi n d \cos\theta}{\lambda} \Delta\theta$,

subscript t and r denote the transmission-type and the reflection-type holograms, respectively, Δn and n are the refractive index and the amplitude of its variation, d is the length of the fiber, and θ and $\Delta\theta$ are the internal construction angle and its variation. Notice that for weak coupling condition, i.e., $|v| << |\xi|$, η can be approximated as

$$\eta_t \approx v_t^2 \mathrm{Sinc}^2 \xi_t, \tag{4}$$

$$\eta_r \approx v_r^2 \mathrm{Sinc}^2 \xi_r. \tag{5}$$

For example, The typical values of n and Δn for a $LiNbO_3$ crystal would be n = 2.28 and $\Delta n_{max} = 10^{-3} - 10^{-5}$.[17] Since Δn (for each multiplexed hologram) is sufficiently smaller than Δn_{max}, Eqs. (4) and (5) would be good estimations for PR fiber hologram construction under the weak coupling assumption.

We note that the angular selectivity is inversely related to the full width of the Sinc factor, the internal (write-in) angular selectivity would be

$$\Delta\theta_t = \frac{\lambda}{\mathrm{nd}\,\mathrm{Sin}\theta}, \quad \Delta\theta_r = \frac{\lambda}{\mathrm{nd}\,\mathrm{Cos}\theta}. \tag{6}$$

By applying the Snell's law of refraction (see Fig. 6), such as

$$\mathrm{Sin}\alpha = \mathrm{n}\,\mathrm{Sin}\theta, \tag{7}$$

$$\mathrm{Cos}\alpha = \mathrm{n}\,\mathrm{Cos}\theta, \tag{8}$$

where Eqs.. (7) and (8) are for transmission-type and reflection-type holograms, respectively, the angular selectivity (in radian) expressed in terms of the external write-in angle α can be written as[1-2]

$$\left\{\frac{1}{\Delta\alpha}\right\}_t = \frac{\mathrm{Sin}\alpha\mathrm{Cos}\alpha}{\sqrt{\mathrm{n}^2 - \mathrm{Sin}^2\alpha}}\frac{\mathrm{d}}{\lambda}, \tag{9}$$

$$\left\{\frac{1}{\Delta\alpha}\right\}_r = \frac{\mathrm{Sin}\alpha\mathrm{Cos}\alpha}{\sqrt{\mathrm{n}^2 - \mathrm{Cos}^2\alpha}}\frac{\mathrm{d}}{\lambda}. \tag{10}$$

To calculate the wavelength selectivity, **k**-vector diagrams are considered as shown in Fig. 7. By referring to the principle of moment conservation, a change in the diffraction beam angle (θ) by $\Delta\theta$ is needed, if the readout wavelength is deviated from the recording wavelength by $\Delta\lambda$. In view of Fig. 7(a), we have

$$(|\mathbf{k}+\Delta\mathbf{k}|)\mathrm{Sin}\theta + (|\mathbf{k}+\Delta\mathbf{k}|)\mathrm{Sin}(\theta-\Delta\theta) = 2|\mathbf{k}|\mathrm{Sin}\theta, \tag{11}$$

where $|\mathbf{k}| = 2\pi\mathrm{n}/\lambda$. Thus, the wavelength selectivity for the transmission-type fiber hologram (in terms of the internal write-in angle) can be written as

$$\left|\frac{\lambda}{\Delta\lambda}\right| = 1 + \frac{2}{\dfrac{\mathrm{Sin}\theta}{\mathrm{Sin}(\theta-\Delta\theta)} - 1}. \tag{12}$$

It is apparent that, by referring to Eqs..(6)-(8), the wavelength selectivity for the transmission-type is given by

$$\left\{\left|\frac{\lambda}{\Delta\lambda}\right|\right\}_t = 1 + \frac{2}{\dfrac{\mathrm{Sin}\alpha/\mathrm{n}}{\mathrm{Sin}[\mathrm{Sin}^{-1}(\frac{\mathrm{Sin}\alpha}{\mathrm{n}}) - \frac{\lambda}{\mathrm{dSin}\alpha}]} - 1}, \tag{13}$$

which is a more exact solution as compared with the result that we have previously derived.[1]

Similarly, for the reflection-type fiber hologram, as represented by the **k**-vector diagram in Fig. 7 (b), the wavelength selectivity of the fiber hologram can be written as[1]

$$\left\{\left|\frac{\lambda}{\Delta\lambda}\right|\right\}_r = \frac{\sqrt{n^2 - \mathrm{Cos}^2\alpha}}{\lambda} d. \tag{14}$$

To test the validity of the derived results, a comparison with the experiments obtained by Hesselink[7] is made, as tabulated in Table 1. From this table, we see that the calculated results are somewhat consistent with the experimental data obtained by Hesselink, in which we have seen that the rate of changes is in the same order. The consistency of the calculated results may be explained that, by the assumption of two light beams interact along the fiber (due to internal reflections) are behaviors in a similar manner as the light beams interacted along the thickness of the crystal.

The angular and the wavelength selectivities for the transmission-type and the reflection-type PR fiber holograms, as a function of external write-in angle, are shown in Figs. 8 and 9, respectively, in which we have seen that both the transmission-type and the reflection-type hologram offers higher selectivity as the fiber length increases. From these figures, we have also noted that there is a limited range of angular selectivity, i.e., $15^\circ \leq \alpha \leq 75^\circ$ (FWHM) for both the transmission and reflection fiber holograms. On the other hand, the wavelength selectivity for the reflection-type hologram is somewhat higher and more uniform distributed over the entire range of recording angles. Thus we would state that the reflection-type wavelength-multiplexed PR fiber hologram would be a better choice for holographic data storage. Recently, Rakuljic et al[18] and Curtis et al[19] have subsequently have shown that the cross-talk in the wavelength multiplexed volume holograms are not as severe as in the angular multiplexed holograms. In order to have a feeling of number, we assume that the length of PR fiber d = 7 mm, the construction wavelength $\lambda = 0.7\ \mu m$, then the wavelength selectivity for a reflection-type fiber hologram $\{\lambda/\Delta\lambda\}_r$ would be in the order of 10^4.

5. CROSS-TALK NOISE

It is well known that the upper bound storage capacity of an volume hologram can be estimated as V/λ^3 (where V is the volume of the recording material and λ is the wavelength), which is as high as 10^{12} bits for a 1 cm^3 cube.[20] However, there are some obvious limitations that prevent it to achieve this goal in practice. The cross-talk noise is one of the major limiting factors among them. Mention must be made that cross-talk noise for angular multiplexed hologram has been subsequently discussed by Ramberg[21] and Gu et al,[22] in which they have shown that the level of cross-talk noise is proportional to the number of stored patterns. In a more recent paper by Gu et al,[23] the effect due to the wavelength spread is evaluated, in which they have shown that the level of cross-talk noise will approach to a constant when the number of stored patterns is large enough. However, their study is based on the assumption of infinitesimal linewidth of the reading beam. In this section, we shall evaluate the cross-talk effect due to the wavelength spread among the multiplexed hologram channels. Since the reflection-type fiber hologram has been shown to be a better choice for the fiber holographic process, we shall focus our attention primarily on this type of fiber hologram as follows:

If we regard the reconstructed hologram images from a particular multiplexed channel (j) as signal and the intrusion (into this channel) due to wavelength spread as noise, then the signal-to-noise ratio (SNR_{CT}) due to cross-talk can be shown as[24]

$$SNR_{CT}^{(j)} = \frac{\int \text{rect}\left(\frac{\Delta\lambda}{\Delta\lambda_1}\right)\eta d(\Delta\lambda)}{\sum\limits_{\substack{m=-(M-j) \\ m\neq j}}^{M+j} \int \text{rect}\left(\frac{\Delta\lambda - m\delta\lambda}{\Delta\lambda_1}\right)\eta d(\Delta\lambda)}, \tag{15}$$

where we have assumed that a total number of 2M+1 holographic channels is used, $\delta\lambda$ is the separation of holographic channels, and $\Delta\lambda_1$ is the spectral width of the light source. For simplicity, we assume that recording linewidth is negligibly small, by which the evaluation of the cross-talk effect would be primarily due to the reconstruction linewidth. Notice that similar result can also be obtained when the recording linewidth is not negligibly small. In order to minimizing the cross-talk noise between the holographic channels, we assume that the multiplexed channel separation ($\delta\lambda$) is equal to the spectral bandwidth of the channels $\Delta\lambda_{min}$, that is

$$\delta\lambda = \Delta\lambda_{min}.$$

To have a feeling of magnitude, Let the modulated refractive index be $\Delta n = 2x10^{-5}$, the length of the fiber be $L = 5$ mm, the central wavelength be $\lambda_o = 670$ nm, the wavelength-multiplexed channel separation be $\delta\lambda = \Delta\lambda_{min} = 0.02$ nm, and the total number of multiplexed holographic channels be 2M+1 = 501, then the $SNR_{CT}^{(j)}$ as a function of channel number j for $\Delta\lambda_1 = 1.5x10^{-4}$ nm (~ 100

MHz) and $\Delta\lambda_1 = 7.5\text{x}10^{-3}$ nm (~ 5 GHz) are plotted in Fig. 10. In view of this figure, we see that the SNR_{CT} are basically uniformly distributed, and the one using a narrower spectral width offers a 10 dB higher SNR_{CT}. Figure 11 shows the SNR_{CT} as a function of the total number of multiplexed holographic images, in which we see that the noise performance is better for using a narrower linewidth, as expected. As the number of stored images increases, the behavior of the SNR_{CT} tends to be independent of storing numbers. The effects on SNR_{CT} due to the spectral width of the light source are also depicted in Fig. 12. As anticipated, the further the channel separation is used, the better the channel performance (i.e., the SNR_{CT}) will be. And we also see that SNR_{CT} decreases rapidly as the linewidth of the reading beam increases.

Since the cross-talk in a reflection-type wavelength multiplexed fiber hologram is determined by the spectral bandwidth of the channel, it in turn determines its storage capacity. For example, if the laser tuning range is assumed to be in the order of 10 to 10^2 nm, the cross-talk-limited storage capacity (C_{CT}) of the fiber hologram can be determined by

$$C_{CT} = \frac{\text{Tuning} - \text{range}}{\text{channel} - \text{separation}}. \qquad (16)$$

Let us assume that $\Delta\lambda_{min}$ is in the order of 10^{-2} to 10^{-3} nm, then C_{CT} would be in the order of 10^3 to 10^5 pages of holographic data with a read-out SNR_{CT} ranging from 10 to 20 dB.

6. OPTICAL INTERCONNECTION USING FIBER HOLOGRAMS

Refer to the angular selectivity, either from Eq. (9) or Eq. (10), we see that the selectivity can be very high as the length of the fiber increases. As an example, if the refractive index of the PR fiber core is $n_{co} = 2.2$, $\lambda = 0.5$ μm, $\alpha = 45°$, and d = 10 cm, then the angular selectivity can be determined such as $\delta\alpha \approx 10^{-5}$ rad. Notice that the diffraction limitation of the fiber is much greater than its angular selectivity due to small aperture of the fiber. Thus the interconnection density is limited by the numerical aperture of the fiber instead of the Bragg diffraction. The numerical aperture of a fiber can be expressed as $\left(n_{co}^2 - n_{cl}^2\right)^{1/2}$, where n_{co}, n_{cl} are the refractive indices of the core and cladding of the fiber.

Because of the high electro-optic coefficient in the PR fiber, the PR fiber can be used as an electro-optic switches as proposed in Fig. 13. In other words, when a PR fiber hologram is constructed with a given electrical field pattern, the recorded spatial information can only be reconstructed by using the same applied field distribution. It is therefore apparent that a multi-channel switches (e.g., a reconfigurable optical interconnection network) can be constructed by simply manipulating the applied electric field patterns on the fiber.

The preceding fiber hologram construction can be extended to multi-channel fiber bundle interconnection as shown in Fig. 14. For instance, different angular illumination of the object transparencies are sequentially imaged into the input-end of the fiber bundle, using a spatial light modulator (SLM) with a lenslet array. In other words, the object transparencies, illuminated at different directions, are coupled into the fiber bundle by using a SLM and a lenslet array interconnection architecture. The main features of this fiber holographic memory organization (FHMO) is that, a spatial light modulator (SLM) is used to divide the object beam space. The illuminating waves are focused by a condensing lens onto the SLM, which is served as a movable sampling pinhole that allows only one of the focusing beams (i.e., Fourier Spectra) to pass at the illuminated object direction. Thus, different object transparencies can be recorded with different illuminating beams by using a specific pinhole by the SLM. Hence, the object beam information is essentially divided into two planes, namely, the object and the SLM planes, which are considered addressable in 4-dimension. As for the reference beam space, the reference beam is derived from a single mode fiber that can provide different angular deviations by simply displacing the fiber. Due to the degeneration of the Bragg diffraction, it is trivial that the reference beam has only one degree of freedom. Thus, when the fiber bundle holograms are read-out, there would have a 5 degrees of freedom, by which the stored data can be randomly accessed. Notice that one can add one more adjustable parameter (wavelength) if a tunable diode laser is used. In the reference beam space, the reading beams can also be generated by using a linear laser diode array. In other words, each laser diode generates a reading beam, which is conjugated with a specific reference beam during the recording process. Thus, at the object beam space, a set of reconstructed images from the fiber holograms would be projected by one of the reference beams. As can be seen, when the pinhole generated by the SLM is set at a specific position, only one image would appear at the object plane. Thus, the stored data can be randomly accessed. This FHMO can also be employed for optical interconnection application. The interconnections between the laser diode array and the object plane can be reconfigured using a programmable SLM.

We note that the FHMO can address massive information storage capability. If the aperture of the object plane is AxA and the size of the SLM is BxB, then the spatial-bandwidth product of FHMO would be $(AB/\lambda f)^2$, where f is the focal length of the condensing lens. Thus, by choosing appropriate values of A, B, and f, a large space-bandwidth product (in the order of 10^{10}) of the PR fiber bundle can be fully utilized. Since the number of reference beams is limited by the spatial multiplexing capacity of the PR fiber, which is in the order of 10^5

order (i.e., the number of the mode in the fiber), the overall information storage capacity can be as high as 10^{15} bits.

In practice, the spatial multiplexing capacity of the PR hologram is limited by the maximum index modulation Δn of the crystal. Accordingly, the number of holograms that can be recorded in a material is given by[25]

$$N = \frac{\Delta n}{n} \frac{d}{\lambda} \frac{\pi}{\eta^{1/2}}, \qquad (17)$$

where η is the diffraction efficiency and d is the effective length of the material. Since the effective length of the fiber is expected to be one or two orders of magnitude longer than the bulk material, the limitation imposed by Eq. (17), on the spatial multiplexing capacity for the PR fiber, would be far more relaxed.

We further note that the proposed FHMO can be used either for optical reconfigurable interconnection or massive memory. One of the major advantages is that the stored data can be accessed randomly at high speed without using any mechanical intervention or beam deflecting devices.

7. EXPERIMENT

For experimental demonstration, we have used a Ce: Fe: doped $LiNbO_3$ crystal fiber to construct a fiber hologram, as shown in Fig. 15. The crystal fiber was grown by LHPG system in the Material Research Laboratory at Penn. State. The fiber growing temperature was about 1260 °C. The doping levels of Ce: Fe: in the PR crystal fiber was measured to be about 300 ppm and 500 ppm, respectively. The annealing temperature for the fiber was set at 700° C for 5 hours in open air environment. The size of the PR fiber used for the experimental demonstration is about 7 mm in length and 0.7 mm in diameter, as depicted in Fig. 16. Notice that the Ce:Fe: doped $LiNbO_3$ PR fiber has a higher photosensitivity in the red-light region, which is suitable for using a low-power visible-light tunable diode laser.[26-27] We also note that the PR fiber has a high curie point at about 1200° C, which is rather robust against the ambient temperature variation.

A 5 mw tunable laser diode (manufactured by New Focus Inc. model 6102, $\lambda \approx 670$ nm, with a tuning range of 12 nm) is used as the light source for the construction of the PR fiber hologram. The object beam is measured to be about 2 mw/cm^2 with a write-in angle of $\alpha \approx 90°$. By using the wavelength tuning step of $\Delta\lambda = 0.1$ nm per step with about 5 second exposure time, a wavelength multiplexed PR fiber hologram is constructed. Needless to mention that, by blocking the object beam and tuning the laser diode, a sequence of holographic images can be reconstructed as depicted (in part) in Fig. 17. In view of these fiber holograms, we have seen the interference fringes which are scanning lines of the monitor while the distortions are primarily due to the non smooth surfaces of the crystal fiber. We believe that these effects can be further improved if the PR fiber has been smoothly polished. At any rate, from these experimental results, the potential application of PR fibers to optical memory, interconnections, switchings, signal processing, and many others are anticipated in the near future.

8. CONCLUDING REMARKS

In summary, we have briefly reviewed some recent works done on PR fibers and their applications at the Pennsylvania State University. We have proposed using a phase conjugate holographic construction architecture to compensate the modal distortion in the PR fiber. We have also illustrated that higher and more uniform selectivity can be achieved by using the reflection-type wavelength multiplexed fiber hologram. Channel cross-talk noise for a wavelength multiplexed fiber hologram has been evaluated, in which we have shown that lower cross-talk performances can be achieved by using a narrower linewidth of the light source. A 5-dimensional holographic addressable memory based on PR fiber bundle is proposed, in which we have shown that an all-optical high speed large-capacity random addressable memory can be realized in practice. The major advantages of the fiber bundle memory organization is that no mechanical intervention or beam deflecting devices are needed. In concluding the paper, we provided a set of fiber holographic images, in which we have demonstrated the feasible applications of the fiber holograms to large capacity memory, interconnections, switchings and many others. In short, PR fiber devices are at the threshold of the technical reality. However, much research remains to be done before it would become wide spread practical applications.

9. ACKNOWLEDGEMENT

The authors wish to acknowledge the support of the Defense Advanced Research Projects Agency through the U.S. Army Research Office under Contract No. DAAL 03-92-G-0328.

10. REFERENCES

1. S. Wu, Q. Song, A. Mayers, D. Gregory, and F. T. S. Yu, Appl. Opt. **29**, 1118 (1990).
2. F. T. S. Yu and S. Jutamulia, *Optical Signal Processing, Computing and Neural Networks*, Wiley-Interscience (1992), Ch. 7.
3. P. A. Yeh, A. E. T. Chiou, and J. Hong, Appl. Opt., **27**, 2093 (1988).

4. L. Hesselink and S. Redfield, Opt. Lett. **13**, 877 (1988).
5. F. Ito and K. Kitayama, Opt. Lett., **17**, 1152 (1992).
6. Y. Qiao and D. Psaltis, Opt. Lett., **17**, 1376 (1992).
7. L. Hesselink, International Journal of Optoelectronics J, **5**, 103 (1990).
8. K. Kitayama and F. Ito, Multidimensional Systems and Signal Processing **2**, 401 (1991).
9. J. K. Yamamoto and A. S. Bhalla, Mater. Res. Bull., **24**, 761 (1989).
10. H. Yoshinaga, K. Kitayama, and H. Oguri, Appl. Phys. Lett. **56**, 1728(1990).
11. Y. Sugiyama, I. Yokohama, K. Kubogera, and S. Yagi, IEEE Photo. Tech. Lett., **3**, 744 (1991).
12. S. Wu, A. Mayers, S. Rajan, and F. T. S. Yu, Appl. Opt. **29**, 1059 (1990).
13. H. Kogelnik, The Bell Sys. Tech. J. **48**, 2909 (1969).
14. R. S. Feigelson, *Crystal Growth of ELectronic Materials*, E. Kalkis ed., P. 127. North-Holland, Amsterdam (1985).
15. M. Saifi, B. Dubois, E. M. Vogel and F. A. Thiel, J. Mater. Res. **1**, 452 (1986).
16. W. A. Gambling, D. N. Payne and H. Matsumura, Appl. Opt. **14**, 1538 (1975).
17. P. Gunter and J.-P. Huiguard, *Photorefractive Materials and Their Applications I*, Springer-Verlag, Berlin (1988), P. 53.
18. G. A. Rakuljic, V. Leyva, and A. Yariv, Opt. Lett. **17**, 1471 (1992).
19. K. Curtis, C. Gu, and D. Psaltis, Opt. Lett. 18, 1001 (1993).
20. P. J. Van Heerden, Appl. Opt. **2**, 393 (1963).
21. E. G. Ramberg, RCA Rev. **33**, 5 (1972).
22. C. Gu, J. Hong, I. McMichael, R. Saxena, and F. Mok, J. Opt. Soc. Am. A, **9**, 1978 (1992).
23. C. Gu and J. Hong, in OSA Annual Meeting Technique Digest Series, **23**, 1992 (OSA, Washington, DC., 1992), paper WT2.
24. F. T. S. Yu, F. Zhao, H. Zhou, and S. Yin, "Cross-talk noise in a wavelength-multiplexed reflection-type photorefractive (PR) fiber hologram," submitted to Opt. Comm.
25. L. Solymer and D. J. Cooke, *Volume Holography and Volume Gratings*, P. 30, Academic Press, 1981.
26. S. Yin and F. T. S. Yu, IEEE Photonics Technology Lett. **5**, 581 (1993).
27. S. Yin and F. T. S. Yu, in OSA Annual Meeting Technical Digest Series, **23**, 1992 (OSA, Washington, D. C., 1992), paper MOO7.

recording angel(2α)	Angular selectivity calculated results (degree)	Angular selectivity experimental data (degree)
0.7	2.41	1.43
3	0.64	0.86
5	0.39	0.57
10	0.19	0.44
20	0.098	0.26

Table 1 calculated and experimental data of the angular selectivity for the fiber hologram $\lambda = 515$ nm $n = 2.28$, $d = 4$ mm

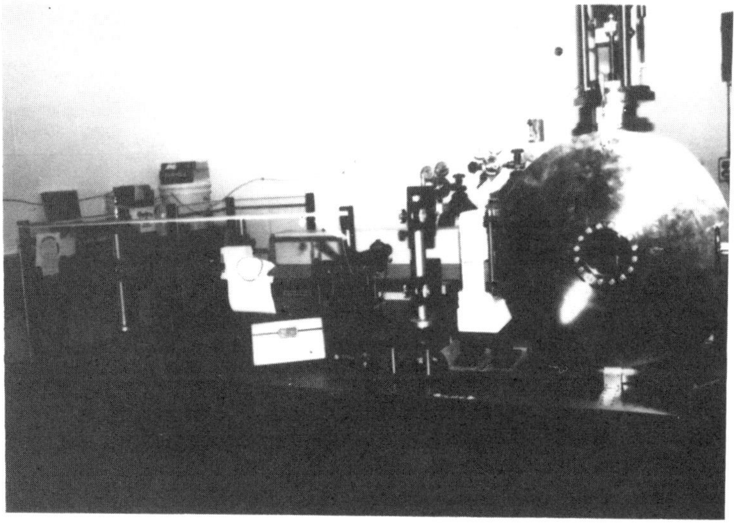

Fig. 1 The Laser Heated Pedestal Growth (LHPG) System

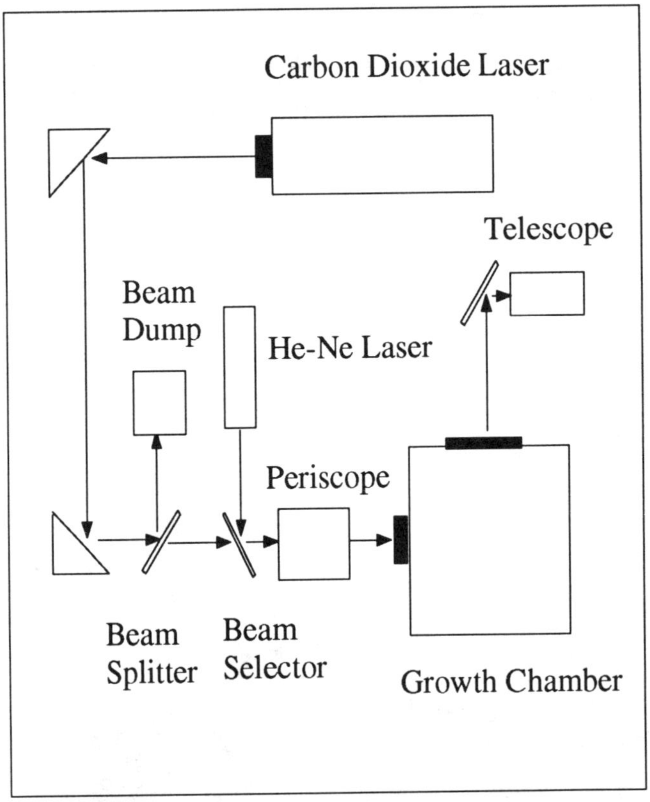

Fig. 2 Schematic diagram of (LHPG) system

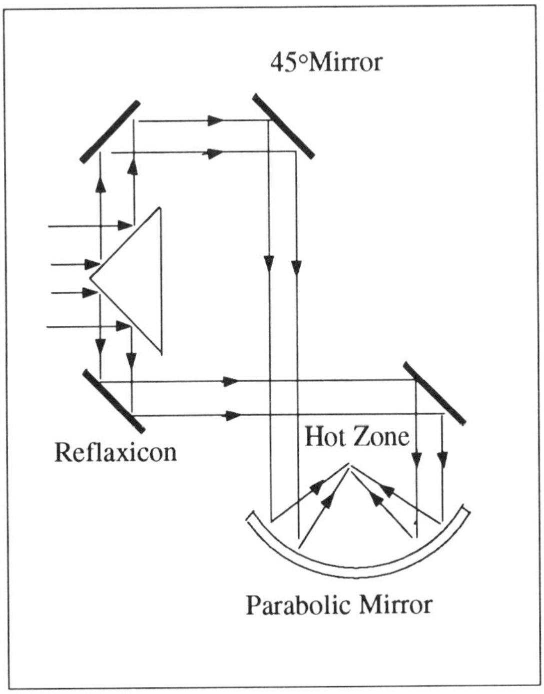

Fig. 3 The growth chamber.

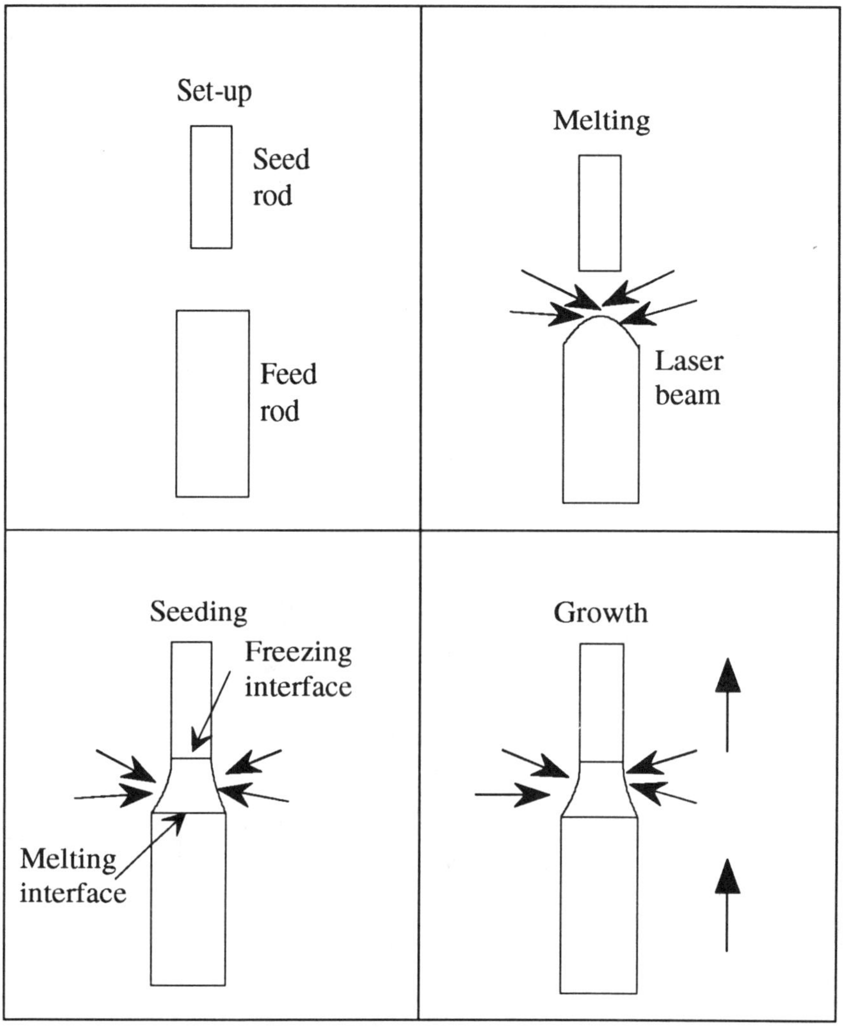

Fig. 4 Schematic diagram of the melt zone

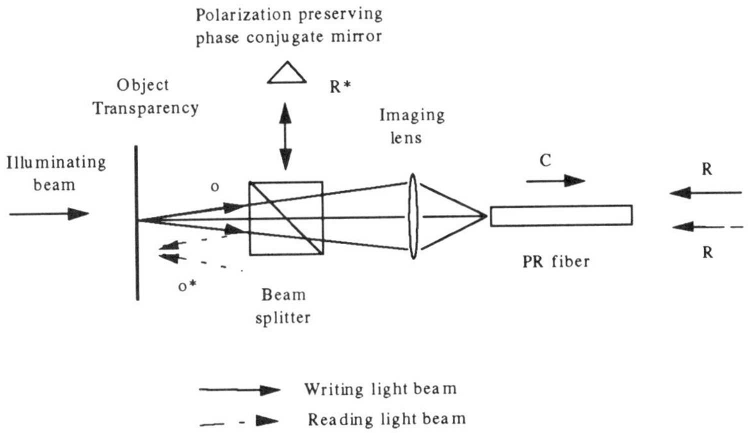

Fig. 5 Construction of a phase conjugate PR fiber hologram

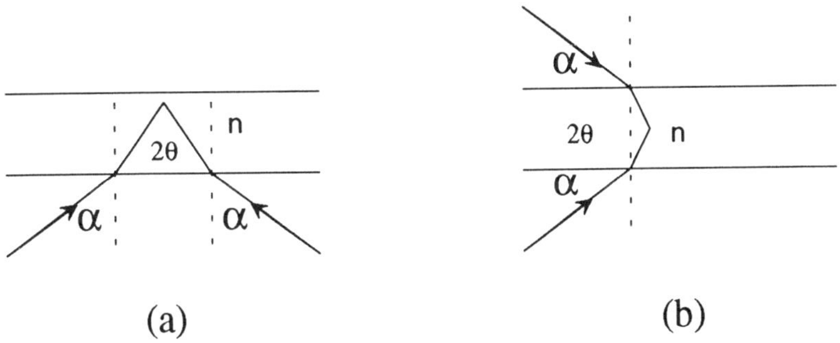

Fig. 6 Unslanted volume holographic construction
a). Transmission-type hologram;
b). Reflection-type hologram.

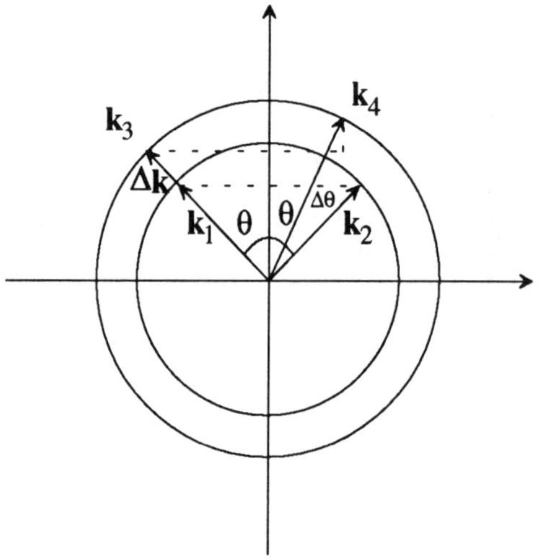

(a)

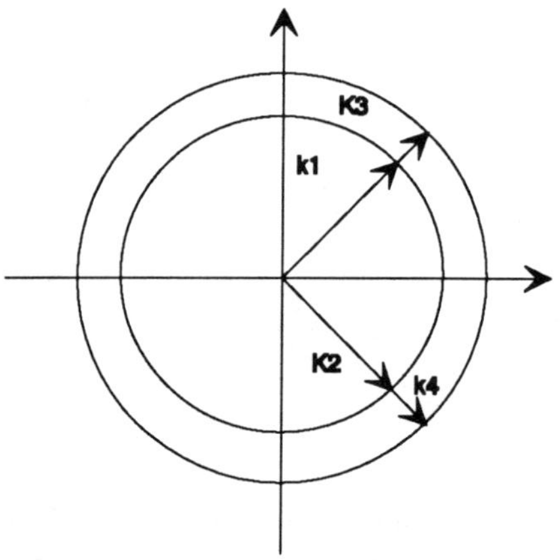

(b)

Fig. 7. Momentum conservation principle. $\mathbf{k}_1$, $\mathbf{k}_2$; writing wave vectors. $\mathbf{k}_3$; readout wave vector. $\mathbf{k}_4$; diffracted wave vector. $\mathbf{\Delta k}$ reading wavelength deviation.
a). for transmission-type hologram,
b) for reflection-type hologram.

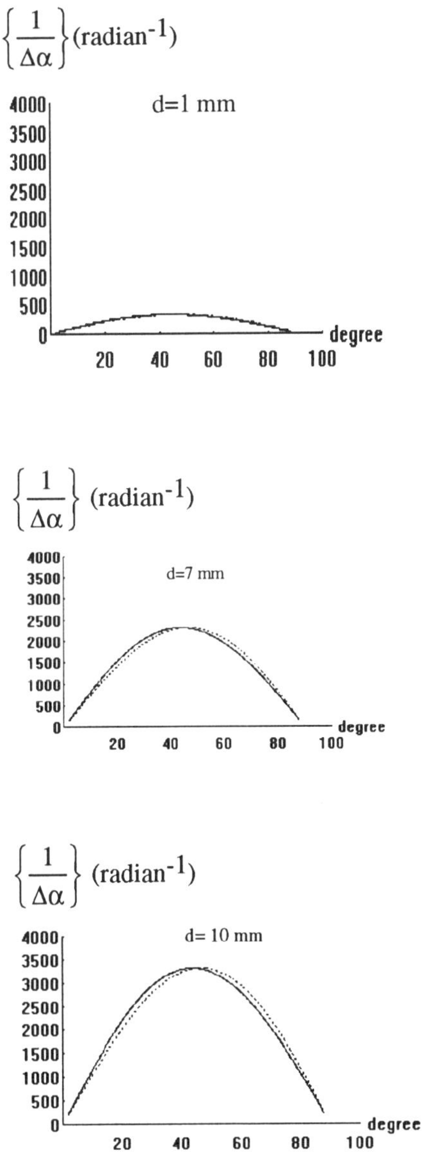

Fig. 8. Angular selectivity (dashed line for transmission-type hologram and solid line for reflection-type hologram), $\lambda = 0.7\ \mu m$

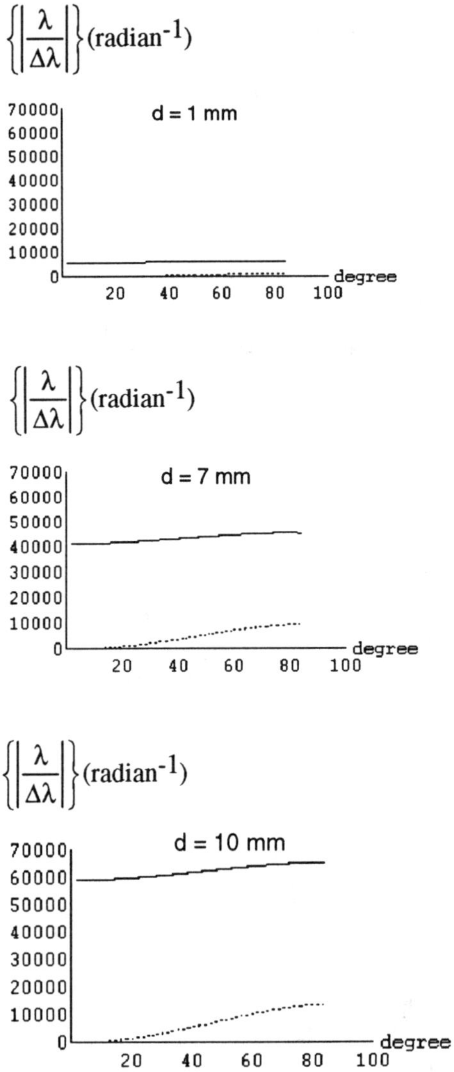

Fig. 9. Wavelength selectivity (dashed line for transmission-type hologram and solid line for reflection-type hologram), $\lambda = 0.7\ \mu m$

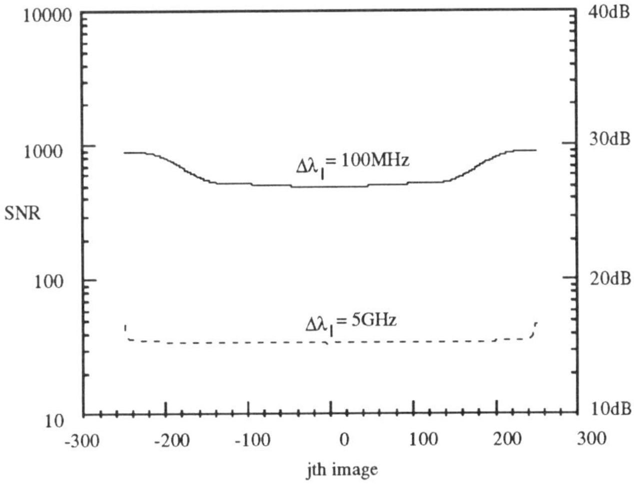

Fig. 10 Distribution of SNR_{CT} versus the jth holographic channel

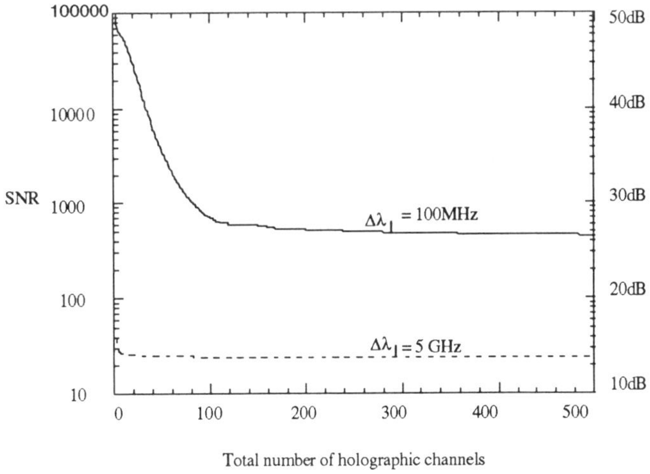

Fig. 11 SNR_{CT} versus the total number of holographic channels

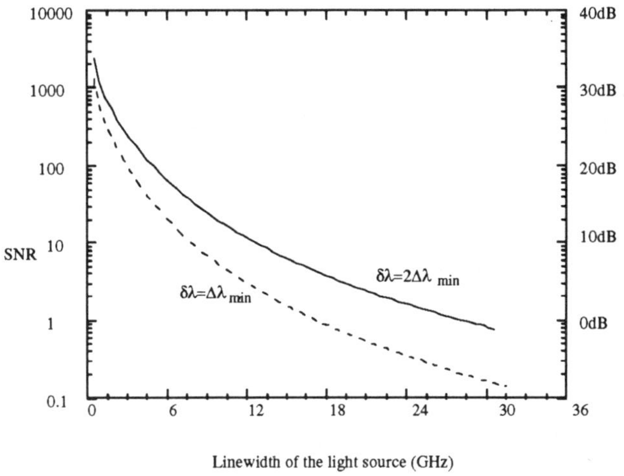

Fig. 12 SNR_{CT} versus the linewidth $\Delta\lambda_1$ for different channel separations.

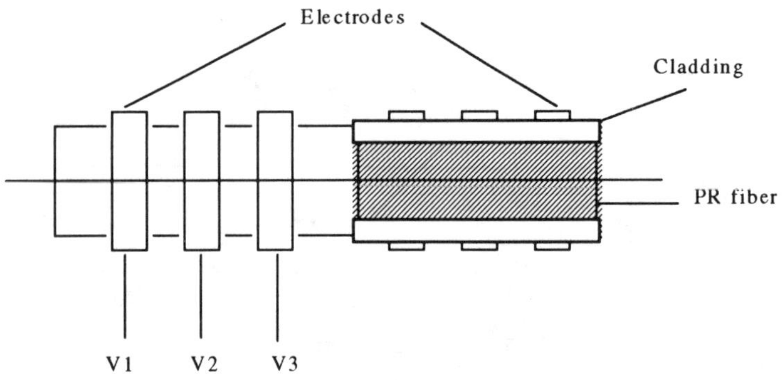

Fig. 13 A PR fiber switches

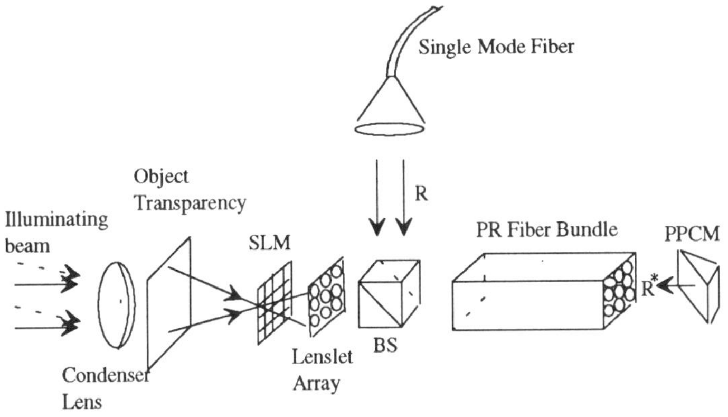

Fig. 14 5 dimensional addressable holographic memory

(a)

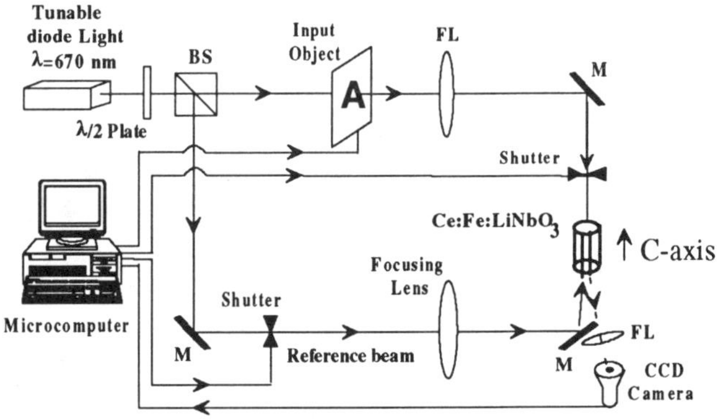

(b)

Fig. 15. Experimental set-up for the construction of a wavelength-multiplexed reflection-type PR crystal fiber hologram.
(a) picture of the experimental set-up
(b) schematic diagram of the system

(a) (b)

Fig. 16 Picture of a c-axis Ce: Fe: doped $LiNbO_3$ PR crystal fiber
(a) side view
(b) cross section view

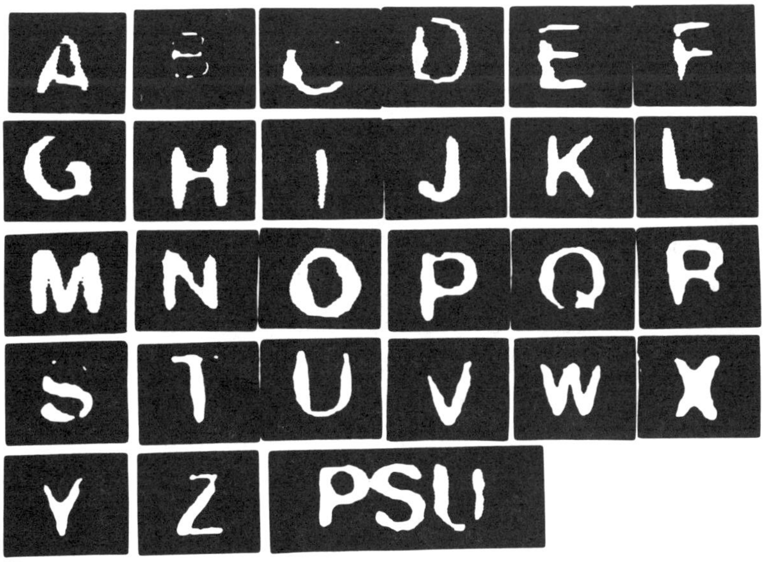

Fig. 17. Holographic images obtained from a Ce: Fe: doped $LiNbO_3$ PR fiber hologram.

Adaptive optical networks for image recognition

Demetri Psaltis, Yong Qiao, and Sergei Orlov

California Institute of Technology
Department of Electrical Engineering
Pasadena, California 91125

ABSTRACT

Periodic copying and electrical fixing are presented as two practical methods for controlling the decay rate of volume photorefractive holograms in the implementation of optical image processing systems.

1. INTRODUCTION

In this paper we examine issues that arise when we use 3-D photorefractive holograms in the implementation of optical image processing systems such as correlators and neural network architectures. There is strong motivation to use volume holograms in such systems in order to increase their computational capability and hence its competitiveness versus electronic implementations. For example, let us consider the classic VanderLugt correlator. Assume that the input device is an SLM with 10^6 pixels and operating at 30 frames per second. In computing the 2-D correlation of the input image, the optical system performs the equivalent of $10^6 \times 30 \times \log_2(10^6) \approx 6 \times 10^8$ multiplications per second that are required for an FFT electronic implementation. This speed can be achieved by relatively standard digital signal processing circuits even though the size and power consumption of such digital circuits may currently exceed the requirements of the optical implementation. Therefore, optical correlators can be limited to applications where size and power consumption are crucial. Continued improvements in device technologies promise to increase the speed of the SLM to 1,000 frames per second, leading to an increase in the processing speed to 2×10^{10} multiplications per second which is more competitive with digital electronics. While there are very few applications that require processing of input images at this high rate, there are applications that require high-speed correlation of the same input image against many references that could make use of the speed of optics. However, the classic correlator configuration places a large burden on the electronic memory and associated interface electronics that are necessary to read this high-speed memory. An optical correlator implemented with 3-D holograms can simultaneously provide increased processing speed and the capability to correlate an input image against a large library of references. The storage medium for the reference images is the hologram itself and therefore the need for auxiliary memory and high speed interfaces is eliminated. Such a system produces a 1-D slice of the 2-D correlation between the input image and each of the stored references. Since a single volume hologram can store 1,000 references we can calculate the processing speed (assuming that the speed of the SLM is 30 frames per second) to be $10^3 \times log_2(10^3) \times 10^3 \times 10^3 \times 30 \approx 3.3 \times 10^{11}$ multiplications per second. This dual benefit in processing speed and storage capability is also realized when optical neural network architectures are implemented with volume holograms.

The price we pay for using volume holograms, however, is the reduced diffraction efficiency. The reduction in efficiency is due to the fact that 3-D holograms must be recorded with multiple monochromatic exposures and/or multiple wavelengths [1]. A fully developed argument for this assertion is contained in Ref. 1, but a simple argument for the trade-off between 3-D storage and diffraction efficiency goes as follows: Information is recorded in the holograms by shining light through the surface of the crystal. Since the surface is 2-D, only 2-D patterns can be used to illuminate the volume of the crystal at any one time. In order to use 3-D storage it is necessary to expose the crystal multiple times (or with multiple wavelengths). The diffraction efficiency of a hologram recorded with M exposures falls as $1/M^2$ [1]. Therefore, as more exposures are used, the volume of the recording medium is utilized more effectively to store data but at the same time the diffraction efficiency rapidly falls off. When the recording medium is photorefractive, the $1/M^2$ dependence of the diffraction efficiency comes out of the exposure schedule formalism [2], in which early exposures are made longer to compensate for their erasure by subsequent exposures.

The loss in diffraction efficiency that accompanies the superposition of many holograms in the same volume generally limits the read-out speed since as M increases, the detector requires a longer integration period to detect the light diffracted by the hologram reliably. In practice this has not been a major limiting factor since the parallelism of the optical system is so high that I/O speed, rather than the detector speed, sets the limit in overall processing speed. A more serious problem arises from the following two considerations: 1. In learning systems such as neural networks and adaptive filters, the number of exposures, M, can be indefinitely large and is not generally known a priori; 2. Photorefractive holograms are erased when exposed to light during the read-out period. One solution to both problems is periodic copying [3-7]. In this case the stored holograms are copied onto a second medium, amplified, and then copied back to the original crystal. This approach provides the necessary re-normalization to arrest the $1/M^2$ drop-off in diffraction efficiency when a large number of exposures are used to form the hologram, and it also naturally leads to non-volatile storage. The second method we will discuss is electrical fixing in SBN crystals. Fixing addresses the non-volatile storage issue directly but it can also impact adaptive image processing systems.

2. PERIODIC COPYING

In the simplest case (no applied or photovoltaic field), the dynamics of hologram formation in a photorefractive crystal can be approximated by a simple first order differential equation:

$$\tau \frac{dw_{ij}}{dt} + w_{ij} = \alpha x_i y_j, \tag{1}$$

where w_{ij} is the amplitude of one of the sinusoidal grating components that make up the hologram, τ is the time constant of the crystal (inversely proportional to the incident light intensity), x_i and y_j are the amplitudes of the two plane wave components that interfere to record the grating w_{ij}, and α is determined by the crystal parameters and the total illuminating intensity. For short exposure times we can approximate Eq. (1) by

$$\Delta w_{ij} \approx (\frac{-w_{ij}}{\tau} + \frac{\alpha}{\tau} x_i y_j)\Delta t. \tag{2}$$

The above equation clearly shows the weight decay problem. As new data is recorded on the hologram the existing hologram is erased in proportion to the amplitude of each grating. Stronger gratings decay faster. Weight decay can be a useful mechanism that has been used in neural network algorithms before. Specifically in hardware implementations, the limited dynamic range of the devices that implement the weight require a mechanism that normalizes the weights during the learning phase to prevent saturation. Weight decay can accomplish this normalization task. However, it is necessary to have accurate control of the rate of weight decay. Otherwise the weights can decay too rapidly leading to small average weight values (and hence small diffraction efficiency) and more importantly the learning dynamics will be negatively affected since the effect of early examples will be forgotten.

It is possible to control the relative strengths of the learning and forgetting terms by changing the overall intensity I_o. The parameters τ and α in Eq. (2) both are proportional to $1/I_o$. Therefore, as I_o decreases, the forgetting term in Eq. (2) becomes smaller whereas the learning term remains unaffected. Even though this method works and it has been used in most of the learning experiments that have been done with photorefractive holograms, it has important shortcomings. In order to keep the forgetting term sufficiently small, the intensity must be kept at a very low level which slows the learning rate. Moreover, there is a minimum exposure that is required to record a hologram to overcome the thermally generated carriers in the photorefractive crystal. For large networks that need to be trained with large training sets, the minimum weight decay rate that can be realized by decreasing the intensity is not small enough.

Periodic copying allows us to control the weight decay rate while working with relatively high intensities. The basic copying setup is shown in Figure 1. Holograms are recorded by interfering the signal beam from the input to that particular layer with a set of plane-wave reference beams whose amplitudes are typically controlled by the error signals during the training phase of the network. In addition to this basic recording mechanism a scanning plane wave beam is used to read out the stored hologram. At each angular setting of the reference a different set of weights are reconstructed from the hologram and after the reference beam scans (in one dimension) over all the angles within the numerical aperture of the optical system used in the recording, all the stored weights are read-out. Let w_{ij} be one of the weights reproduced by the current setting of the read-out beam, as a plane wave whose amplitude is w_{ij}. This plane-wave component is optically amplified and reintroduced at the hologram where its interference with the read-out beam is formed. In this way, a hologram of w_{ij} is copied back onto itself, reinforcing the existing grating and arresting its decay due to the read-out. We can modify the expression for the modification of the weights (Eq. (2)) by introducing a second recording term that is proportional to w_{ij}:

$$\Delta w_{ij} \approx (\frac{-w_{ij}}{\tau} + \frac{\alpha}{\tau} x_i y_j + \frac{\beta w_{ij}}{\tau R})\Delta t. \tag{3}$$

In the above equation, R is the "rank" of the hologram which is the number of beams that are needed to read out all the stored weights. The coefficient β is determined by the gain of the optical amplifier and the amplitude of the reference beam. If we combine the two terms in Eq. (3) that contain w_{ij}, we obtain an expression identical to Eq. (2) except that the decay time constant is equivalently

$$\tau_{eq} = \tau \frac{R}{R - \beta}. \tag{4}$$

Notice that if the feedback gain (β) is large enough, the decay time constant can be made very large and the decay mechanism slowed down. Several versions of this copying method have been demonstrated in recent years, mostly with holographic memory applications in mind [3-7]. In what follows we will describe the method presented in Ref. 7 applied to the training of a single perceptron.

A schematic diagram of the experimental apparatus is shown in Figure 2. The input images are presented to the system through a 2-D spatial light modulator (SLM). The Fourier transform of the SLM illuminates the photorefractive crystal ($BaTiO_3$) where a hologram stores the weights that interconnect all the input pixels to the output detector. The signal from the detector is the output signal of the perceptron and it should be above the threshold if the input belongs to category #1 and below threshold if the input is from category #2. The objective of the training is to select a set of weights in the hologram that classifies the training patterns correctly. The training is implemented by forming a hologram of the current input with a plane wave reference beam and adding it to the previously recorded holograms. The perceptron algorithm for training the weights requires the following modification of the weights for each presentation of an input training pattern:

$$\Delta w_{ij} = c_j f_i. \tag{5}$$

In the above equation f_i is the ith pixel of the input image and c_j is +1 (-1) if the input is from category #1 (#2) but the system produces a response below (above) threshold. c_j is 0 if the input is classified correctly. The introduction of an electrically controlled liquid crystal phase retarder in the reference beam implements the bipolar c_j in the setup of Figure 2. In other words, a hologram of the input image f_i is recorded either in phase or out of phase if the input is classified incorrectly. No holographic exposure is done if the input is correctly classified. Since we can readily set the product $x_i y_j$ equal to $f_i c_j$ (there is only one output neuron in this case and therefore j=1), the difference between Eqs. (3) and (5) is the decay term present in Eq. (3). The feedback loop shown in Figure 2 partially compensates for this decay. The reconstructed hologram is Fourier transformed through the beam splitter onto the photosensitive side of an optically addressed SLM (a Hughes liquid crystal light valve). The SLM is read out from the opposite side by illuminating it with a plane wave. The modulated light is then Fourier transformed onto the crystal where a hologram is recorded using the read-out beam as a reference. The copying of the hologram onto itself compensates for the decay as described above, if the feedback gain is large enough. In this set up the feedback gain is provided by the optically addressed SLM since the read-out light intensity can easily be much higher than the illuminating light intensity. Therefore, the addition of the feedback loop reduces the decay term in Eq. (3) and the optical system performs a much better approximation of the true perceptron as the feedback gain is raised. The nonlinear transfer function of the optically addressed SLM, however, makes the system different from the true perceptron by essentially performing a thresholding of the weights each time the hologram is copied. In many respects this thresholding is desirable since it normalizes the weights and reduces noise. It has been previously shown that perceptrons with binary weights have approximately the same capacity as perceptrons with real weights, and that such binary perceptrons can be readily trained even though there is no convergence proof for them.

In Figure 3 we show the error rate on the training set of the optically trained perceptron with and without copying. In this experiment the training set consisted of 20 pictures of the faces of two individual people (10 each). The training procedure converged in both cases but the copying significantly improved the convergence speed. This speed-up is expected to increase when larger training sets are used.

3. ELECTRICAL FIXING

The second method we will discuss for controlling the rate at which the holograms decay is electrical fixing. Holograms recorded in SBN:75 were first electrically fixed in 1973 by Micheron and Bismuth [8]. These experiments were not successfully duplicated until recently [9]. These recent experiments showed the same basic behavior as the original work of Micheron and Bismuth, except that the diffraction efficiency of the fixed gratings is smaller. Two-wave mixing may have been responsible for the strong gratings reported in Ref. 8. Several new modalities have been devised for recording and using fixed gratings in SBN:75, most significantly the use of revealed gratings that are typically much stronger than the fixed gratings. Fixing of holograms in SBN:75 without the application of an electric field but with relatively strong light intensities has also been reported recently [10].

The optical setup used to perform the experiment is shown in Figure 4. An SBN crystal grown at the Rockwell Science center with its c-axis parallel to the grating vector was used. An ordinary polarized plane wave from an argon laser (λ = 488 nm) is split into three beams, two of which interfere to record a grating in the crystal, and the third is used as a non-Bragg-matched erasing beam. The diffraction efficiency η is monitored with a weak, extraordinary polarized HeNe laser beam incident at the Bragg angle. An electric field can be externally applied along the c axis of the crystal, and it is called positive (negative) if its direction is the same as (opposite to) that of the initial poling field.

First a photorefractive grating with a grating spacing Λ = 11.6 μm was recorded in the completely poled crystal without any applied field. The recording beams were blocked after the diffraction efficiency η reached its saturation value of 11%, and a negative voltage pulse with amplitude V = -1 kV and duration t = 0.5 s was then applied to the crystal. The diffraction efficiency fell quickly during the voltage pulse, and recovered a portion of its initial value after the pulse was removed. Then the crystal was illuminated with the non-Bragg-matched erasing beam, and η decreased until it reached a steady state of η = 0.06%, which could not be further erased optically. Finally the erasing beam was blocked and a positive voltage pulse with amplitude V = + 2 kV and duration of a few seconds was applied to the crystal. During the positive voltage pulse $\eta = 1.8\%$, and after the pulse a grating with $\eta = 0.4\%$ was revealed. This revealed grating can be optically erased. The experimental results of recording, fixing, revealing, and erasing are shown in Figure 5, where the diffracted signal is plotted as a function of the time during the different stages of the experiment.

In this experiment, the electrical fixing reported in Ref. 8 was duplicated, except that the fixed grating we obtained was weaker than in Ref. 8. An important difference is our observation of a revealed grating upon applying a strong positive voltage across the crystal at relatively large grating spacings.

In general, the strengths of the fixed and revealed gratings can be greatly enhanced if we apply a constant negative voltage during the recording of the initial grating. With an applied constant voltage of V = - 500 V, we recorded a grating with $\eta = 11.6$ μm using an exposure time of 50 s. Then the recording beams were blocked, the applied voltage was removed, and the grating was allowed to settle down to a steady level. Upon illuminating the crystal with the erasing beam, η decreased first and then rose back to a steady state, indicating a fixed grating of $\eta = 0.8\%$. Finally,a positive voltage pulse same as the one used in the previous experiment was applied, which revealed a grating of $\eta = 0.5\%$. Figure 6 shows the experimental result.

Electrical fixing with an applied constant negative voltage during recording was repeated for several different grating spacings, and the results are summarized in Figure 7. Under the reported experimental conditions, the maximum diffraction efficiency of the fixed grating is found to be 0.8% at $\Lambda = 11.6$ μm. The revealed grating is observed only at relatively large Λ and its strength increases drastically with the increase in Λ, achieving about 17% diffraction efficiency at $\Lambda =$ 29 μm.

We were also able to record and fix an image hologram in the SBN:75 crystal. The reconstructions of the initial and fixed holograms are shown in Figure 8.

Our experimental observations can be explained with the polarization grating formation mechanism suggested in Ref. 8. With the aid of the negative applied field, the electronic space-charge field established during the holographic recording creates a spatial modulation of the ferroelectric polarization. Under the illumination of the erasing beam, the electronic grating is erased until a balance between the polarization grating and the electronic grating is reached. The observed fixed grating is attributed to the polarization grating that is partially compensated by the electronic grating. A strong positive voltage applied to the crystal erases the polarization grating (i.e., re-aligns the ferroelectric domains), revealing the compensating electronic grating which of course can be erased optically. A physical model describing this process has been developed based on the band transport model, and the theoretical prediction fits very well with the experimental results.

4. DISCUSSION

Volume photorefractive holograms are attractive for the implementation of optical processing systems due to their ability to store, access, and modify information in 3-D. The two important issues that arise in such applications are the control of dynamic behavior of photorefractive holograms and non-volatile storage. Periodic copying and electrical fixing have provided effective solutions to these problems. Optical processing systems based on such techniques promise a very large number of adaptations and nondestructive read-out.

REFERENCES

[1] D. Brady and D. Psaltis, "Control of volume holograms," J. Opt. Soc. Am. A **9**, 1167 (1992).

[2] D. Psaltis, D. Brady, and K. Wagner, "Adaptive optical networks using photorefractive crystals," Appl. Opt. **27**, 1752 (1988).

[3] D. Brady, K. Hsu, and D. Psaltis, "Periodically refreshed multiply exposed photorefractive holograms," Opt. Lett. **15**, 817 (1990).

[4] Y. Qiao, D. Psaltis, C. Gu, J. Hong, P. Yeh, and R. R. Neurgaonkar, "Phase-locked sustainment of photorefractive holograms using phase conjugation," J. Appl. Phys. **70**, 4646 (1991).

[5] H. Sasaki, Y. Fainman, J. E. Ford, Y. Taketomi, and S. H. Lee, " Dynamic photorefractive optical memory," Opt. Lett. **16**, 1874 (1991).

[6] S. Boj, G. Pauliat, and G. Roosen, "Dynamic holographic memory showing readout, refreshing, and updating capabilities," *Opt. Lett.*, **17**, 438 (1992).

[7] Y. Qiao and D. Psaltis, "Sampled dynamic holographic memory," *Opt. Lett.*, **17**, 1376 (1992).

[8] F. Micheron and G. Bismuth, "Field and time thresholds for the electrical fixation of holograms recorded in $Sr_{0.75}Ba_{0.25}Nb_2O_6$ crystals," *Appl. Phys. Lett.*, **23**, 71 (1973).
[9] Y. Qiao, S. Orlov, D. Psaltis, and R. R. Neurgaonkar, "Electrical fixing of photorefractive holograms in $Sr_{0.75}Ba_{0.25}Nb_2O_6$," *Opt. Lett.*, **18**, 1004 (1993).
[10] A. Kewitsch, M. Segev, A. Yariv, and R. Neurgaonkar, "Selective Page-Addressable Fixing Of Volume Holograms In $Sr_{0.75}Ba_{0.25}Nb_2O_6$ Crystals," *Opt. Lett.*, **18** (15), 1262 (1993).

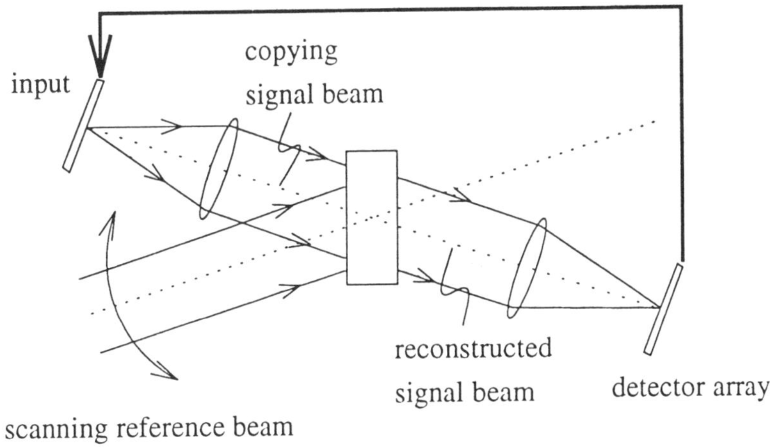

Figure 1. Basic setup for period copying.

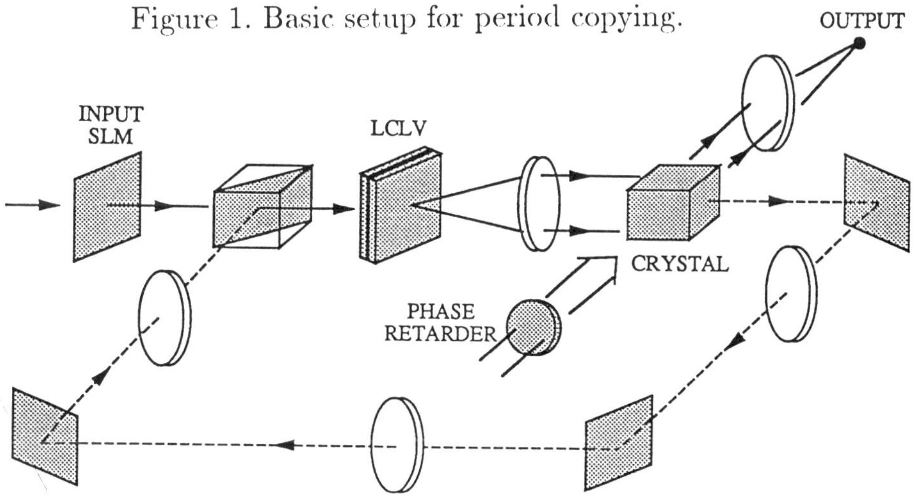

Figure 2. Periodic-copying optical perceptron network

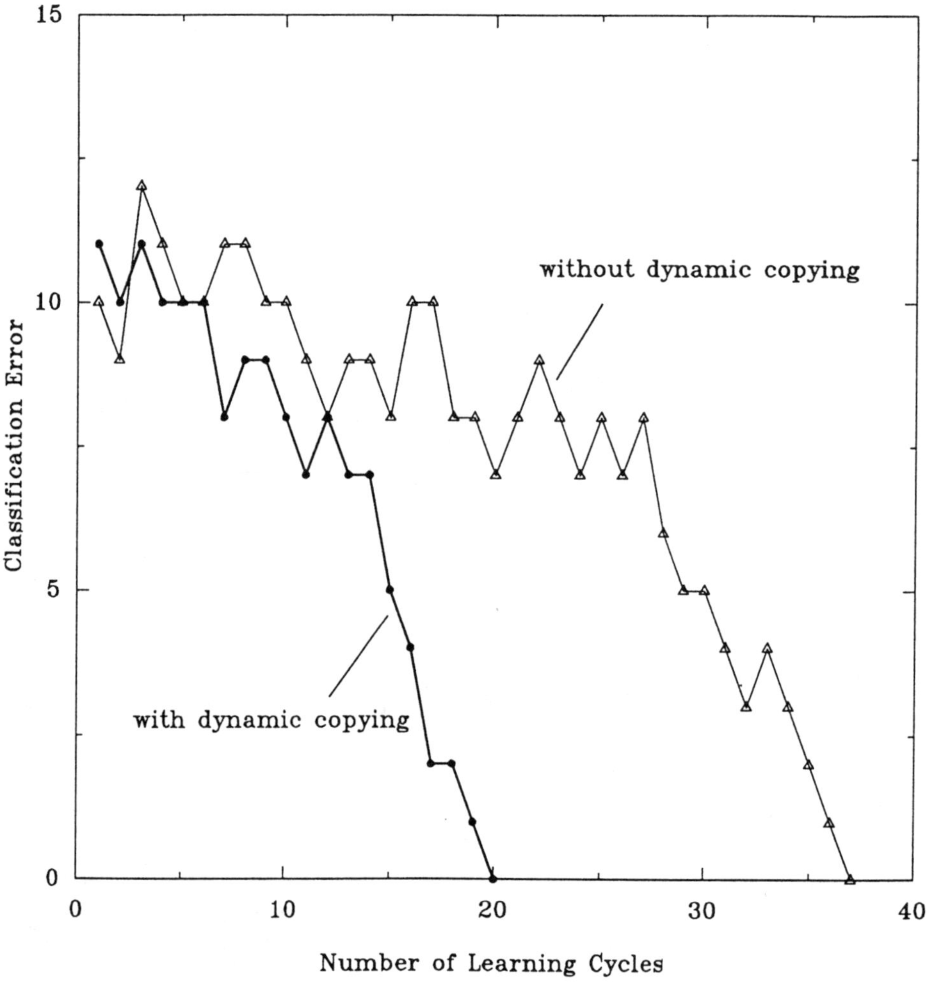

Figure 3. Error rates of the experimental perceptron

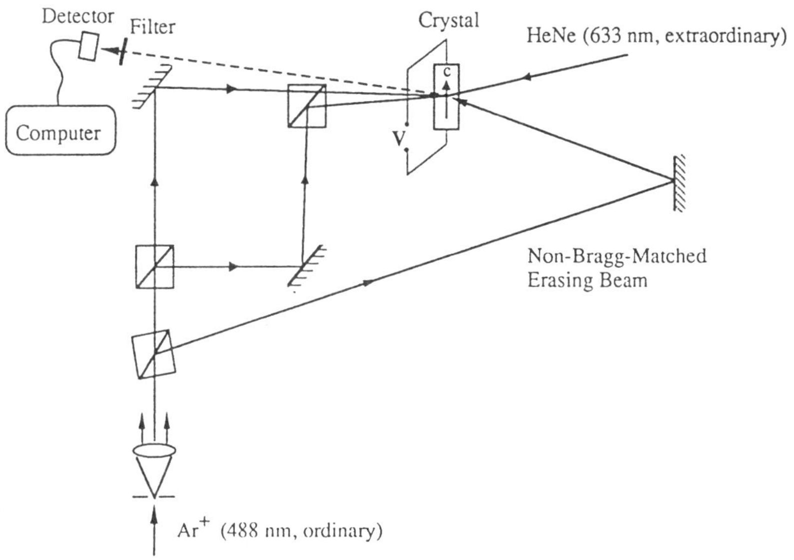

Figure 4. Optical setup for electric fixing

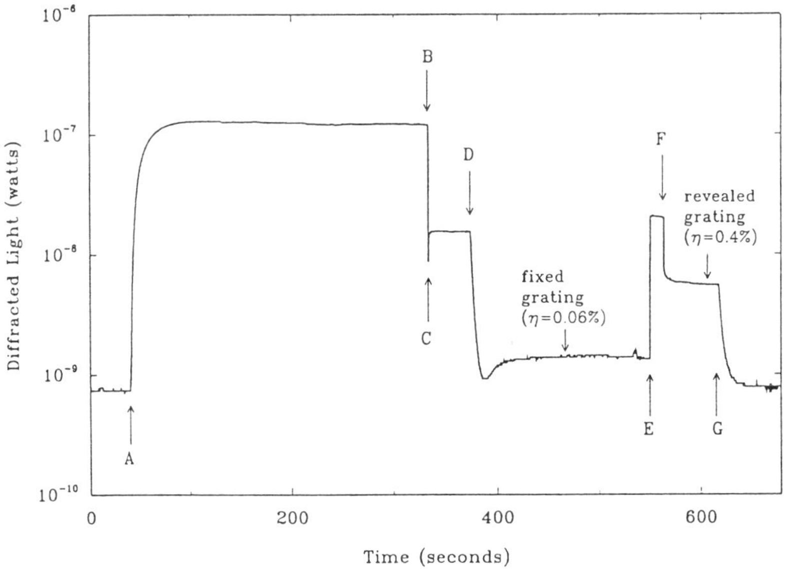

Figure 5. Diffracted light as a function of time for fixing with a negative voltage pulse ($\lambda = 11.6\ \mu m$). The transmitted light power in the absence of the grating is 1.1 μW. (A) Hologram recording begins. (B) Negative pulse is applied. (C) End of negative pulse. (D) Optical erasure begins. (E) Erasing beam is blocked and positive voltage pulse is applied. (F) End of positive pulse. (G) Optical erasure begins.

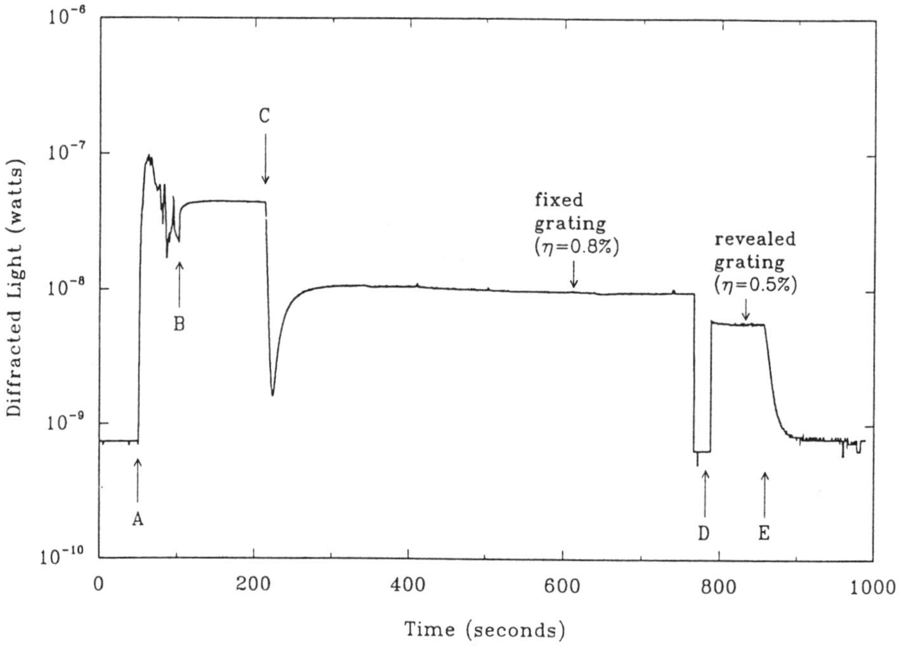

Figure 6. Diffracted light as a function of time for fixing with a constant negative voltage during recording (λ = 11.6 μm). The transmitted light power in the absence of the grating is 1.1 μW. (A) Hologram recording begins and negative voltage is applied. (B) Negative voltage is removed and recording beams are blocked. (C) Optical erasure begins. (D) Positive voltage pulse is applied (Probe beam is blocked). (E) Optical erasure begins.

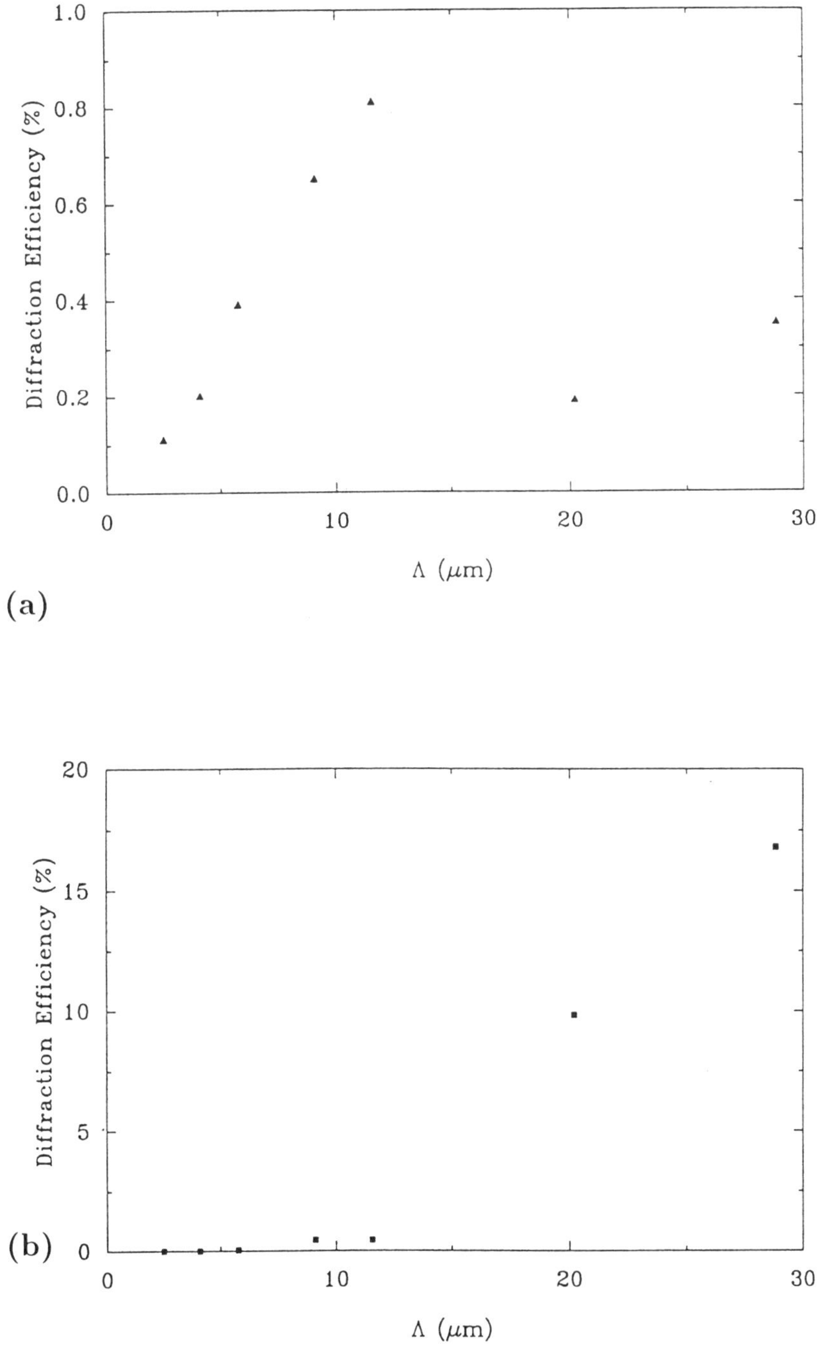

Figure 7. Diffraction efficiency as a function of grating spacing for (a) the fixed grating and (b) the revealed grating.

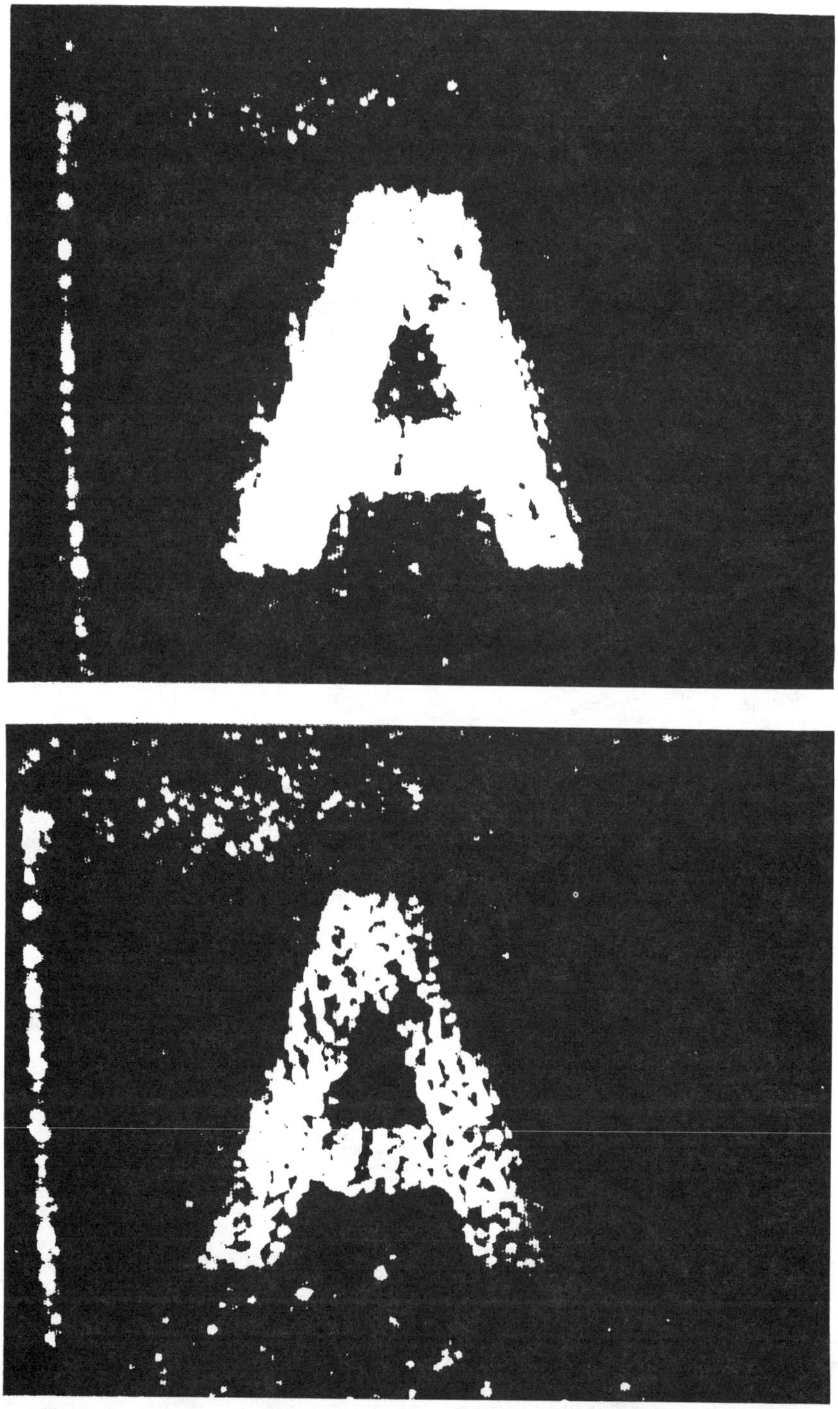

(a)

(b)

Figure 8. (a) Reconstruction of the recorded hologram. (b) Reconstruction of the fixed hologram.

Applications of Photorefractive Materials for Optical Interconnects and Storage

Sing H. Lee, Jian Ma, Yeshaiahua Fainman, Joseph E. Ford
Department of Electrical & Computer Engineering
University of California, San Diego, La Jolla, CA 92093-0407

Abstract

Photorefractive materials are useful for optical interconnection and data storage in optoelectronic computing systems because of their unique properties (e.g., write/read/erase capability, real time wave mixing, and high storage capacity). In this paper we will review the recent advances in these two application areas that make use of these properties and the techniques for improving their real-time and storage performance.

I. Introduction

Photorefractive materials have distinct characteristics making them unique among optical materials, namely (1) write/read/erase response, (2) real-time wave mixing, and (3) volume storage. These features have made photorefractives useful for many applications including optical interconnects and storage.

The photorefractive effect is a light-induced change in the local index of refraction occurring in materials which are both electro-optic and photoconductive. Consider two plane waves crossing in a photorefractive crystal (PRC). The light generates charge carriers through photoabsorption and conduction, so that the spatial intensity distribution of the interference between these two waves generates a corresponding spatial charge distribution. The charge distribution creates a space charge electric field, which in turn modulates the refractive index via the electro-optic effect, creating a phase grating. When the recording light is turned off, the phase grating remains stored in the material until thermal or optical excitation redistributes the charge carriers. If the crystal is uniformly illuminated, the index modulation is erased by the same process. The cycle can be infinitely repeated. This unique ability of write/read/erase cycling can be used for reconfiguring optical interconnection [1-2], updating interconnection weights for neural networks [3-4] and updating information in a parallel access optical memory [5-7].

Real-time wave mixing in PRC offers several useful functions such as energy coupling and phase conjugation. In two-beam mixing, propagating waves interact with the self-induced grating, diffract into each other and causes energy exchange between the two beams. This effect can be applied to image amplification [8], incoherent-to-coherent optical conversion [9-10], 2-D optical logic [11], matrix inversion [12] and optical interconnects [13]. In four-wave mixing, three input waves stimulate the radiation of a fourth wave. Under certain conditions, the fourth wave is a phase conjugated (sometimes called time reversed) version of one of the input waves. Four-wave mixing is useful for phase-conjugation [14], optical correlation [15] and optical interconnection [16].

The large achievable thickness of PRC allows volume holographic storage. The ultimate theoretical storage capacity limit of such holograms is V/λ^3 bits [17], so that a crystal of rectangular area A and thickness L can store L/λ planes of information each consisting of A/λ^2 resolvable elements. The multiple superimposed images can be differentiated by the recording wavelength [18-19], beam angles [20-21] and relative phases [22], allowing image multiplexing with high diffraction efficiency, high signal-to-noise ratio (SNR) and low crosstalk.

We found that it is useful to separate photorefractive applications into two categories [23]: real-time operation that requires large energy coupling and fast response time, and storage operation that requires large storage capacity and long storage time.

II. Real-Rime Operation of Photorefractive Materials

2.1 Real-Time Operation Characteristics

The real-time operations are characterized by response time, sensitivity, operation wavelength, exponential gain coefficient (Γ) for two-wave mixing, and phase-conjugate reflectivity (R) for four-wave mixing. Γ and R are related to coupling strength γ through [24]

$$\Gamma = 2\,\mathrm{Re}(\gamma)\ , \qquad (1)$$

$$R = \left| \frac{\sinh\left(\frac{\gamma L}{2}\right)}{\cosh\left(\frac{\gamma L}{2} + \frac{\ln r}{2}\right)} \right|^2 , \qquad (2)$$

where

$$\gamma = \frac{i\,\omega\,E_{sc}\,e^{i\phi}\,r_{eff}}{2\,n\,c\cos\theta}\ , \qquad (3)$$

and L is the thickness, r is the beam ratio, ω is the optical frequency, n is the refractive index chosen according to the polarization of the beam, c is the speed of light, E_{sc} is the space-charge field, ϕ is the phase shift between light intensity and E_{sc}, r_{eff} is the effective Pockels coefficient and θ is the half angle of the two writing beams inside the crystal. The product of E_{sc} and r_{eff} is proportional to the steady-state refractive index modulation Δn_{ss}.

Table.1: Figures of merit of some of the existing materials for real-time operation

Materials	Sensitivity (cm^3/J) $\Delta n/\alpha It$	Δn_{ss}	Time response @ $1W/cm^2$	Spectral response (nm)	λ (μm)	Comments
$BaTiO_3$ [25]	3×10^{-3}	1.5×10^{-4}	60 ms	400-1100	0.5	} *Large gain but slow*
SBN:60:Ce [26]	6×10^{-4}	9.6×10^{-5}	80 ms	400-850	0.5	
BSO [25]	1×10^{-3}	5.0×10^{-6}	1 ms	400-700	0.5	
GaP [27, 28]	2×10^{-3}	1.5×10^{-6}	0.4 ms	600-1000	0.63	} *Fast but small gain*
InP:Fe [29]	2×10^{-2}	2.1×10^{-6}	0.1 ms	900-1300	1.06	
GaAs [30]	4×10^{-2}	3.4×10^{-6}	80 μs	900-1800	1.06	
CdTe:V [31]	5×10^{-1}	7.0×10^{-6}	8 μs	900-1500	1.5	*Longer operating wavelength*
MQW [32, 33]	2	1.0×10^{-3}	0.5 μs	836	0.836	*Limited by thickness*

For achieving a large coupling gain or large phase-conjugate reflectivity, a large steady-state index modulation Δn_{ss} is always preferred. In addition, fast response (i.e., on the order of nsec with a moderate (1 W/cm^2) beam intensity), and near infrared (IR) spectral response are desirable for most applications due to the recent advances of tunable semiconductor laser operating at near IR.

Table 1 lists the comparison of the existing most commonly used materials for real-time operation. All the data were obtained without using the enhancing techniques described later in this paper. These data might not be the most accurate because of the differences in results reported by different authors. Some data were obtained by estimation based on the available experimental result. However, by searching through the table, it is clear that no single material is perfect. Ferroelectric oxides ($BaTiO_3$ and SBN) have large achievable Δn_{ss} (providing large gain or phase conjugation) but they are too slow and are operated at unfavorable wavelengths. Most of the semiconductors respond quickly but have small Δn_{ss}. CdTe is a most attractive material for its relatively large Δn_{ss} and high sensitivity, however its optimum operating wavelength is beyond the wavelength of semiconductor microlasers. The characteristics of CdTe at 1.06μm does not show much difference with GaAs [34]. Multiple Quantum Well (MQW) microstructures provide the most favorable characteristics but the achievable material thickness is limited to a few microns. Although the large value of Δn_{ss} in MQW provides a large exponential gain coefficient Γ ($\approx$200 cm^{-1}), the net energy coupling gain, which is determined by the product of Γ (or γ for phase conjugation) and the thickness L, is small. In order to overcome these material limitations, several techniques were developed for enhancing real-time operation.

2.2 Techniques for Enhancing Real-time Operation

Most of the important efforts to enhance real-time operation have been demonstrated using two-beam coupling, i.e., by increasing the exponential gain coefficient Γ. By inspecting Eqs. (1) and (3), it is found that Γ can be enhanced through optimizing three parameters: (i) the effective electro-optic coefficient r_{eff}, (ii) the space charge field E_{sc}, and (iii) the grating phase shift ϕ. The corresponding techniques for optimizing these three parameters are discussed in the following three sub-sections.

2.2.1 Optimizing r_{eff} by 45°-cut $BaTiO_3$

The geometric configuration for two-beam coupling in a conventional 0°-cut $BaTiO_3$ is shown in Fig. 1. Both beams are linearly polarized in the extraordinary direction, with $\hat{e}_1$ and $\hat{e}_2$ lying in the plane of incidence. Their interference creates an index grating of wave vector $\mathbf{K_g}$ with an angle β to the crystal **c** axis. The effective Pockels coefficient r_{eff} can be expressed as

$$r_{eff} = \frac{1}{2}[n_o^4\, r_{13} \left(\cos 2\theta - \cos 2\beta\right) + 4\; n_e^2\, n_o^2\, r_{42} \sin^2\beta + n_e^4\, r_{33} \left(\cos 2\beta + \cos 2\theta\right)] \cos\beta \quad , \tag{4}$$

where n_o and n_e are the ordinary and extraordinary index of refractive respectively, r_{ij} are the elements of the linear electrooptic tensor, θ and β are defined in Fig.1. The nonzero linear electro-optic coefficients are $r_{13} = 8$, $r_{33} = 28$, and $r_{42} = 820$, in units of 10^{-12} m/V. The exceptionally large value of r_{42} will yield a high value of r_{eff} when the

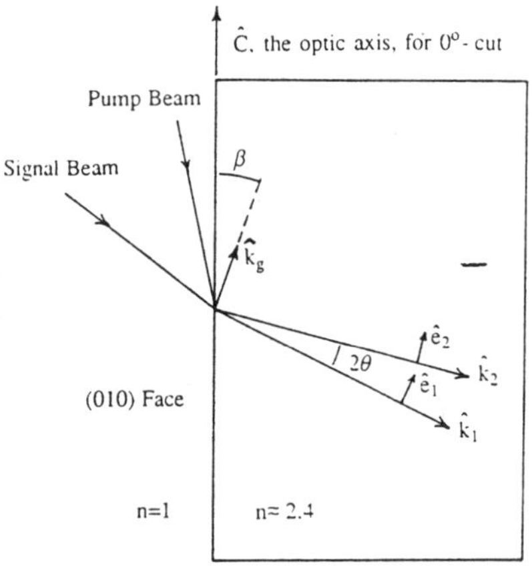

Fig.1. Geometric configuration for two-wave mixing in $BaTiO_3$ [8, 35].

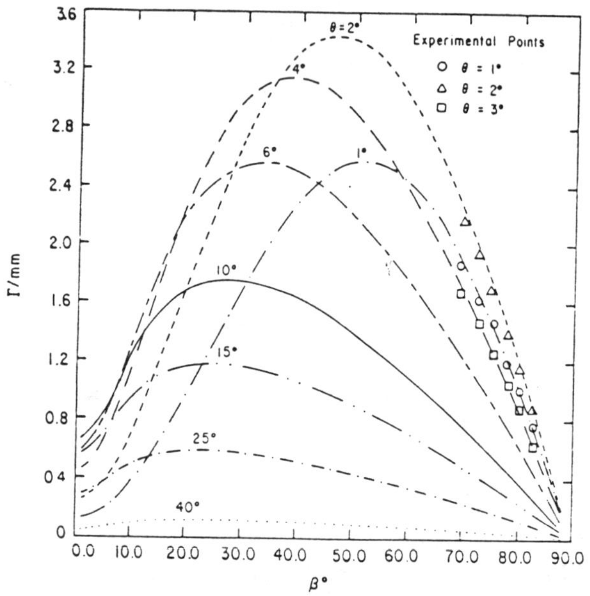

Fig.2. Exponential gain coefficient Γ for two-wave mixing in $BaTiO_3$ as a function of externally controlled angles θ and β (measured inside the crystal). Experimental points were taken with a conventional 0°-cut crystal [8, 35].

two beams interfere to generate the slanted grating of nonzero β. Fig. 2 shows the theoretically calculated dependence of Γ on β and θ. It is seen that maximum gain occurs for the grating vector angle β set at approximately 30° to 50° to the optical axis **c**.

Because the index of refraction of $BaTiO_3$ is high (≈ 2.4) compared with air, the input beams are strongly refracted at the input face. This limits the range of β accessible without index matching to $0° < \beta < 23°$ or $67° < \beta < 90°$, depending on the input face used

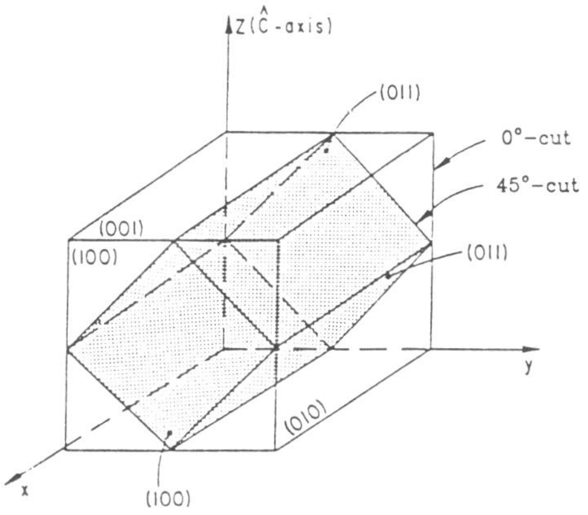

Fig.3. Orientation of the 45°-cut $BaTiO_3$ crystal [with faces on the (100), (011), and (011) crystallographic planes] relative to the 0°-cut crystal [8, 35].

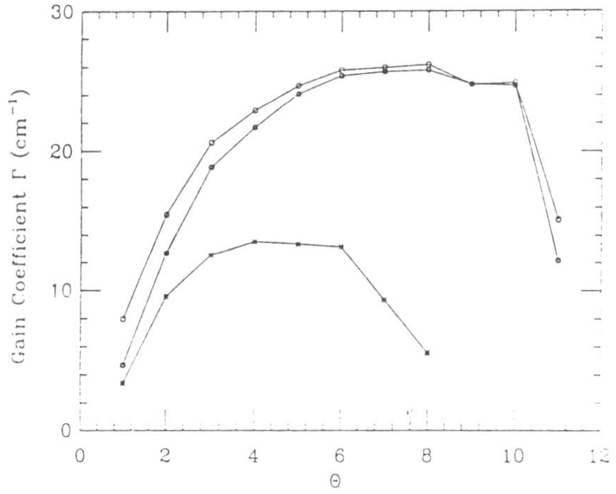

Fig.4. Experimental result of Γ vs θ for both conventional (represented by starts) and 45°-cut (represented by circles) $BaTiO_3$ [35].

(see experimental points in Fig. 2). Even with a index matching fluid (n = 1.624) and acute beam incident angles, the peak Γ is only marginally accessible from the side. Another problem associated with these acute beam incident angle is the reduction in the effective input aperture of the crystal and an incomplete utilization of the crystal volume.

A straight forward solution to these problems, first proposed in Ref. [8] and then experimentally performed and characterized in Ref. [35], is to cut the $BaTiO_3$ crystal with its input faces at 45° to **c**, as showed in Fig. 3. This new orientation effectively adds a bias angle of 45° to β, so that beam pairs with near normal incidence on the crystal face will access the broad peak of the Γ curve. The maximum (calculated) Γ of 36 cm^{-1} is potentially achievable with the full front surface area of the crystal as the input aperture. Because $\mathbf{K_g}$ is now perpendicular to the endfaces, it becomes possible to apply a uniform external electric field parallel to $\mathbf{K_g}$, which is otherwise difficult if not possible to achieve. Fig. 4 shows the experimental result of Γ as a function of θ for 0°-cut as well as 45°-cut $BaTiO_3$ samples. The Γ produced by the 45°-cut crystal was substantially higher than that of the 0°-cut crystal and was available over a broader range of θ. The peak value of 26 cm^{-1} was reached at $6° < \theta < 8°$. This deviates from the theoretical calculation (Fig. 4) which predicted a sharp peak (34 cm^{-1}) at $\theta = 2°$, followed by a gradual decline as θ increases. The deviation is apparently a result of the fanning of the signal beam.

2.2.2. Optimizing E_{sc} and ϕ by DC Field and Moving Grating

The most attractive materials for real-time operation are semiconductors which have fast response. However, their low refractive index modulation provide only a small coupling gain. The refractive index modulation can be improved by enhancing the modulus of space-charge field through applying a DC electric field [25],

$$E_{sc} = E_q \left[\frac{E_d^2 + E_o^2}{E_o^2 + (E_d + E_q)^2} \right]^{\frac{1}{2}} \tag{5}$$

where E_q is the maximum sustainable space-charge field, E_d is the diffusion field and E_o is the applied field. However, the optimum phase condition will be violated when an external DC field is applied. The beam coupling in PRC is due to a phase mismatch ϕ between the holographic grating and light fringes. The maximum energy transfer [Γ in Eqs.(1) and (3)] is obtained when this phase mismatch equals $\pi/2$. ϕ is a function of electric fields expressed by [25]

$$\tan\phi = \frac{E_d}{E_o}\left(1 + \frac{E_d}{E_q} + \frac{E_o^2}{E_d E_q}\right) \tag{6}$$

A $\pi/2$ phase shift exists when the space charge field is generated via diffusion mechanism (no "internal" or external applied electric field, i.e. $E_o = 0$).

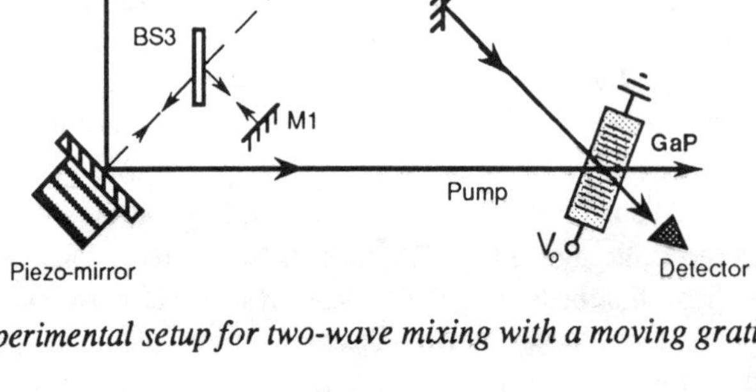

Fig.5. Experimental setup for two-wave mixing with a moving grating [27].

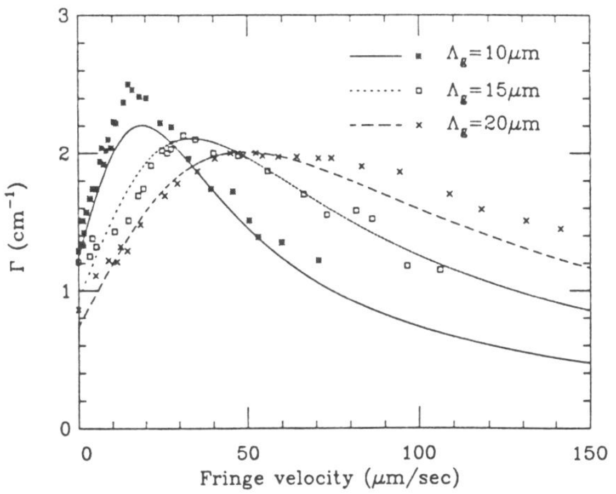

Fig.6. Experimental results of two-beam coupling gain coefficient as a function of fringe velocity for different spacings in GaP. The curves are theoretical prediction [27].

Table 2. Comparison of Γ at dc field and moving grating

Material	Γ (cm^{-1}) No applied field	dc field	moving grating
BSO [39]	4.0	9.0	12.0
GaAs [38]	0.6	2.6	7.0
GaP [27]	0.3	1.0	2.5

In order to optimize both Δn_{ss} (via enhancing E_{sc}) and ϕ, a moving grating technique was proposed by Huignard et al in 1981 for enhancing beam-coupling in BSO crystals [36]. The solution of Kukhtarev's equations with a moving grating [37] has shown that there is a resonance effect so that at a particular optimum velocity, a maximum space charge field which is $\pi/2$ phase shifted with respect to the moving fringes is created, allowing efficient beam coupling. This technique was applied to enhancing real-time operation of semiconductors including GaAs [38] and GaP [27]. The experimental setup for two-beam coupling is shown in Fig.5. The recording fringes were made to move during recording by reflecting one beam from a moving piezo-mirror, driven by a sawtooth voltage. The velocity of the fringe was detected in real time using a Michelson interferometer formed by beam splitter BS3, mirror M1 and the piezo-mirror.

An example of gain enhancement in GaP using a moving grating is demonstrated in Fig. 6 which shows Γ as a function of fringe velocity with an external applied field of E_0 = 16.7 kV/cm and several different grating periods Λ_g. Table 2 shows a comparison of the beam coupling gain coefficient Γ in BSO, GaAs and GaP using DC field and moving grating techniques.

Table 3. Comparison of Γ (cm^{-1}) in GaAs for various mechanism and techniques [40]

	Nonlinear mechanism		
Technique	EOPR	ERPR	EOPR+ERPR
No applied field (diffusion)	0.6	0	0.6
dc field	2.6	2.8	7.6
Moving grating	7.0	...	16.3

2.2.3. Optimizing E_{SC} by Band-Edge Electroabsorption

A technique for enhancing photorefractive gain using band-edge electroabsorption was proposed and demonstrated by Partovi et al on GaAs [40] and more recently by Millerd et al on InP:Fe [41]. A large electric field will induce changes in absorption (electroabsorption) near the band edge in semiconductors through the Franz-Keldysh effect. The Franz-Keldysh effect is an electric field-induced change of the complex dielectric constant of a direct band-gap semiconductor, occurring at optical energies close to the band-gap energy. The Franz-Keldysh effect has two parts, "electroabsorption" (changes in the optical absorption coefficient of the material $\Delta\alpha$) introduced by a field-induced tunneling between valence and conduction band states, and "electrorefraction" (changes of the refractive index of the material Δn) introduced by the Kramers-Kroning relationship between Δn and $\Delta\alpha$. Large beam-coupling energy transfer has been observed by combining electrorefractive photorefractive (ERPR) and the conventional electro-optic photorefractive (EOPR). A comparison of beam-coupling gain coefficient in GaAs using different mechanisms and techniques is shown in Table 3 [40]. A gain coefficient of Γ = 16.3 cm^{-1} was obtained at λ = 939 nm and E_o = 10 kV/cm. It will be more interesting if band-edge photorefractive effect were to occur at the wavelengths where maximum laser diode powers are available, such as λ = 830 nm.

In summary, techniques for enhancing PRC real-time operation have been discussed through a few examples, including special crystal orientations, applying an external DC field and using a moving grating, and through band-edge effects. There are other techniques which are also interesting and important. These techniques include applying an AC field [42], applying a pulsed field [43] and especially by using temperature stabilization [44]. The result of Γ = 19 cm^{-1} obtained in InP:Fe by combining temperature stabilization and band-edge effect was quite exciting [41]. It will be interesting to see if the technique of temperature stabilization can be applied to other semiconductor materials.

III. Storage Operation of Photorefractive Materials

The difference between real-time operation mode and storage operation mode is that in real-time operation all of the recording beams (references and signal) are simultaneously present in the material and interact on each other, while in storage operation recording and reconstruction are separated into two stages, as depicted in Fig.7.

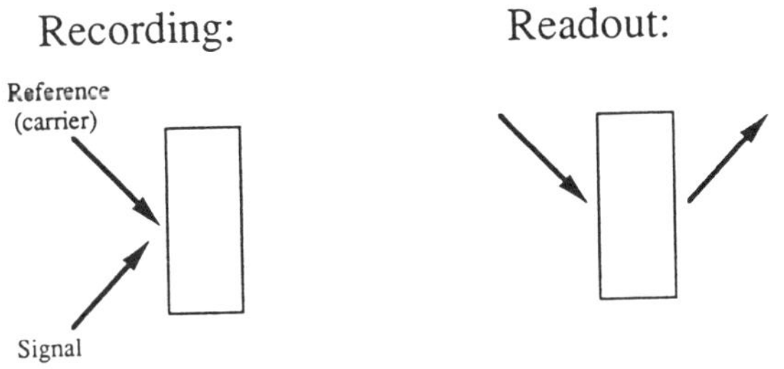

Fig.7. Two stages of storage operations: recording and readout.

3.1 Storage Operation Characteristics

The limitation of the photorefractive dynamic range on storage capacity is expressed by

$$N = \frac{\Delta n_{ss}}{\Delta n_i} \frac{\tau_e}{\tau_w} \,, \tag{7}$$

where Δn_{ss} is the steady-state refractive index modulation, τ_e and τ_w are the erasing and writing time constants respectively, and Δn_i is the required diffraction efficiency of the hologram calculated by Kogelnik's equation [45] for the diffraction efficiency:

$$\eta = \sin^2\left(\frac{\pi \, \Delta n_i \, L}{\lambda \cos \theta}\right) \,. \tag{8}$$

For practical applications, we wish the storage time be as long as possible and to limit the hologram decay to a minimum during reconstruction. Spectral response is another important consideration. Currently special attention is being paid to development of photorefractive materials that operate at the wavelength of semiconductor lasers, in the near IR range [18, 47-48]. Table 4 compares the figures of merit of the existing most popular materials for storage. Research on improving storage operation has been conducted to meet the practical application requirements.

3.2 Techniques for Enhancing Storage Performance

Work on enhancing storage performance fall into two categories: enhancing material performance and enhancing system performance. The first category includes applying

Table 4. The figures of merit of the some of the existing materials for storage:

Materials	Δn_{ss}	τ_e/τ_w	Sensitivity (cm^3/J)	λ	Dark Storage time	Techniques
$LiNbO_3$:Fe [48-49]	1×10^{-3}	9	5×10^{-6}	0.5 μm	months-1yr	
$LiNbO_3$:Fe [50]	4×10^{-4}	-	4×10^{-6}	0.5 μm	>1yr	fixed by heating
SBN:60:Ce [21]	9.6×10^{-5}	7	6×10^{-4}	0.5 μm	days-months	10kV/cm field
$BaTiO_3$ [25]	4.8×10^{-5}	1	3×10^{-3}	0.5 μm	days	

DC field for enhancing capacity [21], moving grating for enhancing diffraction efficiency [51] and special doping/processing for enhancing near IR operation [18, 46-47], while the second category includes orthogonal phase encoding for fast random access and incremental recording for signal uniformity [22].

3.2.1. Applying DC Field for Enhancing Capacity

Eq.(7) indicates that the storage capacity (the number of hologram which can be superimposed) can be increased by enhancing the steady-state refractive index modulation Δn_{ss} and/or by enhancing the ratio of erasing/writing time constant τ_e/τ_w. It

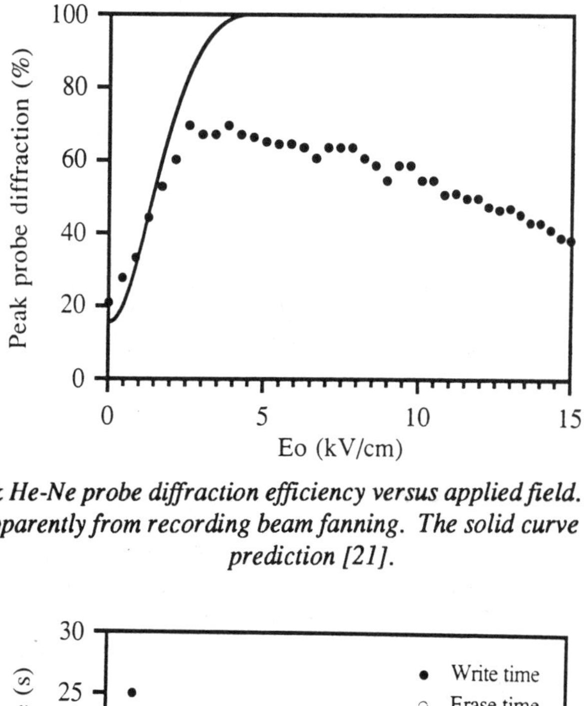

Fig.8. Peak He-Ne probe diffraction efficiency versus applied field. The drop after 5kV/cm is apparently from recording beam fanning. The solid curve is the theoretical prediction [21].

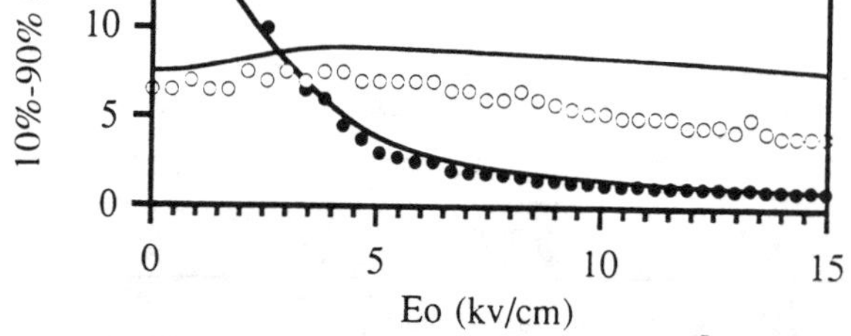

Fig.9. Write and erase time response times versus applied field. The write time is the time to rise from 10% to 90% of peak diffracted intensity, and the erase time is the time to decay from 90% to 10%. The solid curves are theoretical calculations [21].

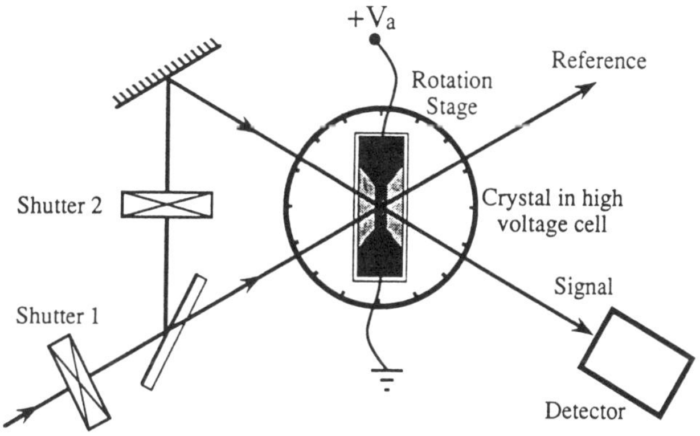

Fig.10. Experimental setup for hologram angle multiplexing. Shutter 2 controls exposure time. The crystal is held on a computer-controlled rotation stage. During readout, shutter 2 is closed, and the stage rotates continuously as the diffraction is recorded[21].

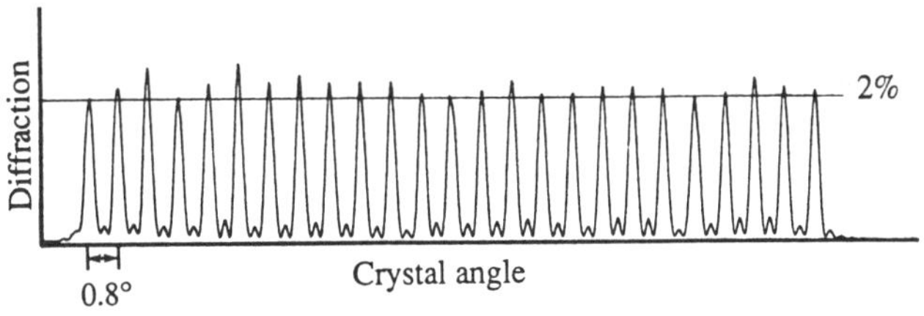

Fig.11. Diffraction as a function of crystal angle for 25 holograms recorded and read at 8.2kV/cm with 2% minimum diffraction efficiency. Angular separation was 0.75°[21].

has been shown that in Ce-doped (0.02 wt.%) SBN:60, an external DC electric field can be used to enhance both Δn_{ss} and τ_e/τ_w [21]. Fig. 8 shows the peak probe diffraction efficiency as a function of the applied field. The calculated peak probe diffraction reaches 100% at $E_o = 5$ kV/cm.

The time required for recording and erasing the gratings is shown in Fig.9. The 10%-90% writing response time is defined as the time during writing for the grating diffraction efficiency to rise from 10% to 90%, similarly the 90%-10% response time is defined as the time during erasing for the grating diffraction efficiency to decay from 90% to 10% of its peak value. When the external field is applied, τ_e/τ_w increases from 0.3 at $E_o = 0$ to 3 at $E_o = 5$ kV/cm.

The experimental setup for recording superimposed angle-multiplexed hologram is shown in Fig. 10. The crystal is mounted upon a computer-controlled rotation stage. A series of plane-wave holograms were superimposed, using shutter 1 to control the exposure time. Shutter 2 was closed during readout. The holograms were reconstructed by continuously rotating the crystal and measuring the diffraction efficiency on a chart recorder. Fig. 11 shows an example of the result for 25 holograms recorded with diffraction efficiency ≥ 2% with $E_o = 8.2$ kV/cm applied to the 1.1-mm thick SBN:60 crystal.

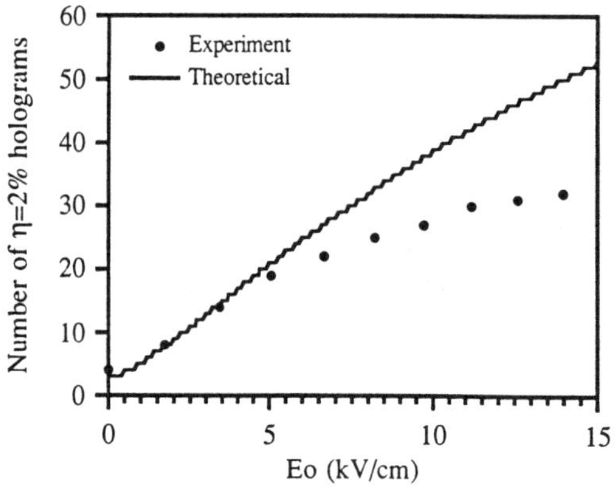

Fig.12. Hologram capacity for 2% minimum diffraction efficiency holograms as a function of applied field. The number rises from 4 with no applied field to 32 at 14kV/cm [21].

The multiplex hologram capacity increased substantially as field was applied to the crystal. Fig. 12 shows the hologram capacity when a minimum of 2% diffraction is required. The number rises from 4 with no applied field to 32 at 14 kV/cm.

3.2.2 Moving Grating for Enhancing Diffraction Efficiency

While energy coupling in PRC is useful for real-time operation as discussed in Section 2, it is often an unfavorable factor when the photorefractive materials are used in storage mode. Energy transfer between the recording beams tends to reduce fringe contrast and limit the achievable index modulation depth. For example, in a 5-mm thick crystal of SBN:Ce with a gain coefficient of 30 cm^{-1}, energy coupling alone would cause the beam ratio to drop from an input value of 1 to 3×10^{-7} at the exit face. This greatly limits the usable crystal thickness.

It was shown in Section 2.2.2 that a moving grating can be used to enhance coupling gain via optimizing the phase angle between the intensity distribution and the induced photorefractive grating. In this section we show that a moving grating can also be applied to cancel the coupling between the recording beams, obtaining an enhanced diffraction efficiency [51].

In general, the complex value of the space-charge field E'_{sc} is composed of its imaginary component $Im(E'_{sc}) = E_{sc}\sin\phi$ and its real component $Re(E'_{sc}) = E_{sc}\cos\phi$. $Im(E'_{sc})$ has a phase shift of $\pi/2$ with respect to the light intensity distribution, and contributes directly to the energy coupling; while $Re(E'_{sc})$ has zero-phase shift and has no contribution to energy coupling. For real-time operation, the moving grating can be used to optimize $Im(E'_{sc})$, while for storage operation, the moving grating can be used to minimize $Im(E_{sc})$ but to optimize $Re(E'_{sc})$. By solving Kukhtarev et al's equations for the case of a moving grating [37, 52-53], the real, imaginary, and the modulus values of the space-charge field E_{sc} as a function of grating velocity were obtained for several applied voltage V_o in SBN:60 (see Fig. 13). At the optimum fringe velocity υ_{opt}, $Im(E'_{sc})$ becomes zero, while the real part and the modulus of E_{sc} reach their maximum values. The highest diffraction efficiency is expected at υ_{opt}. Energy coupling should be canceled at that velocity, and an enhanced hologram with more uniform index modulation along the direction of propagation can be recorded.

The experimental setup for simultaneously measuring the beam-coupling gain and diffraction efficiency with a moving grating is shown in Fig. 14. An argon-ion laser at wavelength 514.5 nm was collimated and split into two extraordinarily polarized beams of equal intensity as the recording beams. The fringes were made to travel by reflecting beam 1 from a moving piezo mirror. A weak Bragg-matched He-Ne probe beam (λ = 632.8 nm) was used to monitor the grating strength independently. Detector D1 and D2 measured the transmitted intensities of beam 1 and 2, while detector D3 measured the diffracted probe-beam intensity. The output for 3-kV applied voltage with stationary (υ = 0) and moving ($\upsilon = \upsilon_{opt}$ = 0.048 µm/sec) grating are compared in Fig. 15. The energy coupling effect was almost canceled and the peak diffraction efficiency increased by a factor of 1.6 when the grating velocity was adjusted to the optimum value.

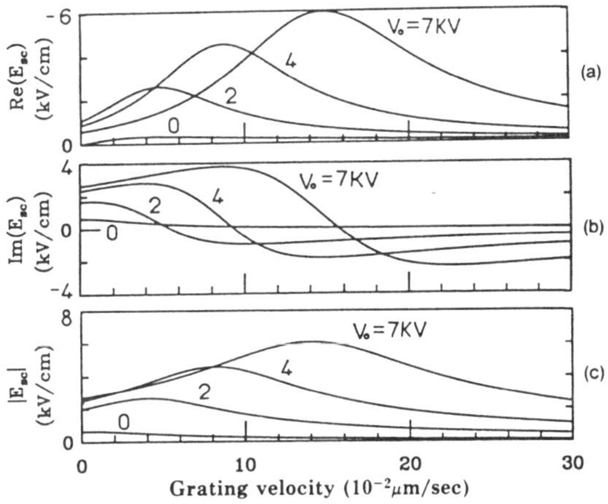

Fig.13. Theoretical results. (a) The real part of the space-charge field, (b) the imaginary part of the space-charge field, and (c) the modulus of the space-charge field versus the moving grating velocity for different applied voltage V_O ($E_O = V_O/d$, with electrode spacing d = 0.57 cm) [51].

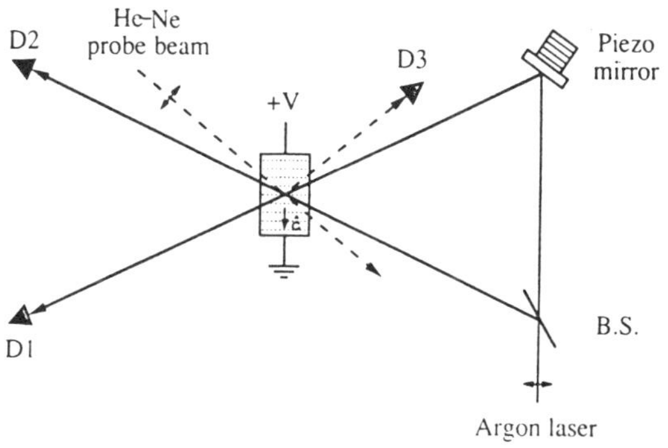

Fig.14. Schematic of the experimental system for simultaneously measuring the beam-coupling gain and diffraction efficiency with the moving grating [51].

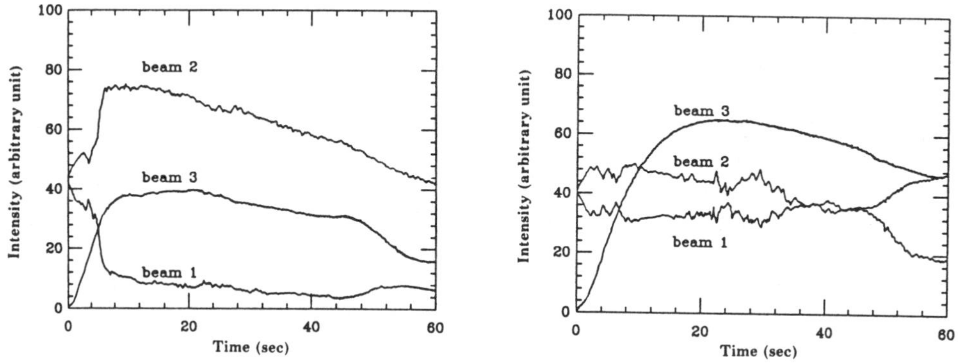

(a) (b)

Fig.15. Experimental curves of the intensities of the transmitted recording beams and diffracted probe beam versus the recording time. The applied dc field E_o = 5.3kV/cm. (a) Without the moving grating. The peak diffraction efficiency is 58%. (b) With optimum moving grating velocity v_{opt} = 0.048 μm. The peak diffraction efficiency is 93% [51].

Moving grating recording has been applied to hologram multiplexing in SBN:60. For 5 superimposed holograms, the average diffraction efficiency was increased from 16% with no moving grating to 18% with moving grating. The improvement was not substantial as our originally expected. The reason is that, in hologram multiplexing using scheduled recording, only the first hologram tends to reach the peak diffraction efficiency and suffer from beam coupling, while the majority of the holograms are recorded in the small index modulation region with short exposures. The coupling effect therefore becomes less important as the number of multiplexed holograms increases.

3.2.3 Special Doping/Processing for Enhancing Near IR Operation

Recent advances in developing tunable semiconductor lasers operating in the near IR [54-55] make applications of PRC using these wavelengths attractive. Although IR sensitive photorefractive materials for real-time operation are more or less available, the lack of IR materials suitable for storage operation is a matter of an important concern. Cobalt-doped $BaTiO_3$ [56] and Ta-doped $KNbO_3$ [57] have been investigated in near IR. For storage applications, $LiNbO_3$ has been shown to be the most promising material for its large storage capacity. Catanzaro et al [47] reported that the near IR response in $LiNbO_3$ can be substantially enhanced by annealing Fe-doped sample in argon atmosphere or by a special Nd doping. Three different samples of $LiNbO_3$, 0.05% (mole) Fe-doped (standard visible sample), argon annealed 0.1% (mole) Fe-doped and 0.05% (mole) Nd-doped, were tested and compared in the near IR.

Fig. 16 shows the comparison of recording dynamic at λ = 790 nm with a beam intensity of I_o = 100 mW/cm^2. An order of magnitude improved sensitivity of the argon annealed Fe-doped and Nd-doped samples over the standard visible Fe-doped sample were measured.

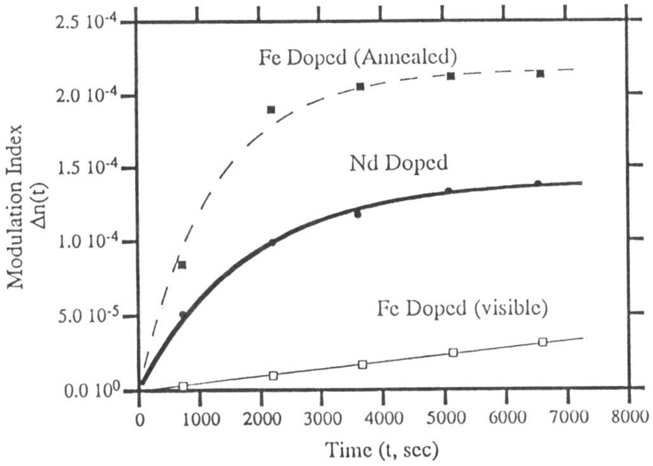

Fig.16. Comparison of holographic recording dynamics at 790nm for three different $LiNbO_3$ samples [47].

Table 5. Comparison of figures of merit for three different $LiNbO_3$ samples at $\lambda = 790$nm

Materials	Δn_{ss}	absorption (cm^{-1})	sensitivity ($\times 10^{-6} cm^3/J$)
$LiNbO_3$:Fe (0.05% mol) (visible)	2.5×10^{-5}	2.3	0.018
$LiNbO_3$:Fe (0.1% mol) (argon annealed)	2.2×10^{-4}	2.3	0.04
$LiNbO_3$:Nd (0.05% mol)	1.4×10^{-4}	0.25	2.2

Table 5 summaries comparison of some of the figures of merit measured for the three samples. The steady-state index modulation were improved by one order of magnitude and the sensitivity were improved by one to two orders of magnitude with the new IR sensitivity samples.

3.2.4 Orthogonal Phase Encoding for Fast Random Access

Storage of 5000 holograms in $LiNbO_3$:Fe has been reported using angle-multiplexing [58]. For fast access of angle-multiplexed holograms, the incidence angle of the reference beam is controlled by an acoustic-optic device (AOD). The access speed is limited by the speed of the AOD, which is about 10 μs/page. For achieving even faster access, an orthogonal phase encoding method has been studied and demonstrated [22, 59], which is expected to achieve random access speed of 0.1 μsec/page.

Fig. 17 shows the schematic of experimental setup. The input image is provided by the input-SLM and is Fourier transformed into the crystal. The reference beam, modulated by a 1-D phase-SLM, consists of a set of unit amplitude plane waves, each separated from the others by more than the Bragg selectivity requirement. Each phase code reconstructs only its corresponding hologram (to first order). That is,

$$\sum_{j=1}^{N} R_{ij}^{*} R_{kj} = \begin{cases} N & (i=k) \\ 0 & (i \neq k) \end{cases} , \qquad (9)$$

where R_{ij} represents the coded reference beam of the jth pixel of the ith code. We need to determine the phase codes which will allow crosstalk-free reconstructions. For the case of binary (0 or π) phases, the solution of this equation is the Walsh function which can be generated by a simple binary tree logic structure from $\log_2 N$ bit address. Using this binary phase reference beam allows the same memory addressing capacity as the conventional angle-multiplexed scheme, but allows implementations which are potentially two orders of magnitude faster because of the faster speeds of binary phase SLMs compared to accousto-optic modulators. An access speed of 0.1 μsec/page can be achieved with a SLM operated at 10 MHz, such as a PLZT SLM [60].

The initially implemented experimental system is similar to Fig. 17 except that for simplicity, the input images were made on transparencies and the phase codes recorded as fixed patterns rather than using SLMs. The initial demonstration was intended to test the phase coding only, rather than a combination of algorithm, material, and SLM performance. The resulting memory output for 2 of 16 stored images is shown in Fig. 18. The measured average signal to noise ratio was 100 to 1.

3.2.5 Incremental Recording for Enhancing Signal Uniformity

Superposition of multiple holograms in real-time responding material requires using some type of recording schedule which takes into account the partial erasure of each hologram present in the material while the other holograms are being recorded. Two holographic recording techniques have been used to obtain a set of uniform superimposed holograms: scheduled recording [3, 61] and incremental recording [22]. Schedule

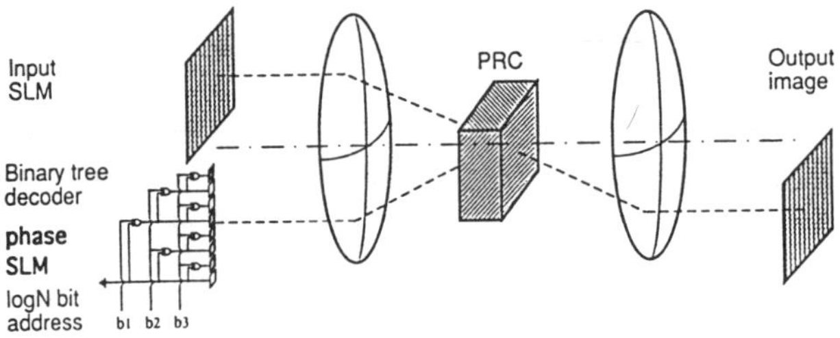

Fig.17. Experimental scheme of orthogonal phase encoding for random access.

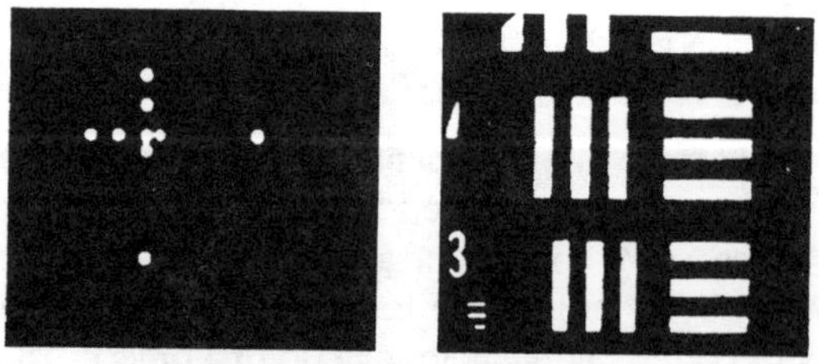

Fig.18. Reconstructed images, which indicate good resolution and low cross talk, with an average signal-to-noise ratio of 100:1 [22].

recording normally involves a somewhat complicated recording procedure with a long first exposure, followed by shorter and shorter exposures. The schedule is calculated from the material's response times and the maximum attainable index modulation. A small error in material characterization (also nonuniformities or changes in material characteristics) can result in highly nonuniform diffraction efficiencies in the final set of holograms. This problem can be avoided in incremental recording. Incremental approach was first introduced by Anderson, et al [62] and Paek, et al [63] for neural network applications, and was later investigated by Taketomi, et al [22] for holographic storage applications.

In incremental recording, each of N holograms is recorded with a series of exposures Δt, which are extremely short compared with the material's response time. During recording, each image and reference pair is sequentially displayed, repetitively cycling through all N images. The first hologram, initially recorded with an exposure Δt, will experience erasure during the time $(N-1)\Delta t$ of the following N-1 exposures, while the other holograms are recorded, as shown in Fig. 19. However, the material's writing and erasing sensitivity both depend strongly on the index-modulation depth. Near the foot of the recording curve, the writing sensitivity is greatest and the erasing sensitivity is lowest. Therefore some of the first recorded hologram will remain after incrementing the other N-1 holograms (see Fig. 20). As this process is repeated, all the holograms' diffraction efficiencies gradually increase. After many cycles, the recording process approaches equilibrium, where the increase of the recording increment is exactly matched by the decrease during the (N-1) erasure increments.

At equilibrium, the hologram diffraction efficiency η shows a small variation ε because holograms incremented more recently are stronger. This variation is controlled by the recording increment time Δt. If we require that all holograms satisfy $\eta_{eq} \leq \eta \leq \eta_{eq}(1+\varepsilon)$, then the maximum Δt is determined to be

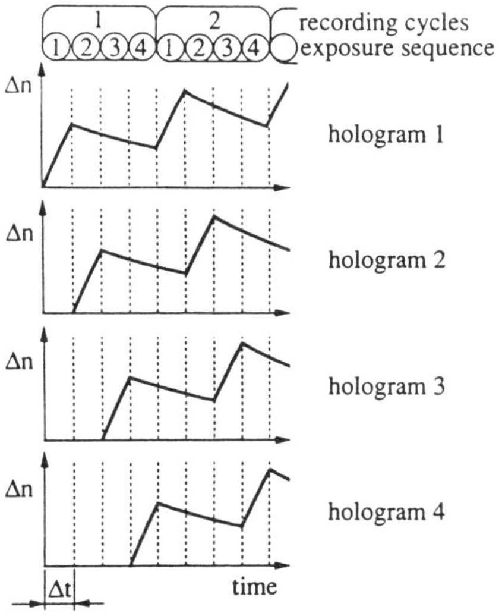

Fig.19. Incremental evolution of the superimposed holograms. The diffraction efficiency of all the holograms increases as the cycle is repeated, gradually approaching the saturation value [22].

$$\Delta t = \frac{\tau_e}{2(N-1)} \ln(1+\varepsilon) \ . \tag{10}$$

The signal uniformity can be made to satisfy the system requirements by using the single parameter Δt. Its value depends only on the erase response time τ_e, which can be measured accurately in any crystal. Therefore, with this technique, control of signal uniformity is simple and accurate.

In incremental recording, the system must repeatedly display the input and reference pairs, reproducing the interference pattern to within much less than one fringe width. To achieve this, the orthogonal phase encoding method discussed in the last section was used. The combination of incremental recording and binary orthogonal phase codes was demonstrated experimentally with a system similar to that shown in Fig. 17. The signal uniformity for the incremental recording was investigated by varying the uniformity parameter ε. The experimental results are shown in Fig. 21 for $\varepsilon = 0.5$ and $\varepsilon = 0.1$. As expected, the signal uniformity was significantly improved with a reduced recording increment.

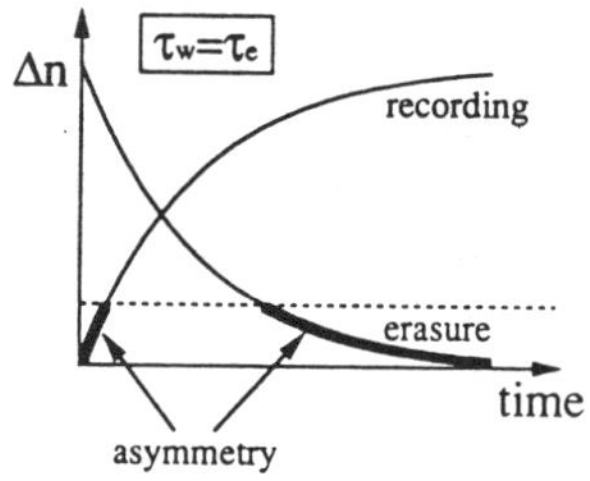

Fig.20. Index modulation Δn versus time, which shows that the slope of the foot of the exponential rise in the recording curve is much steeper than that of the erasure curve, even when the write and erase response times are identical [22].

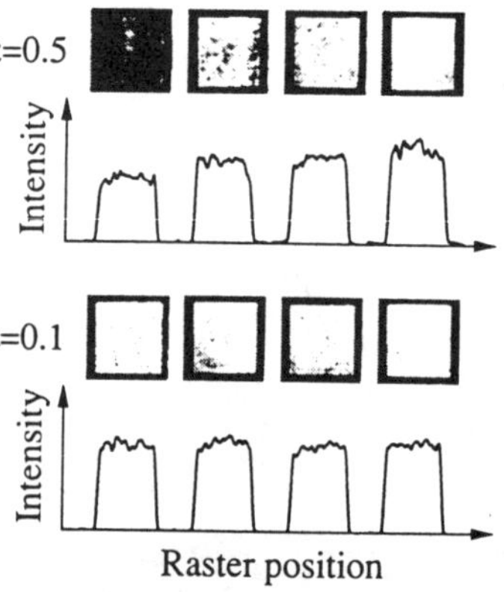

Fig.21. Reconstructed images and the raster scan results for signal uniformity control. The signal uniformity was improved from $\varepsilon = 0.5$ to $\varepsilon = 0.1$ by reducing the recording incremental Δt from $\tau/15$ to $\tau/63$ [22].

IV. Applications

Applications of photorefractive materials have been reviewed in a variety of publications [23, 64-66]. In this paper, we mainly concentrate on the recent application accomplishments for optical interconnection and storage. The discussion will be divided into three parts. Section 4.1 and 4.2 describe the applications of PRC for optical interconnects using real-time and storage operations respectively, and Section 4.3 describes the application of PRCs in dynamic memory.

4.1 Applications of Real-Time Operation for Optical Interconnects

Optical interconnects can be used to remove two important communication bottlenecks: communication between processing elements in a parallel processing array and communication between these processing elements and memories [67]. Reconfigurability is particularly important because it can be used to address such needs as network contention and fault tolerance [68]. Using photorefractive real-time operation, the reconfiguration is achieved by updating the interconnection holograms (the refractive-index gratings) in real-time.

4.1.1 Matrix-Vector Multiplication Using Two-Beam Coupling

Optical matrix-vector multiplication can be used to implement optical crossbar switch for arbitrary interconnections between two 1-D arrays [69-71]. However, this implementation is limited by an intrinsic fan-out energy loss, leading to an energy efficiency of 1/N. Yeh et al proposed and demonstrated that this energy loss can be minimized using beam coupling in PRC [13, 72].

Fig. 22 shows the schematic on implementing the crossbar switch based on a matrix-vector multiplier with a photorefractive amplifier. A small portion of the input beamlets v is split off to be used as signal beam, while the majority of the energy is pass through the beam splitter and illuminates the crystal as a pump beam. The signal beam is fanned by the cylindrical lens and then passes through the spatial light modulator (SLM), which carries the interconnection pattern. After passing through the SLM, the signal beam is imprinted spatially with the interconnection pattern. The pump beam does not pass through the SLM and thus suffered no energy loss due to the fanout. The signal beam which contains the interconnection information is then recombined with the pump beam in a PRC. The energy of the pump will be coupled into the signal beam via

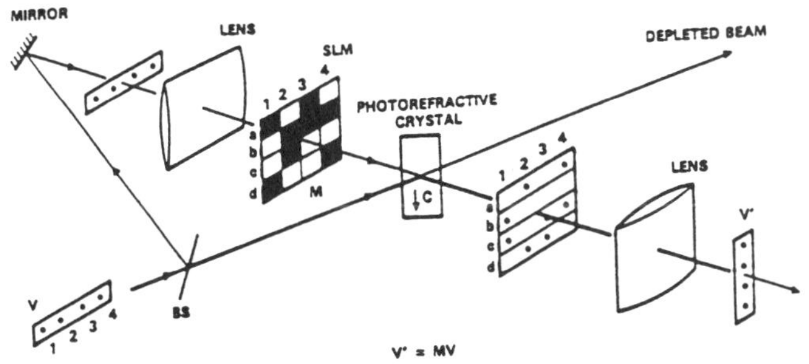

Fig.22. Schematic on implementing the crossbar switch based on a matrix-vector multiplier with a photorefractive amplifier to compensate for fan-out losses [13].

photorefractive two-beam coupling, yielding an extremely high energy efficiency. Reconfiguration is achieved by varying the transmission pattern of the SLM combining with the photorefractive real-time operation nature of simultaneously recording and readout. The elegant idea involved in this approach has been used recently to photorefractive memory in which a weak beam was used to read the memories and the reconstruction was amplified by a photorefractive two-beam coupling, achieving a long term readout [73].

The energy efficiency of the interconnection system shown in Fig. 22 was optimized by completely overlapping the signal and pump beams inside the PRC using two extra Fourier transform lenses (not shown in the figure). In this approach, high energy efficiency is attained for data modulation speeds in excess of the photorefractive response time only at a cost in the contrast ratio. The data modulation depth decreases as the theoretically achievable efficiency increases, since the data signal is only a portion of the transmitted energy. Coherent addition of the data and amplifier signal helps maintain a significant modulation depth even for efficient systems. An energy efficiency of 30-40% was obtained [72].

A matrix-vector multiplier provides N^2 parallelism, allowing only 1-D input array (N) to 1-D output array (N) interconnection. However, a full optical interconnects of an N x N input signal array to an N x N output array requires N^4 parallelism [74]. A matrix-tensor multiplication is required to achieve this N^4 parallelism.

4.1.2 Phase-Coded Correlation Matrix-Tensor Multiplication Using Four-Wave Mixing

Methods of N^4 parallel interconnection were demonstrated to use space multiplexed interconnect weight matrix [75] and angle-multiplexed hologram [3]. In the space-multiplexed scheme, the interconnect tensor contains N x N sub-matrixes, and each sub-matrix contains N x N pixels providing N to N interconnection. A large size of SLM is required to implement the interconnection. Another drawback associated with this scheme is that the dimension scale of input and output arrays are different by a factor of N x N. The angle-multiplexing approach suffers crosstalk between holograms arising from Bragg degeneracy, reducing the parallelism to N^3. Ford et al proposed and demonstrated a novel 2-D array interconnection approach, called phase-coded photorefractive Correlation Matrix-Tensor Multiplier (CMTM) [2, 16, 76-77]. In CMTM, matrix-tensor multiplication was implemented by a correlator via four-wave mixing in PRC. Phase-encoding was used to improve the SNR, and the interconnect tensor was compressed for reducing its size and SBP and producing the same dimension scale for both input and output arrays.

In CMTM, the required output array c_{pq} is the product of the input array g_{ij} with an interconnect tensor W_{ijpq} containing the N^4 complex-valued interconnect weights,

$$c_{pq} = \sum_{i,j=1}^{N} W_{ijpq}\, g_{ij}\,, \qquad\qquad p, q, = 1, 2, ..., N\,. \tag{11}$$

This product is contained in the correlation of an input image with a properly constructed control image. The interconnection tensor is arranged into a 2-D image of N x N array of N x N pixels sub-array W_{pq}. Each sub-array contains the interconnection information for a single output position. In effect, the $(p, q)^{th}$ sub-array is a mask which if superimposed over the input image would transmit only those input pixels connected

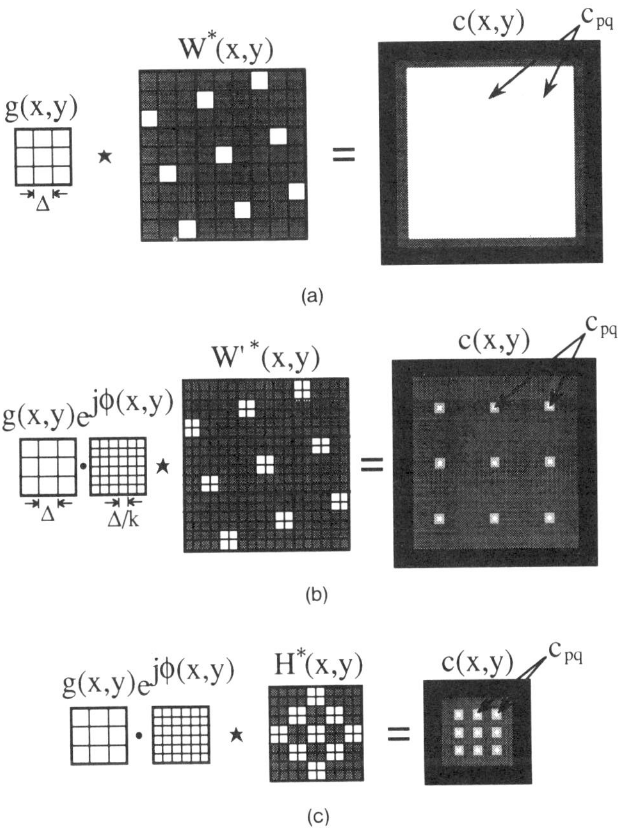

Fig.23. Conceptual construction of the correlation matrix-tensor multiplier algorithm in three steps. (a) Correlation of the input array with the control image produces the connected output at sites embedded in a field of noise. (b) Phase coding of the input and control image reduces background noise relative to the signal. (c) The final control image H(x,y) is produced by compressing (overlapping) W'(x,y), reducing the control SBP at the cost of superimposing noise onto the signal sites [2].

$(p, q)^{th}$ output position. Fig. 23 shows an example 3 x 3 input image g(x, y) and 1-to-1 interconnection pattern W(x, y). The output image c(x, y) is the cross correlation of g(x,y) with the complex conjugate of W(x, y)

$$c(x,y) = g(x,y) * W^*(x,y)$$
$$= \int_{-\infty}^{+\infty}\int_{-\infty}^{+\infty} g(\alpha,\beta) \sum_{p',q'=1}^{N} W_{p'q'}(\alpha - x, \beta - y) d\alpha d\beta \quad . \tag{12}$$

Each point of this correlation represents the product of the shifted input image with the control image. The points $x_p = N\Delta - pN\Delta$, $y_p = N\Delta - qN\Delta$ in the correlation plane correspond to where the shifted input image exactly overlaps the control image's (p, q)th sub-array. The output sites are spaced N times father apart than the input pixels, which is similar to that of the space-multiplexed sub-matrixes implementation [75]. Between these sites the correlation produces side-lobes from the neighboring correlation peaks. For enhancing the SNR of the CMTM output, a unit-amplitude random phase code $\exp[j\phi(x, y)]$ composed of Nk x Nk pixels of size Δ/k was introduced. A phase-coded control image W'(x, y) can be made by multiplying each sub-array W_{pq} by $\exp[j\phi(x, y)]$. If the input is also multiplied by the same phase code before correlating with the conjugate of W'(x, y) [see Fig. 23 (b)], then on the c_{pq} sites the two additional phase factors cancel and there is no effect on matrix-tensor products. At the off-site positions, however, the added phase factor does not cancel. The noise becomes a sum of random phasors adding in a 2-D random walk, so that the average noise amplitude grows only as the square root of the number of contributions. The matrix-tensor products c_{pq} would then lie in a field of lower intensity background noise. Introduction of the phase code increases the SBP of the control image by a factor of k^2, i.e., k^2N^4. For reducing the SBP of the control image, the control image was spatially compressed, as shown in Fig. 23 (c). The final control image, H(x, y), is given by

$$H(x,y)=\sum_{p,q=1}^{N} W_{pq}(x-p\Delta,y-q\Delta)\exp[-j\phi(x-p\Delta,y-q\Delta)] \ . \tag{13}$$

Essentially, the phase-coded sub-arrays have been shifted so that their centers are separated by exactly Δ, then summed together. The correlation of the phase-coded input with the conjugate of the compressed interconnection tensor,

$$c(x,y)=\{g(x,y)\exp[j\phi(x,y)]\}* \left\{\sum_{p,q=1}^{N} W^{*}_{pq}(x-p\Delta,y-q\Delta)\exp[j\phi(x-p\Delta,y-q\Delta)]\right\} \tag{14}$$

once again produces the required matrix-tensor products c_{pq}, now located at $x = l\Delta$, $y = m\Delta$.

This compression produces two advantages; it reduces the size and bandwidth of the connection tensor, and it rescales the output pixel locations to the original (input) spacing. The penalty invoked is a reduction in SNR, since noise arising from the overlapped phase-coded subarrays now lands on the output matrix-tensor product locations. The number of pixels of the control image is reduced from k^2N^4 of the uncompressed image to $[(2N-1)k]^2$ of the compressed image. As k decreases, the theoretically infinite SNR of the interconnection can be trade in for a reduction of control image SBP. Note that although the sub-arrays W_{pq} now overlay one another, the element are summed, not multiplied, so that the transmission of each does not affect the others, and arbitrary weighted interconnections are still possible.

The first implementation of CMTM algorithm was carried out using real-time four-wave mixing in PRCs on the system shown in Fig. 24 [16]. In four-wave mixing, three input waves incident upon a crystal generate a fourth waves. By using the Fourier transforms of three input images, an optical triple product is formed. In the geometry shown in Fig. 24, the output wave $U_3(x, y)$ is given by [15]

$$U_3(x,y)=KU_1(-x,-y)\otimes U_2(-x,-y)*U_4(-x,-y) \ , \tag{15}$$

where U_1, U_2 and U_4 are the input amplitudes, K is a constant depending on the material properties and operational parameters, ⊗ denotes convolution, and * denotes correlation. Taking

$$U_1(x,y) = H(x,y)$$
$$U_2(x,y) = g(-x,-y)e^{j\phi(-x,-y)} \quad . \qquad (16)$$
$$U_4(x,y) = \delta(x,y)$$

we obtain the required correlation output, $U_3 = g(x,y)e^{j\phi(x,y)} * H^*(x,y)$.

The phase-code and control images were made using surface-relief etched and amplitude patterned glass plates, while the input was a simple transparency. The PRC used was a 5 x 5 x 2.5 - mm^3 45°-cut $BaTiO_3$ [35]. All inputs were presented and interacted simultaneously to produce the output wavefront. For this experiment, the uncompressed control image of the center of Fig. 23 was used. The output from two different inputs in a 3 by 3 array are shown in Fig. 25. The interconnection was a randomly distributed half-filled pattern with binary weights. That is, each input connected to 5 of the 9 outputs. The average SNR for all the interconnection was 25 (the worst SNR was 2), while the average signal to background intensity (between pixels) was 200. The overall input to output efficiency was approximately 2.5%.

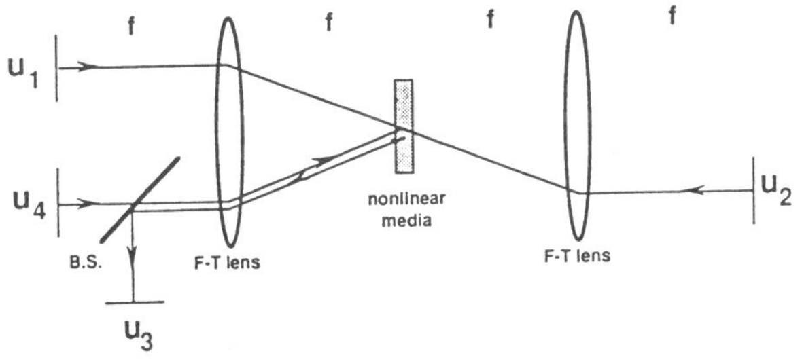

Fig.24. Photorefractive correlator system. Inputs U1, U2 and U4 are Fourier transformed into the crystal to generate the output U3 [76].

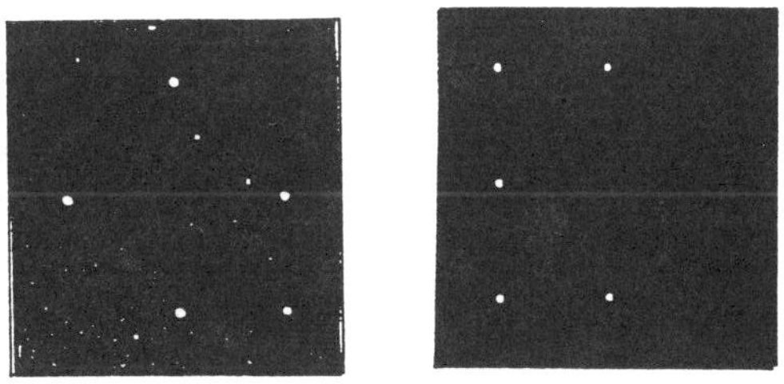

Fig.25. Real-time CMTM system output. The output for two input is shown [76].

4.2 Application of Storage Operation for Optical Interconnects

The main limitation of real-time operation came from the slow photorefractive response time. The response time for real-time operation in currently available materials may vary from microseconds to seconds depending on the material and incident light intensity (see Table 1 in Section 2.1). However, most opto-electronic computing applications demand interconnection switching at MHz rates or faster. Although photorefractive semiconductor materials offer high speed with relatively low optical input intensities, operation is limited by the complicated techniques and critical conditions for optimizing photorefractive effects (see Section 2.2). In certain cases, however, storage operation can be used to decoupled the interconnection reconfiguration time from photorefractive response time.

Many interconnection patterns can be superimposed in a single crystal using a proper multiplexing technique (wavelength, angle or phase-encoding). The interconnection could be reconfigured by changing the readout conditions (wavelength, angle or phase-code) to access different pre-stored patterns. Recent advances in developing tunable semiconductor laser array support random access times of 100 nsec [54], while discrete devices can select one of twelve wavelengths with a latency of 1 nsec [55]. These lasers make wavelength multiplexing a very attractive method. Two examples of storage operation for optical interconnection will be given in this section. Both of them used wavelength multiplexing.

4.2.1 Correlation Matrix-Tensor Multiplication Using Storage Operation

Instead of using "real-time" four-wave mixing, reconfigurable interconnection with CMTM was implemented by wavelength-multiplexed holographic storage operation [76-77]. The heart of the experimental system for preprogrammed interconnection is shown in Fig. 26. Reflection holograms are suitable for reconfigurable interconnection application because they simultaneously offer maximum wavelength selectivity (≈ 1 Å with 1-mm thick crystals) and minimum angular selectivity (≈ 1° with 1-mm thick crystals). A series of control images are recorded in a z-cut $LiNbO_3$ at different wavelengths. The correlation of the input with the desired control-image is achieved by tuning to the corresponding wavelength. The input passively reads the stored holograms without changing the index modulation. The time to respond to a new input and the switching time between prerecorded interconnection patterns is limited only by the speed at which you can modulate the input and the switch readout wavelength, respectively.

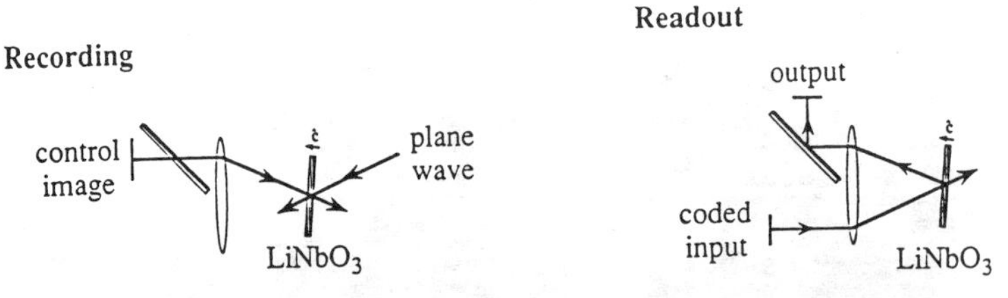

Fig.26. Storage CMTM operation using reflection holograms recorded in z-cut $LiNbO_3$. Several interconnection are superimposed by recording a series of hologram, each with a unique wavelength address. During operation, the input coded by wavelength (and by unchanging phasecode), reads one reference pattern [76].

(a) (b) (c)

Fig.27. Readout of 3 wavelength multiplexed interconnections. Three interconnections were stored at 476 (broadcast), 488 (irregular), and 496 (broadcast) nm. Interconnected output from a single input pixel is shown at each wavelength. No crosstalk is visible [76].

The laser used for the experiments was an argon ion laser, which has discrete output wavelength distributed over the blue-green spectrum. Because these lines are separated by several manometers, it was impossible to test theoretical predictions of sub-Angstrom selectivity. However, this selectivity was demonstrated in our next example of the Reconfigurable Memory Distribution Network (Section 4.2.2) in which a tunable Ti-Sapphire laser was used. Fig. 27 shows the individual pixel readouts for three different array interconnections stored by wavelength multiplexing. Interconnections were recorded at three strong argon blue lines, 470, 488 and 496 nm. All connections had the same phasecode. Fig.27 (a) and (c) were full broadcast connections while Fig.27 (b) was a randomly half-filled connection where each input connected to 12 of the 25 outputs. The figure which shows the system output for a single input pixel as the wavelength is scanned through the three recording wavelengths. No crosstalk is visible between the different wavelength multiplexed holograms.

CMTM algorithm performance was also investigated in large array interconnection using fully compressed phase-only interconnection control images. The scaling of output SNR with the phasecode density k was tested. At constant array size, the intensity and uniformity of the output increases with k, as predicted. For a constant N = 16, the average output SNR increases from 14.5 with k = 8 to 319 with k = 32. Finally, the scaling performance with increasing array size was measured. Fig. 28 shows the output for different input array size (N^2) and phase-code density k. The interconnection control image SBP was held constant by reducing k linearly with N. The control image size was 1000 x 1000 pixels (25 mm x 25 mm), approaching the realistic maximum for an optical system to resolve with good accuracy. The intensity and uniformity of the output decreases as N is increased, as expected. The average output SNR decreases from 319 with N = 16 and k = 32 to 8.1 with N = 64 and k = 8. The overall results match the theoretical predictions and are encouraging for eventual practical application of CMTM interconnection.

4.2.2 Reconfigurable Memory Distribution Network

The Memory Distribution Network (MDN) is particularly concerned with the parallel interconnection between large arrays of Processing Elements (PEs) and a bit plane

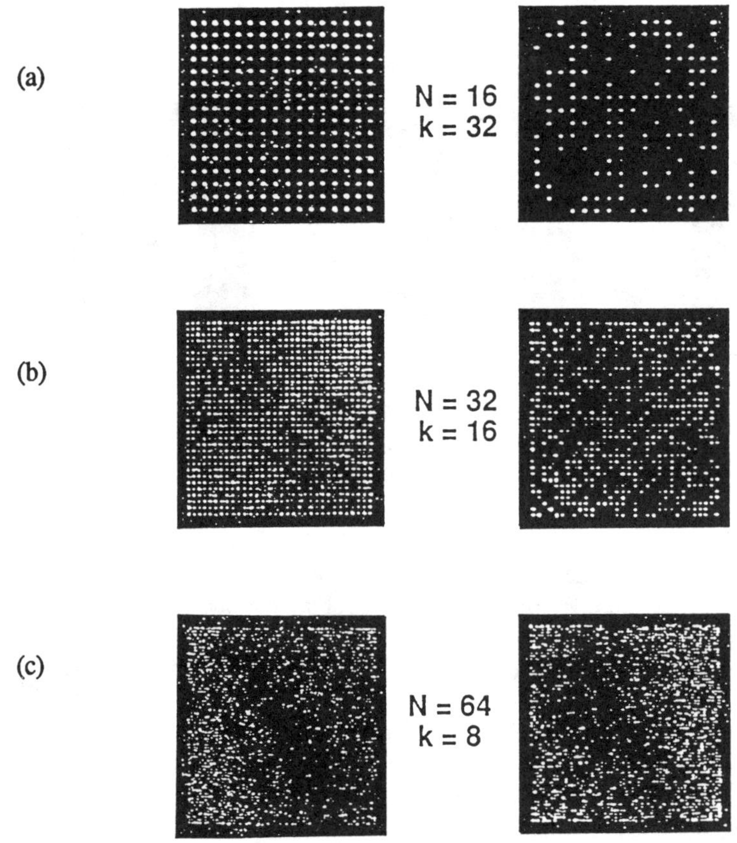

Fig.28. Effect of array size on output SNR. The control image SBP was held at about 1000 x 1000 by decreasing k linearly with N. At left, all inputs on. At right, half the inputs masked. Output SNR decreases as N^2 grows from 256 to 1024 to 4096 [76.

oriented optical secondary storage system. The MDN consists of a page oriented memory, a reconfigurable volume holographic lenslet array for interconnection, an electronic selection unit, and a processing array (Fig. 29). The volume holographic lenslet array is used to image portions of the memory page onto an electronic selection unit via a series of off-axis lenses recorded in a PRC using wavelength multiplexing. By illuminating the hologram with different wavelengths, different lenslets multiplexed in the volume can be accessed to achieve reconfiguration. Once a portion of the memory has been imaged onto an electronic selection unit, the information is latched until a request is made from the processing array. When the processing array sends a request for data, the information is then electronically selected and optically transmitted to the processor.

The optical system diagrammed in Fig. 30 was constructed and studied to implement this system. The system input comes from a memory plane illuminated by a tunable Ti-sapphire laser operated at around 0.8 μm. The memory plane is then imaged to a CCD camera by a wavelength multiplexed reflection holographic lens array.

The key component in this system is the wavelength selective holographic lenslet. We have studied reflection holographic lenses recorded in a 1-mm-thick Fe-doped $LiNbO_3$ crystal [78-79]. The basic characteristics related to these volume holographic lenses are field of view, SBP, wavelength selectivity, aperture limited diffraction resolution, aberration and diffraction efficiency.

Although reflection holograms are used to minimize the Bragg angular selectivity of the volume hologram, we expect a limited field of view in imaging with the holographic lenses. A uniform object will construct a nonuniform output image as shown in Fig. 31. An on-axis lens (focal length = 6 cm, F/# = 3) was simulated and the result is displayed in Fig. 31(a). The lens was then recorded, and the field of view was measured [points in Fig. 31(a)]. This lens was then used to image a 10 mm x 10 mm grid [Fig. 31(b)], demonstrating a field of view. The field of view strongly depends on the off-axis parameters Δx (the lateral image shift) and F-number (F/#) of the lens. Fig. 32 shows the Bragg-limited Field Of View (B-FOV, which is defined to be the maximum lateral dimension of a uniform object that can be imaged with an intensity variation smaller than 50%) as a function of Δx for different F/#'s (or focal lengths). It shows that B-FOV drops quickly as the off-axis parameter increases or as the F/# decreases. This result indicates that the interconnect broadcast distance (Δx) of MDN is limited for uniform imaging.

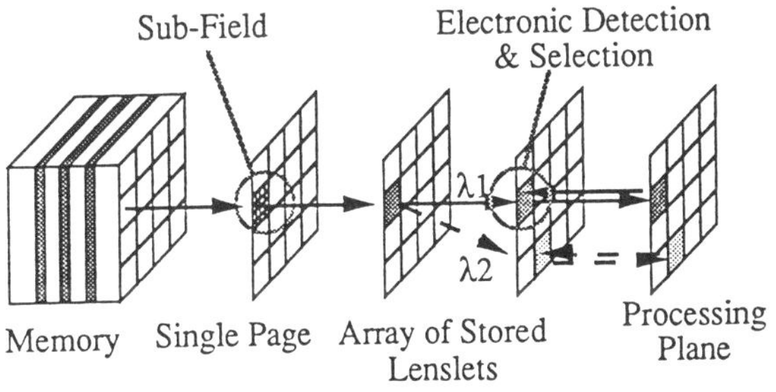

Fig.29. Schematic diagram of the reconfigurable memory distribution network interconnect [80].

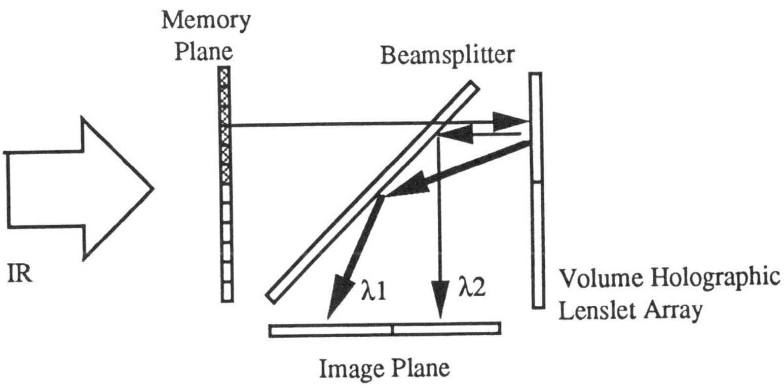

Fig.30. Optical implementation of memory distribution [80].

Because of B-FOV, a small F/# lens might not be always desirable since the throughput cf the lens is not only limited by the conventional SBP of the lens but also limited by the extent of the B-FOV. Two figures of merit were defined to aid in the evaluation of the image capacity of the thick holographic lenses: the SBP of the images limited by the lens aperture, i.e. SBP:Lens; and the SBP of image limited by the B-FOV, i.e. SBP:B-FOV. In general, a small F/# lens will provide a large SBP:lens but a small SBP:B-FOV, and vice versa (see Fig. 33). The optimum condition for a maximum image information throughput is that these two parameters are equal. For example, in the case of $\Delta x = 10$ mm (see Fig. 33), an optimum SBP of 2000 (calculated in one dimension) will be obtained in a F/# = 16 lens. The SBP optimization is important for MDN design.

Wavelength selectivity is maximized in the reflection hologram. Although the B-FOV is strongly dependent on Δx and F/#, the wavelength selectivity $\Delta\lambda$ of the lens is almost independent on these parameters. It is much more strongly dependent on crystal thickness and operating wavelength. $\Delta\lambda = 0.15$ nm was obtained in a 1-mm thick lens recorded in $LiNbO_3$ at $\lambda = 0.8$ µm. Using current technology for tunable surface emitting diode lasers, a range of 26 nm is achievable [54], yielding a total of 173 multiplexed holograms.

The aperture limited diffraction performance of the photorefractive lenses was measured using an on-axis lens (F/# = 3) recorded in a 1-mm thick $LiNbO_3$. A resolution target served as the input object. The image formed by the photorefractive lens was then magnified and imaged to a camera by a F/# = 2 conventional glass lens. Fig. 34 shows the magnified center portion of the image. Group 6-4, or 5.5 µm can be resolved. This is close to the theoretical limitation of 3.8 µm. In both on-axis and off-axis lenses, no noticeable aberration distortion was observed.

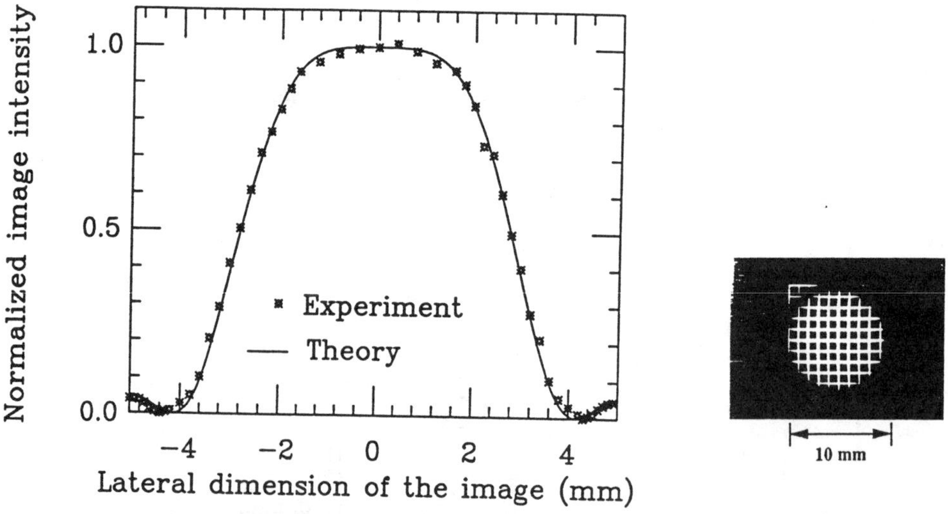

(a) (b)

Fig.31. (a) The experimental and theoretical results on the Bragg angular selectivity on the intensity uniformity of the image formed by volume holographic lenses recorded in a 1-mm-thick $LiNbO_3$. (b) The corresponding image photograph [78, 79].

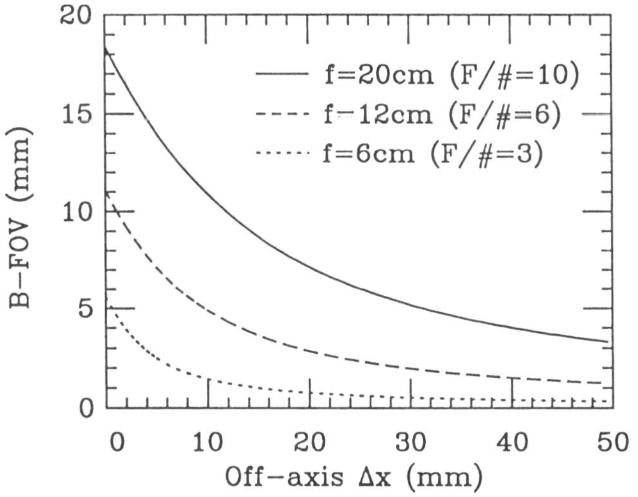

Fig.32. B-FOV as a function of off-axis parameters for different F/# (or focal length) lenses [78, 79].

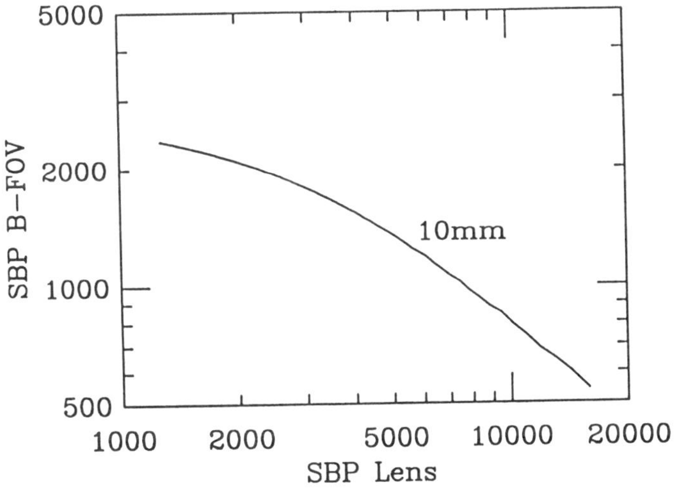

Fig.33. The relationship between SBP:B-FOV and SBP:Lens for off-axis parameter Δx = 10 mm [78, 79].

The diffraction efficiency of a single lens was measured to be 25%. If N holographic lenses are superimposed, the diffraction efficiency for each lens will drop by an approximate factor of N^2. Given a standard MOSIS detector design of 2 μm, the noise equivalent power is 1 μW/MHz. To achieve a SNR of 20:1 (sufficient for digital communication), a signal of 20 μW/MHz is needed. Assuming 1 W input optical power and 5 reconfigurable interconnections, a total number of 500 detectors can be addressed in parallel.

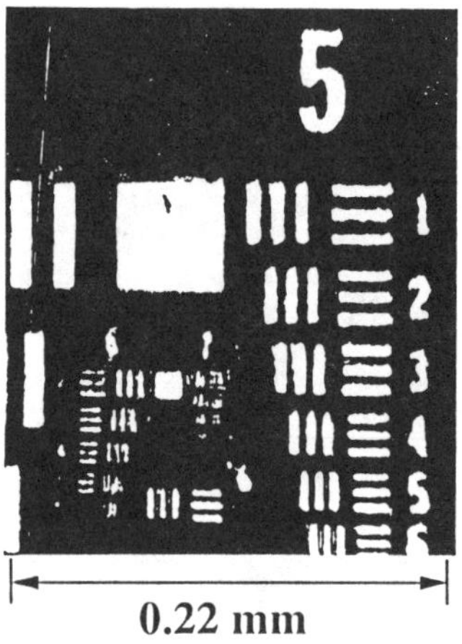

Fig.34. The image of a resolution target formed by a F/# = 3 photorefractive lens. Group 6-4 (5.5 μm) can be resolved [78, 79].

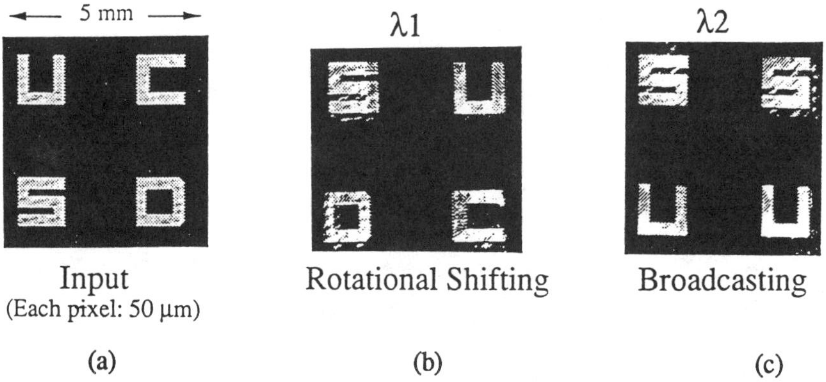

Fig.35. Experimental result of memory distribution network. (a) Input memory page. (b) Output at λ_1, memory permutation. (c) Output at λ_2, memory broadcasting [80].

MDN interconnection was demonstrated by Catanzaro et al [80] using the system shown in Fig. 30. Two superimposed 4 x 4 holographic lens array were recorded in a 1-mm thick IR sensitivity $LiNbO_3$ (see Section 3.2.3 about the material) at near IR with a wavelength separation of 0.15 nm. Memory permutation and broadcasting were implemented. Fig. 35(a) shows the input memory page. The pixel size is 50 μm x 50 μm. Fig.35(b) is the output at λ_1, showing the MDN function of permutation. When the wavelength is changed by 0.15 nm, the result is shown in Fig.35(c), demonstrating the MDN function of broadcasting. The crosstalk was too low to be measured. The ability to permute the memory input page provides PE array with access to a large scope of memory. Arbitrary access can eventually be achieved by cascading several layers. The broadcasting capability essentially provides shared memory access. The fan-out duplicates the memory to several processing elements. Again by cascading several layers of the interconnect, full fan-out can be achieved [81].

4.3 Application of Photorefractive Crystals in Dynamic Memory

The dynamic memory studies were intended to solve requirements which led to contradictory demands on the photorefractive materials. The first problem came from the requirements of memory updating and nondestructive readout. Although the write/read/erase capability of photorefractive material is very useful to memory related applications, such as reconfiguring optical interconnection patterns, updating interconnection weights for neural network or updating information in parallel access optical memory, the stored memory information will be gradually degraded after many readings. Fixation can be used to solve this problem, but it is only suitable for permanent memory without requiring memory updating. An alternative solution of using a dynamic circulating photorefractive memory architecture investigated by Sasaki, et al [5] will be discussed in this section. The second problem came from the requirements on slow erasure for large storage capacity and fast memory updating. It was pointed out in Section 3.1 that an asymmetric erasing/writing time constant τ_e/τ_w is desirable for a large storage capacity. However, in contrast, fast erasing is also preferred in certain applications that require fast memory updating, such as neural networks. This contradiction being resolved by the dynamic memory can be further enhanced by employing selective erasure [82].

4.3.1 Write/Read/Erase and Nondestructive Readout

The objective of the dynamic circulating memory approach is to construct a dynamic photorefractive memory that provides (1) nondestructive readout, so that after many readings the memory can be refreshed by rewriting the still-useful information from one crystal into the other, and (2) write-erase capability, so that while rewriting the useful part of the information from one crystal to the other, we also can replace obsolete information by new information introduced from the memory input. A schematic diagram of the photorefractive circulating memory is shown in Fig. 36. The input images are stored sequentially in one of the two PRCs (say, PRC1) until the total memory capacity of that crystal is reached. After many readouts the stored information in PRC1 will be degraded but still legible. We then transfer the useful part of the information from PRC1 to PRC2.

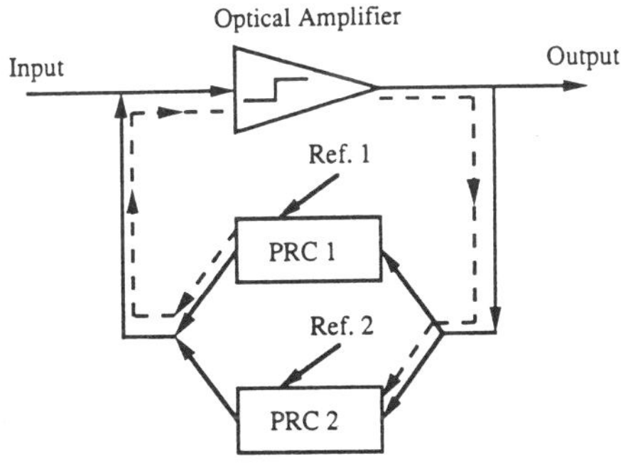

Fig.36. Schematic diagram of the dynamic photorefractive memory. The system consists of an optical amplifier and two functionally identical PRCs. The dashed line is the transfer path from PRC1 to PRC2 [5].

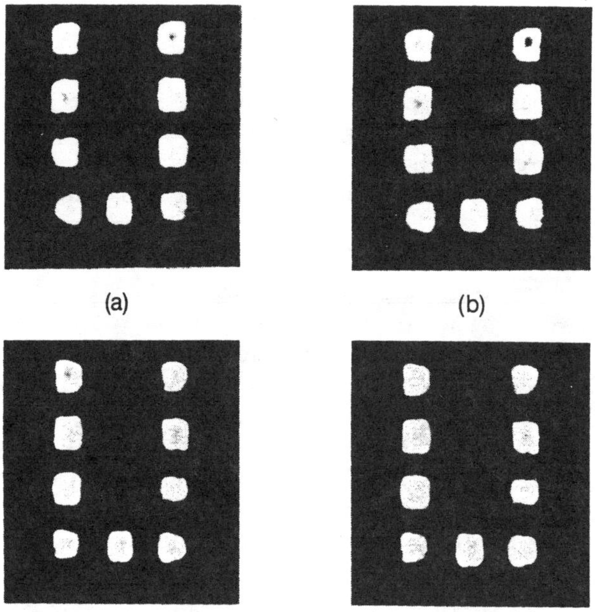

Fig.37. Photograph of transferred images after (a) one, (b) five, (c) ten, and (d) fifteen transfers [5].

If the storage capacity of PRC2 is not reached after the transfer, additional information can be recorded in PRC2. The first crystal is then optically erased to prepare it for the next transfer cycle. The two crystals are functionally identical, exchanging tasks periodically.

An optical amplifier is used to introduce an effective write-erase asymmetry into the system by using a weak readout beam in one crystal and an intense writing beam in the other. The experiment demonstrated transferring a single image between two 1-mm thick crystals of Ce-doped SBN:60. A Hughes liquid-crystal light valve (LCLV) operated in the thresholding regime was used as an optical amplifier. A single image of approximately 1% diffraction efficiency was stored in one of the two crystals and transferred to the second crystal, then back to the first one, and so on. To avoid phase distortion owing to the crystal imperfections, the recording was performed in an image plane. The transferred images are shown in Fig. 37 (a), (b), (c) and (d) after one, five, ten and fifteen transfers, respectively. No noticeable degradation in the pixel (400μm x 400μm) quality was observed after fifteen transfers with the average SNR better than 86. This experiment shows that the circulating memory architecture provides write-erase storage capability with robustness to multiple optical readouts.

4.3.2 Selective Erasure for Fast Memory Updating

The conventional method of incoherent erasing holograms stored in a PRC is to illuminate the crystal with a uniform optical beam. This approach is limited by the slow erasing time due to the nature of the exponential erasing curve. Further, it is incapable of selectively erasing one of the multiple stored memory pages unless a circulating memory architecture (Section 4.3.1) is used. In contrast, the selective erasure process proposed by Huignard, et al [83] has been shown to be much faster than the incoherent erasure process and it is possible to erase only one page out of multiple stored holograms [84]. The dynamics of the selective erasure process was recently studied by Sasaki et al [5, 85].

Fig. 38 shows the principle of selective erasure process based on superimposing a π-phase shifted grating onto the old grating which is selected for erasure. The selective erasure process behaves like an extension to the negative time side of the usual recording exponential curve (see Fig. 39), providing a much faster erasing time than with incoherent erasure. The time to completely erase the old grating from the initial refractive-index modulation Δn_o is given by [82]

$$\tau_e = \tau \ln\left(1 + \frac{\Delta n_o}{\Delta n_{ss}}\right) , \tag{17}$$

where τ is the photorefractive response time.

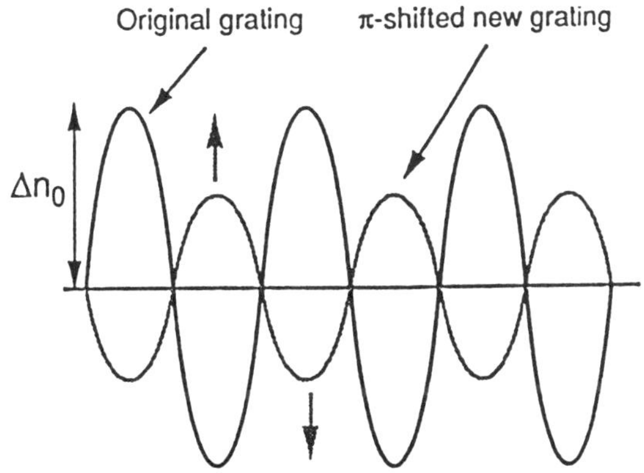

Fig.38. The principle of selective erasure: superimposing a π-phase-shifted image on the old image.

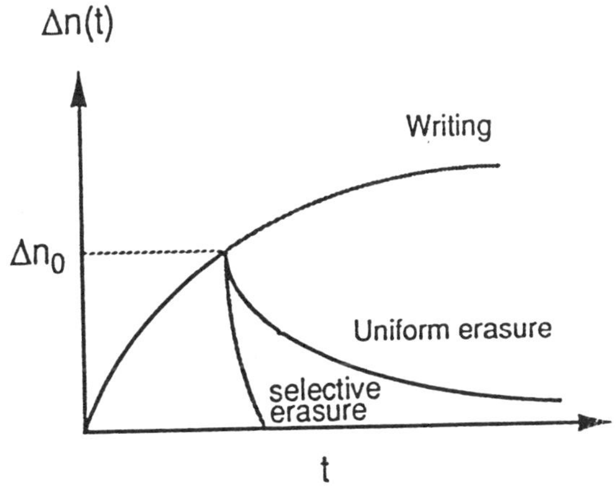

Fig.39. Erasure time comparison shows that selective erasure is much faster than uniform erasure.

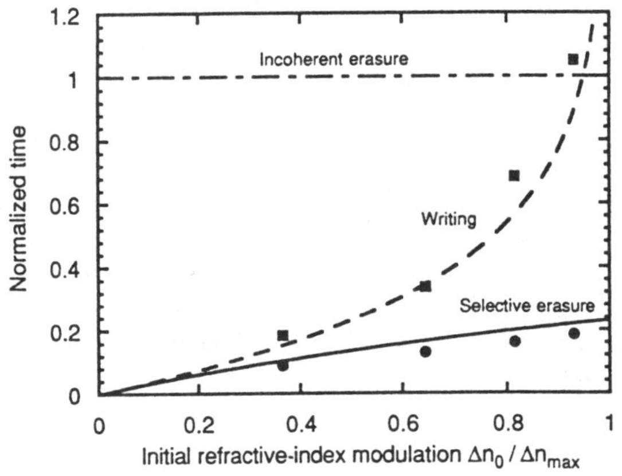

Fig.40. Times for writing, incoherent erasure, and selective erasure of a single grating as functions of the normalized refractive-index modulation $\Delta n_o/\Delta n_{ss}$. Time is normalized by the incoherent erasure time [82].

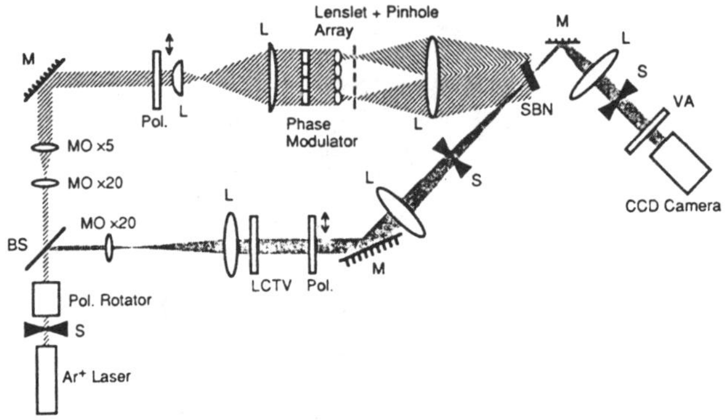

Fig.41. Experimental setup for selective erasure. A combination of the phase modulator and the lenslet and pinhole array is used to generate orthogonal reference beam phase codes [82].

Experiments for verifying the result of dynamic analysis were made by writing a single grating of 2 μm period in a 1.1 mm thick crystal of Ce-doped SBN:60. An ordinarily polarized Ar^+ laser (λ = 0.514 nm) was used for the grating recording and selective erasure process. The diffraction efficiency was monitored by a Bragg-matched extraordinarily polarized weak He-Ne laser beam (λ = 0.633 nm). A π phase shift was induced by a piezo-mirror. Fig. 40 shows the comparison of time for writing, incoherent erasure and selective erasure as a function of the normalized initial refractive-index modulation $\Delta n_o/\Delta n_{ss}$. Using exponential dynamics, the writing and incoherent erasure times were defined as the time required to write a grating from zero to the refractive-index modulation Δn_o and by the time to erase 95% of the original refractive-index modulation Δn_o, respectively. A good agreement between the experimental results and the theoretical curves was found. The selective erasure is much faster than the incoherent

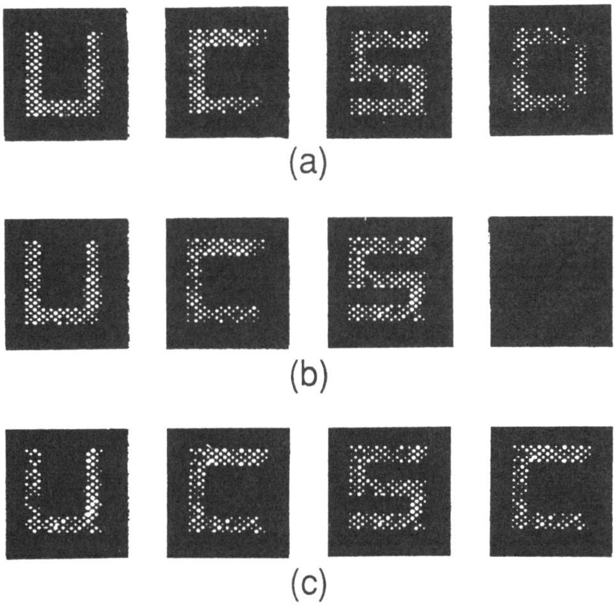

Fig.42 Experimental result of memory updating using selective erasure. Photographs of stored binary images (a) before selective erasure, (b) after selective erasure, and (c) after image update [82].

erasure owing to the coincidence of the previous charge distribution and the π phase shifted new light illumination pattern.

Selective erasure, combined with the incremental recording schedule, (see Section 3.2.5) was applied to construct a dynamic photorefractive memory, The experimental setup is schematically shown in Fig. 41. Images were introduced through a liquid-crystal television (LCTV) and recorded in the Fourier plane. Orthogonal phase codes (see Section 3.2.4) were used for multiplexing and addressing. A one-dimension phase modulation illuminates a lenslet and pinhole array to create a series of phase-modulated point sources. The point sources are Fourier transformed to illuminate the entire crystal surface. A reconstructed image was captured by a CCD camera, thresholded, pixelated electronically and then fed back for display on the LCTV, and recorded back to the PRC during incremental recording cycle. After all the images were stored, selective erasure was used to replace a page of old information by a new one. First the old information page was selectively erased by recording the old image with a π-phase-shifted reference beam. The new information page is overwritten using the same multiplexing address.

Four image pages stored in the PRC were letters **U**, **C**, **S**, and **D** [see Fig. 42(a)]. To demonstrate the selective erasure and overwrite capabilities of the dynamic photorefractive memory, we chose to delete the **D** and replace it with a **C**. The letter **D** was reconstructed from the PRC and stored in the LCTV and then recorded back with a

π-phase-shifted reference phase code at the same memory address so that it could be selectively erased. This selective erasure process was also divided into time increments of Δt to allow circulation of the other three images so that their incoherent erasure could be avoided. After the selective erasure process was completed [see Fig. 42(b)], a new letter C was introduced from the computer and overwritten into the PRC. Incremental recording was used for the new as well as for the old images [see Fig. 42(c)].

V. Conclusion and Future Directions

Photorefractive materials for real-time operation have been applied to reconfigurable interconnections using either two- and four-wave mixing. The difficulty that prevents these applications from becoming actual products is lack of an ideal material. The required material characteristics for real-time operation are large achievable refractive index modulation (i.e., $\Delta n_{ss} > 10^{-5}$), fast response (i.e., nsec) and large crystal thickness (i.e., mm to cm). Ferroelectric materials have large Δn_{ss} but slow response and can only be operated efficiently in the visible wavelength range. Ferroelectrics may only be able to play a major role in "prove of principle" experimental demonstrations. Semiconductor materials have a great chance of success provided that their speed can be improved by at least an order of magnitude and that gain enhancement techniques perform more efficiently and consistently. Recently, special attention has been paid to MQW photorefractive effects for their extremely large Δn_{ss} and nsec response time. However the limited achievable thickness prevents MQWs from becoming useful for immediate applications. The fate of MQW photorefractives may be determined by the possibility of obtaining a thick or multi-layer MQW photorefractive materials (e.g., 230 μm, providing a energy gain of 100 with $\Gamma = 200\ cm^{-1}$).

Applications of photorefractive storage operation have been demonstrated in reconfigurable interconnections and dynamic memories. Large refractive index modulation ($\Delta n_{ss} > 10^{-3}$), large thickness (mm–cm) are always required. In addition, large erasing time constant and small writing time constant are desirable. It seems that only ferroelectric materials have the chance of becoming commercially successful in these applications. $LiNbO_3$ may be the most suitable material for storage applications which do not require fast updating. Although the problem of nondestructive readout in a dynamic memory can be resolved in principle by using a circulating memory system, extensive engineering works need to be done in order to make the idea become a reality. Two-photon photorefractive materials may also become extremely important, because two-photon recording not only prevents destructive readout but also can show better sensitivities. However, most of the research on two-photon was performed in 1970's and early 1980's [86-88], and there is not currently enough energy or attention being paid to this topic. Continuous research on the two-photon effect is absolutely necessary for making the photorefractive dynamic memory an actual device.

VI. Acknowledgments

Contributions made by Hiro Sasaki, Brian Catanzaro, and Yoshinao Taketomi are highly appreciated. The authors are grateful for the research support of the Office of Naval Research under grants N00014-91J-1618 and N00014-86-K0697.

VII. References

[1] A. Marrakchi, "Continuous coherent erasure of dynamic holographic interconnects in photorefractive crystal," Opt. Lett. 14, 326-328(1989).

[2] J.E. Ford, Y. Fainman and S.H. Lee, "Array interconnection by phase-coded optical correlation," Opt. Lett. 15, 1088-1090(1990).

[3] D. Psaltis, D. Brady and K. Wagner, "Adaptive optical network using photorefractive crystals," Appl. Opt. 27, 1752-1759(1990).

[4] Y. Owechko and B.H. Soffer, "Optical neural network based on liquid crystal light valves and photorefractive crystals," Proc. SPIE, 1455, 136-144(1991).

[5] H. Sasaki, Y. Fainman, J.E. Ford, Y. Taketomi and S.H. Lee, "Dynamic photorefractive optical memory," Opt. Lett. 16, 1874-1876(1991).

[6] Y. Qiao and D. Psaltis, "Sampled dynamic holographic memory," Opt. Lett. 17, 1376-1378(1992).

[7] S. Boj, G. Pauliant and G. Roosen, "Dynamic holographic memory showing readout, refreshing and updating capabilities," Opt. Lett. 17, 438-440(1992).

[8] Y. Fainman, E. Klancnik and S.H. Lee, "Optimal coherent image amplification by two-wave coupling in photorefractive $BaTiO_3$," Opt. Eng. 25, 228-234(1986).

[9] A. Marrakchi, " Photorefractive spatial light modulation based on enhanced self-diffraction in sillenite crystals," Opt. Lett,. 13, 654-656(1988).

[10] J. Ma, L. Liu, S. Wu, Z. Wang, P. Mormile, G. Pierattinin and S.D. Nicola, "Photorefractive spatial light modulation by electrocontrolled beam coupling in SBN:Ce crystals," Opt. Comm. 70, 181-184(1989).

[11] Y. Fainman, C.C. Guest and S.H. Lee, "Optical digital logic operations by two-beam coupling in photorefractive material," Appl. Opt. 25, 1598-1603(1986).

[12] H. Rajbenbach, Y. Fainman and S.H. Lee, "Optical implementation of iterative algorithm for matrix inversion," Appl. Opt. 26, 1024-1031(1987).

[13] P. Yeh, A.E. Chiou and J. Hong, "Optical interconnects using photorefractive dynamic holograms," Appl. Opt. 27, 2093-2095(1988).

[14] D.M. Pepper, "Nonlinear optical phase conjugation," Opt. Eng. 21, 156-183(1982).

[15] D.M. Pepper, J. Au Yeung, D. Fekete and A. Yariv, "Spatial convolution and correlation of optical field via degenerate four-wave mixing," Opt. Lett. 3, 7-9(1978).

[16] J.E. Ford, S.H. Lee and Y. Fainman, "Reconfigurable interconnection using the correlation matrix-tensor multiplier," Topical Meeting on Optical Computing, Tech. Dig. 414-415, Kobe, Japan, Apri, 1990.

[17] P.J. van Heerden, "Theory of optical information storage in solids," Appl. Opt. 2, 394-400(1963).

[18] B. Catanzaro, J. Ma, Y. Fainman, S.H. Lee, "Infrared wavelength multiplexed holographic storage in photorefractive $LiNbO_3$," OSA Annual Meeting Technical Digest, 23, p112(1992).

[19] G.A. Rakuljic, V. Leyva and A. Yariv, "Optical data storage by using orthogonal wavelength-multiplexed volume holograms," Opt. Lett. 17, 1471-1473(1992).

[20] F.H. Mok, M.C. Tackitt and H.M. Stoll, "Storage of 500 high-resolution holograms in a $LiNbO_3$ crystals," Opt. Lett. **16**, 605-607(1991).

[21] J.E. Ford, J. Ma, Y. Fainman, S.H. Lee, Y. Taketomi, D. Bize, R.R. Neurgaonkar, "Multiplex holography in strontium barium niobate with applied field," J. Opt. Soc. Am. **A9**, 1183-1192(1992).

[22] Y. Taketomi, J.E. Ford, H. Sasaki, J. Ma, Y. Fainman, S.H. Lee, "Incremental recording for photorefractive hologram multiplexing," Opt. Lett. **22**, 1774-1776(1991).

[23] Y. Fainman, "Applications of photorefractive devices for optical computing," Proc. SPIE Critical Reviews Series, **1150**, 120-141(1989).

[24] B. Fischer, M. Cronin-Golomb, J.O. White and A. Yariv, "Amplified reflection, transmission, and self-oscillation in real-time holograph," Opt. Lett. **6**, 519-521(1981).

[25] G.C. Valley, M.B. Klein, "Optimal properties of photorefractive materials for optical data processing," Opt. Eng. **22**, 704-711(1983).

[26] G.A. Rakuljic, A. Yariv and R. Neurgaonkar, "Photorefractive properties of undoped, cerium-doped, and iron-doped single-crystal $Sr_{0.6}Ba_{0.4}Nb_2O_6$," Opt. Eng. **25**, 1212-1216(1985).

[27] J. Ma, Y. Taketomi, Y. Fainman, J.E. Ford, S.H. Lee and K. Chino, "Moving grating and dc external field in photorefractive GaP at 633 nm," Opt. Lett, **16**, 1080-1082(1991).

[28] K. Kuroda, Y. Okazaki, T. Shimura, H. Okamra, M. Chihara, M. Itoch and I. Ogura, "Photorefractive effect in GaP," Opt. Lett. **15**, 1197-1199(1990).

[29] G.C. Valley, S.W. McCahon and M.B. Klein, "Photorefractive measurement of photoinization and recombination cross section in InP:Fe," J. Appl. Phys. **64**, 6684-6689(1988).

[30] M.B. Klein, "Beam coupling in undoped GaAs at 1.06μm using the photorefractive effect," Opt. Lett. **9**, 350-352(1984).

[31] A. Partovi, J. Millerd, E.M. Garmire, M. Ziari, W.H. Steier, S.B. Trivedi and M.B. Klein, "Photorefractivity at 1.5μm in CdTe:V," Appl. Phys. Lett. **57**, 846-848(1990).

[32] Q.N. Wang, D.D. Nolte and M.R. Melloch, "Two-wave mixing in photorefractive AlGaAs/GaAs quantum wells," Appl. Phys. Lett. **59**, 256-258(1991).

[33] A.M. Glass, D.D. Nolte, D.H. Olson, G.E. Doran, D.S. Chemla and W.H. Knox, "Resonant photodiffractive four-wave mixing in semi-insulating GaAs/AlGaAs quantum well," Opt. Lett. **15**, 264-266(1990).

[34] G.C. Valley, J. Dubard, A. Smirl and A.M. Glass, "Picosecond photorefractive response of GaAs:EL2, InP:Fe, and CdTe:V," Opt. Lett. **14**, 961-963(1989).

[35] J.E. Ford, Y. Fainman and S.H. Lee, "Enhanced photorefractive performance from 45°-cut $BaTiO_3$," Appl. Opt. **28**, 4808-4815(1989).

[36] J.P. Huignard and A. Marrakchi, "Coherent signal beam amplification in two-wave mixing experiments with photorefractive $Bi_{12}SiO_{20}$ crystal," Opt. Comm. **38**, 249-254(1981).

[37] Ph. Refregier, L. Solymar, H. Rajbenbach and J.P. Huignard, "Two-beam coupling in photorefractive $Bi_{12}SiO_{20}$ crystal with moving grating: Theory and experiment," J. Appl. Phys. **58**, 45-57(1985).

[38] B. Imbert, H.Rajbenbach, S. Mallick, J.P. Herriau and J.P. Huignard, "High photorefractive gain in two-beam coupling with moving fringes in GaAs:Cr crystals," Opt. Lett. **13**, 327-329(1988).

[39] H. Rajbenbach, J.P. Huignard and B. Loiseaux, "Spatial frequency dependence of the energy transfer in two-wave mixing experiments with BSO crystals," Opt. Comm. **48**, 247-252(1983).

[40] A. Partovi, A. Kost, E.M. Garmire, G.C. Valley and M.B. Klein, "Band-edge photorefractive effect in semiconductors," Appl. Phys. Lett. **56**, 1089-1091(1990).

[41] J.E. Millerd, S.D. Koehler, E.M. Garmire, A. Partovi, A. Glass and M.B. Klein, "Photorefractive gain enhancement in InP:Fe using band-edge resonance and temperature stabilization," Appl. Phys. Lett. **57**, 2776-2778(1990).

[42] S.I. Stepanov and M.P. Petrov, "Efficient unstationary holographic recording in photorefractive crystals under an external alternating electric field," Opt. Comm. **53**, 292-295(1985).

[43] P. Mathey, G. Pauliat, J.C. Launay and G. Roosen, "Overcoming the trap density limitation in photorefractive two-beam coupling by applying pulsed electric field," Opt. Comm. **82**, 101-106(1991).

[44] P. Gravey, G. Picoli and J.Y. Labandibar,"Stabilization of photorefractive two-wave coupling in InP:Fe under high dc fields by temperature control," Opt. Comm. **70**, 190-193(1989).

[45] H. Kogelnik, "Coupled wave theory for thick hologram gratings," The Bell Syst. Tech. J. **48**, 1909-1947(1969).

[46] S. Yin and F.T.S. Yu, "Specially doped $LiNbO_3$ crystal hologram using low power laser diode," OSA Annual Meeting Technical Digest, **23**, p37(1992).

[47] B. Catanzaro, J. Ma, Y. Fainman, S.H. Lee and L. Xu, "Near infrared holographic recording in lithium niobate," submitted to Opt. Lett.

[48] S. Tao, D.R. Selviah and J.E. Midwinter, "Spatial angular multiplexed storage of 750 holograms in an Fe:$LiNbO_3$ crystal," Opt. Lett. **18**, 912-914(1993).

[49] R.R. Shah, D.M. Kim, T.A. Rabson and F.K. Tittle, "Characterization of iron-doped lithium niobate for holographic storage applications," J. Appl. Phys. **47**, 5421-5431(1976).

[50] D.L. Steabler, W.J. Burke, W. Phillips and J.J. Amodei, "Multiple storage and erasure of fixed holograms in Fe-doped $liNbO_3$," Appl. Phys. Lett. **26**, 182-184(1975).

[51] J. Ma, J.E. Ford, Y. Taketomi and S.H. Lee, "Moving grating for enhanced holographic recording in cerium-doped $Sr_{0.6}Ba_{0.4}Nb_2O_6$," Opt. Lett. **16**, 270-272(1991).

[52] N.V. Kukhtarev, "Kinetics of hologram recording and erasure in electrooptic crystals," Sov. Tech. Phys. Lett. **2**, 438-440(1976).

[53] N.V. Kukhtarev, V.B. Markov, S.G. Odulov, M.S. Soskin and V.L. Vinetskii, "Holographic storage in electrooptic crystals," Ferroelectrics **22**, 949-960, 961-964(1979).

[54] C.J. Chang-Hasnain, J.P. Harbison, C. Zah, M.W. Maeda, C.T. Florez, N.G. Stoffel and T. Lee, "Multiple wavelength tunable surface-emitting laser array," IEEE J. of Quan. Electron. **27**, 1368-1376(1991).

[55] M. Schilling, W. Idler, E. Kühn, G. Laube, H. Schweizer, K. Wünstel and O. Hidebrand, "Interated interferometric injection laser: novel fast and broad-band tunable monolithic light source," IEEE J. of Quan. Electron. **27**, 1616-1624(1991).

[56] D. Rytz, R.R. Stephens, B.A. Wechsler, M.S. Keirstead and T.M. Baer, "Efficient self-pumped phase conjugation at near-infrared wavelengths using cobalt-doped $BaTiO_3$," Opt. Lett. **15**, 1279-1281(1990).

[57] A.E. Clement, G.C. Gilbreath and S.N. Fugera, "Determination of the effective trap density of $Ta:KNbO_3$ and $BaTiO_3$ at 823 nm," technical Digest of Topical Meeting on Photorefractive Materials, Effects and Devices (OSA, Hawaii), 347-350(1992).

[58] F.H. Mok, "Angle-multiplexed storage of 5000 holograms in lithium niobate," Opt. Lett. **18**, 915-917(1993).

[59] Y. Taketomi, J.E. Ford, H. Sasaki, J. Ma, Y. Fainman and S.H. Lee, "Multi-mode operations of a holographic memory using orthogonal phase codes," OSA Topical Meeting on Photorefractive Materials, Effects, and Devices, (Massachusetts, 1991)

[60] S. Esener, M. Title, J.H. Wang, T. Drabik and S.H. Lee, "One-dimensional silicon/PLZT spatial light modulator," Opt. Eng. **26**, 406-413(1987).

[61] E. Maniloff and K. Johnson, "Maximized photorefractive holographic storage," J. Appl. Phys. **70**, 4702-4707(1991).

[62] D. Anderson and D. Lininger, "Dynamic optical interconnects: volume holograms as optical two-port operators," Appl. Opt. **26**, 5031-5038(1987).

[63] E. Paek, J. Wullert and J. Patel, "Holographic on-line learning machine for multicategory classification," Jap. J. Appl. Phys. **29**, 1332-1334(1990).

[64] Y. Fainman, J. Ma, S.H. Lee, "Non-linear optical materials and applications," Materials Science Report, **9**, 53-139(1993).

[65] J.P. Huignard, "Survey of photorefractive nonlinear crystals and their applications," Proc. SPIE, **1127**, 205-215(1989).

[66] "Photorefractive materials and their applications," ed. by P. Günter, J.P. Huignard, (Berlin, New York, Springer-Verlag, c1988-c1989).

[67] J.W. Goodman, F.I. Leonberger, S.Y. Kung and R.A. Athale, "Optical interconnect for VLSI systems," Proc. IEEE **72**, 850-866(1984).

[68] L. Snyder, "Introduction to the reconfigurable, high parallel computer," IEEE Computer **15**, 47(1982).

[69] J.W. Goodman, A.R. Dias and L.M. Woody, Opt. Lett. **2**, 1-3(1978).

[70] R.A. Athale, "Optical matrix processors," Proc. SPIE, **634**, 96-111(1986).

[71] A.A. Sawchuk, B.K. Jenkins, C.S. Raghavendra and A. Varma, "Optical crossbar network, " IEEE Trans. Comput. **C-20**, 56-60(1987).

[72] A. Chiou, P. Yeh, "Energy efficiency of optical interconnections using photorefractive holograms," Appl. Opt. **29**, 1111-1117(1990).

[73] H. Rajbenbach, S. Bann and J.P. Huignard, "Long term readout of photorefractive memories using a storage amplification two-crystal configuration," Opt. Lett. **17**, 1712-1714(1992).

[74] H.J. Caufield, "Parallel N^4 weighted optical interconnection," Appl. Opt. **26**, 4039-4040(1987).

[75] T. Lu, S. Wu, X. Xu and F.T.S. Yu, "Two-dimensional programmable optical neural network," Appl. Opt. **28**, 4908-4913(1989).

[76] J.E. Ford, "Reconfigurable array interconnection by photorefractive volume holography," Ph.D. Dissertation, University of California, San Diego, Jan., 1992.

[77] J.E. Ford, Y. Fainman and S.H. Lee, "Reconfigurable array interconnection by photorefractive correlation," submitted to Appl. Opt.

[78] J. Ma, B. Catanzaro, Y. Fainman and S.H. Lee, "Reflection volume holographic lenses with wide field of view and wavelength multiplexing," OSA Annual meeting Technical Digest, **23**, FQ1, p189(1992).

[79] J. Ma, B. Catanzaro, J.E. Ford, Y. Fainman and S.H. Lee, "Photorefractive holographic lenses and applications for dynamic focusing and dynamic image shifting," submitted to J. Opt. Soc. Am. (A)

[80] B. Cantazaro, J. Ma, Y. Fainman and S.H. Lee, "Wavelength-multiplexed holographic lens array for a memory-distribution network," OSA Annual Meeting Technique Digest, **24**, TuH8(1993).

[81] B. Catanzaro, J. Ma, Y. Fainman, S.H. Lee, "Wavelength multiplexed reconfigurable interconnect using a volume holographic lens array," submitted to Opt. Lett.

[82] H. Sasaki, J. Ma, Y. Fainman, S.H. Lee and Y. Taketomi, "Fast update of dynamic photorefractive optical memory," Opt. Lett. **17**, 1468-1470(1992).

[83] J.P. Huignard, J.P. Herriau and F. Micheron, "Selective erasure and processing in volume holograms superimposed in photorefractive ferroelectrics," Ferroelectrics **11**, 393-396(1976).

[84] G. Kavounas and W.H. Steier, "Fast hologram erasure in photorefractive materials," Digest of Meeting on Photorefractive Materials, Effects, and Devices (OSA, Washington, DC, 1987)pp155-158.

[85] H. Sasaki, J. Ma, Y. Fainman, S.H. Lee and Y. Taketomi, "Dynamics of composite grating in photorefractive crystal for memory application," to be published on J. Opt. Soc. Am. (A)

[86] D. von der Linde, A.M. Glass and K.F. Rodgers, "Multiphoton photorefractive processes for optical storage in $LiNbO_3$," Appl. Phys. Lett. **25**, 155-157(1974).

[87] D. von der Linde, A.M. Glass and K.F. Rodgers, "High-sensitivity optical recording in KTN by two-photon absorption," Appl. Phys. Lett. **26**, 22-24(1975).

[88] H. Vorman and E. Krätzig, "Two-step excitation in $LiTaO_3$:Fe for optical data storage," Soli. Stat. Comm. **49**, 843-847(1984).

Photorefractive crystals for real time image processing

Ken Y. Hsu and Shiuan-Huei Lin

National Chiao Tung University, Institute of Electro-Optical Engineering,
1001 Ta Hsueh Road, Hsinchu
Taiwan 30050, China

AND

Tai-Chiung Hsieh

National Chiao Tung University, Department of Electrophysics,
1001 Ta Hsueh Road, Hsinchu
Taiwan 30050, China

AND

L. J. Hu, T. S. Yeh, H. P. Lin, C. H. Lin, S. L. Tu, S. J. Yang, S. E. Hsu

Chung Shan Institute of Science and Technology,
Material Research and Development Center,
Long-Tan,
Taiwan, China

ABSTRACT

Heavily iron doped lithium niobate photorefractive crystals are grown and characterized. The crystals have high diffraction efficiency and fast temporal response and hence are suitable for real-time image processing. Three applications of thin Fe: $LiNbO_3$ plates are demonstrated: the joint-transform correlator, the optical perceptron, and the hybrid image classifier. In the first and third systems, the heavily doped (0.5% mole) crystals are used. These crystals have fast temporal response (300~400 msec) and thus provide a means for real time optical processing. In the second system, a slightly light doped (0.1% mole) crystal is used. It has relatively long (> 1 minute) response time and is used for recording the dynamic memory in a neural network. Image recognition by these three systems are presented.

1. INTRODUCTION

In this paper we present an experimental study of photorefractive (PR) crystals for real-time image processing. In particular, we describe the growth and characterization of photorefractive $LiNbO_3$ crystals, the preparation of the thin crystal plates, and the are grown and used in this research. The doping concentrations are varied from 0.1% to 1.5% mole. The effect of the doping concentration on the photorefractive properties are experimentally characterized. We then used the thin crystal plates for the image recognition applications. The first system is the joint-transform correlator (JTC) [1][2]. In an ordinary JTC, the crystal volume is used for the square-law detection and thus the shift range of the input patterns is limited by the Bragg condition. The Bragg condition could be released if a thin crystal plate is used. However, a decrease in crystal thickness would result in reducing the wave-interaction length and thus the diffraction signal from the photorefractive crystals will be too weak to detect. In order to overcome this problem, we have grown heavily doped crystals. It is observed that the heavily doped crystal plates produce sufficient signals for obtaining the correlation signal. We operate the crystal plate in the Raman-Nath regime [3] and the correlator is shift-invariant. In addition, the doping concentration provides us a means for controlling the temporal characteristics of the crystals. Two different applications have been designed based on the temporal response of the crystal. The first one, by using a crystal with slightly less doping concentration which produces a relatively-long time constant, is used for the implementation of optical perceptron [4][5]. In this system, the training patterns and the learning algorithm are stored in a personal computer and the dynamic interconnection weights are recorded as holographic gratings in the photorefractive crystal. The system is trained for pattern recognition. It is shown in Section 4 that the dynamic property of the crystal provides the learning capability of the system. However, it is also seen that the learning capacity of the optical perceptron is limited by the finite response time of the crystal. In order to resolve this problem, we have designed the hybrid system which makes use of the fast response of heavily doped crystals. The crystal does not record the dynamic gratings of the neural network. Instead, it simply provides a medium for real-time optical processing. The interconnection weights, the training patterns, as well as the learning algorithm are stored in the personal computer. This system has combined the advantages of parallelism of optical processing with the programmability of electronic computer. Our experimental results demonstrate the potentiality of using heavily doped PR crystals for real-time processing.

2. HEAVILY DOPED Fe:$LiNbO_3$ CRYSTAL PLATES

We now describe the preparation of Fe:$LiNbO_3$ crystal plates for real-time image processing [6]. First, $LiNbO_3$ single crystals with Fe_2O_3 dopant concentrations varying from 0.1% to 1.5% mole are grown. Then, it is x-cut and polished to a thickness from 30 μm to 50 μm. Each thin plate is adhered on an optical glass (BK7) with optical adhesive. The structure of the thin plate is show in Fig. 1. The photorefractive properties of the plates are characterized by measuring the temporal response of holographic grating formation and erasure. In order to operate the sample in the planar hologram regime, we let two extra-ordinarily polarized argon laser beams of 514.5 nm intersect in the crystal plate with the intersection angle of 3.3 degrees outside the crystal. The hologram parameter Q is calculated to be 0.83 and thus the operation is in the Raman-Nath regime. A weak He-Ne laser beam of 632.8 nm is incident from the back side of the crystal to probe this photorefractive grating at an arbitrary incident angle. Up to five orders of diffraction have been observed behind the sample when the intensity of write beam is 4 W/cm^2 [7].

We have measured the diffraction properties and temporal response of this photorefractive grating by monitoring the first order diffraction of the He-Ne probe beam. First, we measured the diffraction power at different probe angles of the He-Ne laser beam. The result is shown in Fig. 2. It is seen that the diffracted power remains the same level at the probing angle 0~4.5 degrees. This confirms the planar hologram property of the Raman-Nath diffraction. Next, we recorded the temporal behavior of the grating formation and dark decay. The result is shown in Fig. 3. It is interesting to see that the build-up time of the grating is about 300 msec and that the dark decay time constant is 480 msec. Comparing this with that of the conventional Fe: $LiNbO_3$ crystals which have long time constant of dark storage (hours ~ days) and slow temporal response, we see that the crystal plates are suitable for real-time optical processing.

3. JOINT-TRANSFORM CORRELATOR BY USING THIN Fe:$LiNbO_3$ PLATES

Fig. 4 shows the generic diagram of a photorefractive joint-transform correlator (JTC). The Fe: $LiNbO_3$ plate is placed in the Fourier plane of the system. Two input patterns u_1 and u_2 are Fourier transformed and interfered in the crystal plate. In order to obtain the shift-invariant property, we need to operate the plate in the Raman-Nath regime. In terms of the Q-parameter of the holograms, it requires $Q < 1$. It was derived in [6] that the amplitude of the m-th order diffraction from the photorefractive gratings is proportional to the m-th order Bessel function:

$$E_{dm} \cong A_3 J_m\left[\frac{\omega}{c}\cdot\frac{n_1 d}{\cos(\theta)}\right], \qquad m = \pm1, \pm2, \pm3, \ldots \tag{1}$$

where A_3 is the amplitude of the read beam, d is the plate thickness, θ is the incident angle of the read beam, and n_1 is the index perturbation induced by the interference pattern of the writing beams in the photorefractive plate. From the photorefractive theory, n_1 can be expressed as:

$$n_1 \cong A_1^* A_2 + A_1 A_2^* \tag{2}$$

where A_1 and A_2 are the amplitudes of the two writing beams in the crystal, and * represents complex conjugation. Under our thin plate conditions, the arguments of the Bessel functions in Eq.(1) are much smaller than 1, hence the first order diffraction can be approximated as:

$$E_{d1} \cong A_3 J_1\left[\frac{\omega}{c}\cdot\frac{n_1 d}{\cos(\theta)}\right] \cong A_3\,\frac{\omega}{c}\cdot\frac{n_1 d}{\cos(\theta)} \qquad (3)$$

By Eqs. (2) and (3), the diffraction from the first order is obtained as:

$$E_{d\pm1} \cong A_3\left(A_1^{*}A_2 + A_1 A_2^{*}\right) \qquad (4)$$

If the A_3 is a plane wave, A_1 is the Fourier transform of the target pattern, $a_1(x, y)$, and A_2 is the Fourier transform of the reference pattern, $a_2(x, y)$, then Eq.(4) represents an operation of spatial filtering. The joint transform correlator is obtained by performing a Fourier transform on the output of the +1 order diffraction of Eq.(4):

$$C(x', y') \cong a_1(x, y) \otimes a_2(x, y), \qquad (5)$$

where $\otimes$ represents cross-correlation operation.

From the above analysis, we know that a JTC system can be obtained by using a thin photorefractive plate that operates in the Raman-Nath diffraction regime. This system provides the shift-invariant capability owing to the planar grating characteristics of the Raman-Nath diffraction. It means that both of the target and reference images can be shifted in the input plane without degrading the correlation signal and also that the reading beam can be incident into the planar hologram from an arbitrary angle. In addition, the dynamic grating erasing/recording properties of the photorefractive grating provides a real-time medium for detecting the joint transform interference pattern of the inputs.

The schematic diagram of our JTC system is shown in Fig. 5. A collimated extra-ordinarily polarized laser beam is derived from the argon laser at 514.5 nm. The beam is split into two beams and each beam passes through a LCTV for carrying the input images into the system. One LCTV shows the reference object and the other one shows the target object. The two beams intersect in the crystal plate after passing through the

Fourier transforming lenses and the interference pattern of the joint transform spectrum is recorded in the Fe-$LiNbO_3$ plate. The focal length of this lens is 40 cm and the intersection angle outside the crystal plate is about 3.3 degree. Therefore, the parameter Q is about 0.83 and the condition of Raman-Nath diffraction is satisfied. The correlation signal was read out by a plane wave of the He-Ne probe beam at wavelength of 632.6 nm and incident from the other side of the crystal plate at an arbitrary angle. The output was imaged on the CCD camera by a Fourier transforming lens.

Fig. 6 shows the photographs of the experimental results. Fig. 6(a) shows the input images to the JTC. The left part is the input image which consists of three Chinese words ". 交通大學 ， 光電所，光子計算， (Chiao Tung University, Institute of Electro-optics, Optical Computing) and the right part is the reference image which contains a Chinese character " 光 " (means "Light"). Fig. 6(b) shows the correlation output. The appearance of two correlation peaks demonstrates the shift invariant property of the correlator. In addition, if we update the input image at the rate of 300 msec per frame, then the correlation output follows the change at the same rate. Thus the system is capable of real-time operation.

4. OPTICAL PERCEPTRON BY USING THIN PHOTOREFRACTIVE PLATES

In the previous joint-transform correlator, if the input pattern is rotated, scaled, or distorted, then the correlation signal degrades rapidly. The synthesis of a matched filter for distortion tolerant image recognition has been an interesting subject to the researchers in optical computing. Recently, a neural network approach has been proposed and demonstrated [8][9][10]. In that approach, the dynamic characteristics of PR crystals has been used for implementing the learning behavior of the neural network. The matched filter obtained by this method was shown to be capable of rotation- and distortion- tolerant image recognition. However, the shift range of the input was very limited because of the Bragg condition of the volume hologram. In order to resolve this restriction we present a method which uses the thin PR plates for recording the dynamic memory of the neural network.

The optical system that we have studied for image recognition is based on the perceptron network [11]. We first explain how perceptron learning algorithm is used for designing the matched filter. The perceptron is originally proposed as a pattern classifier for a set of training patterns. The patterns in the set are specified as either belong to class C1 or class C0. C1 produces high output and C0 produces low response. In the training process, each input pattern vector $\underline{X}$ with N elements is presented to the network, then the perceptron responses to $\underline{X}$ by generating an inner product $\underline{W} \cdot \underline{X}$ where

$\underline{W}$ is the interconnection weight vector stored in the memory. The perceptron is trained to generate a correct response corresponding to each $\underline{X}$.

For the purpose of image recognition, we specify the reference image and its variants as class C1 and the image to be distinguished from the reference image as class C0. Then after the training is complete, the perceptron will produce a high inner product if the input image is similar to the reference image. In optical implementation, high inner product means a high central peak at the correlation plane. In other words, C1 image will produce a high correlation peak and C0 image will not. Therefore, the interconnect weight vector $\underline{W}$ is effectively the matched filter for recognizing the C1 images.

The schematic diagram of the optical perceptron is shown in Fig. 7. It consists of a double Mach-Zehnder interferometer (BS1, M2, M3, BS2), which is for the implementation of grating addition and subtraction based on the Stoke's principle, and a VanderLugt correlator for implementing optical inner product. The a Fe:$LiNbO_3$ plate is placed at the focal plane of two lenses [12]. In the experiments, the training patterns for class C1 are English letter **C** and its variants {**C**, *C*, **C**} and C0 are the patterns {**N**, U, **U**} that we want to distinguish from **C**. All these patterns are stored in a personal computer. In training, the patterns are presented one by one into the system through the LCTV. Light passing through the LCTV makes inner product with the interconnection gratings stored in the Fe:$LiNbO_3$ plate and is diffracted to the photodetector. The detected signal is thresholded by the computer and then compared with the specified class of that pattern. If the pattern is classified correctly then next pattern is presented to the system for further check. On the other hand, if the inner product is too low for a pattern in class C1 then an error signal is generated to turn ON shutters S1 and S3 at the same time shutter S2 is turned OFF. Such beam paths produce an interference pattern that is in-phase with the original grating and thus the holographic gratings are reinforced. On the other hand, if the inner product is too high for a pattern in C0 then shutters S2 and S3 are turned ON and S1 is turned OFF to produce an interference pattern which is π-phase different with the original grating and thus the grating strength is reduced. The training process iterates until all the patterns are correctly classified.

Fig. 8 shows the training patterns. After the training we use the PR plate as the matched filter for image correlation. Fig. 9 shows the experimental result. It shows that the input image can be recognized even it is distorted or in a different shape. In addition, the correlation peak follows the shift of the input pattern. This shows that the system is shift invariant.

The key for the successful implementation of the optical perceptron lies in using the PR plates for recording the dynamic interconnection weight $\underline{W}$. Since each PR crystal has a finite response time, hence it takes time to modify $\underline{W}$. Suppose that the updating time for each pattern is t_1, then the quantity of the weight change is $\alpha(p)$. $[1-\exp(-t_1/\tau)].\underline{x}(p)$, where τ is the time constant of the crystal. Furthermore, since all the holographic gratings are recorded in the same volume of the crystal plate, hence the recorded gratings start to decay whenever the crystal is illuminated for either adding or subtracting the weights. Therefore, the updating behavior of the PR plate is described by the following dynamic equation:

$$\underline{W}(p+1) = \underline{W}(p)e^{-t_1/\tau} + \alpha(p)(1-e^{-t_1/\tau})\underline{X}(p) \tag{6}$$

where $\alpha(p)$ is the updating factor and is given by:

$$\alpha(p) = \begin{cases} 0 & \text{if } \underline{x}(p) \text{ is correctly classified} \\ 1 & \text{if } \underline{x}(p) \in \text{C1 but } |\underline{w}(p)\cdot\underline{x}(p)| < \theta \\ -1 & \text{if } \underline{x}(p) \in \text{C2 but } |\underline{w}(p)\cdot\underline{x}(p)| > \theta \end{cases} \tag{7}$$

In Eq.(6), if $\tau \rightarrow \infty$ then $\underline{W}(p+1) \rightarrow \underline{W}(p)$. This means that if the crystal does not response to light then the system has no learning capability. On the other hand, if $\tau \rightarrow 0$, then $\underline{W}(p+1) \rightarrow \underline{X}(p)$, which means that the network has no memory. In either case, the optical system can not be trained. Therefore, the learning capability of the optical perceptron is governed by the temporal behavior of the PR crystal. On the other hand, it was derived in [12] that the total updating time of the training is on the order of the crystal response time τ, otherwise the training process may not converge. This is physically meaningful because after total exposure time τ then the feature of the first exposure $\underline{X}(1)$ that was previously recorded would have been erased. Thus the perceptron may never learn to recognize $\underline{X}(1)$ unless other patterns have common features with $\underline{X}(1)$ and these features have been reinforced in succeeding training iterations. In other words, finite response time of the PR crystals in fact puts an limit on the learning capacity of the perceptron.

In the next section, we shall present a hybrid perceptron network which has an improved learning capacity. In that system, the fast response PR plates are used for providing a medium for real-time optical processing. There is no recording in the RR plates. Rather, the interconnection weight $\underline{W}$ is stored in a personal computer. Thus the learning capacity is not limited by the temporal behavior of the PR material.

5. A HYBRID PERCEPTRON NETWORK FOR IMAGE RECOGNITION

The schematic diagram for the hybrid perceptron is shown in Fig. 10. It is basically the joint-transform correlator except that the CCD camera is replaced by the combination of a pinhole and a photo-detector. The photodetector is placed at the central position of the correlation plane. Thus, the output signal is the inner product of the two input patterns. For the hybrid perceptron operation, both the training patterns $\underline{X}$ and the interconnection weight $\underline{W}$ are stored in the personal computer. During the training phase, pattern $\underline{X}$ is present to one arm through LCTV1 and the $\underline{W}$ is input to the other arm through LCTV2. The inner product of the two patterns is detected and then thresholded. The result is compared with the specified value of that pattern to produce the updating signal. If, at the p-th iteration, the interconnection $\underline{W}(p)$ is too weak then the $\underline{W}(p)$ is reinforced by adding to it the pattern $\underline{X}(p)$. On the other hand, if $\underline{W}(p)$ is too high, then it is reduced by subtracting from it the $\underline{X}(p)$.The addition and subtraction are performed in the personal computer. The iterations continues until all the patterns are correctly classified.

In the perceptron operations, we consider the case that $\underline{X}$ are patterns with either zero or positive elements, whereas $\underline{W}$ can be bipolar. We operate the LCTVs in amplitude modulation mode and only unipolar patterns can be displayed. To resolve this problem the bipolar $\underline{W}$ is split into two positive unipolar patterns:

$$\underline{W} = \underline{W}^{+} + \underline{W}^{-} \tag{8}$$

where $\underline{W}^{+}$ is the vector of which its non-zero elements are equal to the corresponding positive elements of $\underline{W}$ and other elements are zero which corresponds to the negative and zero elements of $\underline{W}$. Similarly, $\underline{W}^{-}$ has non-zero elements which has the magnitude of the corresponding negative element of $\underline{W}$ and it has zero elements at the positive and zero elements of $\underline{W}$. In the training procedure there are two steps for each iteration. In the first step, $\underline{W}^{+}$ and $\underline{X}$ are presented on the LCTVs and $\underline{W}^{+}\cdot\underline{X}$ is detected by the optical detector. In the second step, $\underline{W}^{-}$ and $\underline{X}$ are displayed and $\underline{W}^{-}\cdot\underline{X}$ is detected. Then the two signals are subtracted in the personal computer. Since

$$\begin{aligned} &\underline{W}^{+}\cdot\underline{X} - \underline{W}^{-}\cdot\underline{X} \\ &= [\underline{W}^{+} + (-\underline{W}^{-})]\cdot\underline{X} \\ &= \underline{W}\cdot\underline{X} \end{aligned} \tag{9}$$

hence inner product $\underline{W}\cdot\underline{X}$ is obtained, and the updating factor $\alpha(p)$ can be produced by comparing $\underline{W}\cdot\underline{X}$ with the threshold value. This completes one iteration.

Fig. 11 shows the training patterns, of which { A,C,S,V } are specified as class C1 and { 2,3,4,5 } are class C0. Fig. 12 shows the training curve. It is seen that the patterns are correctly classified after two iterations. We then use the final W for image recognition. Fig. 13 shows the experimental result. Again the patterns that are similar to C1 produce high inner product values. They are correctly recognized.

There are several advantages of this hybrid perceptron. First, it combines the parallelism of optics and the programmability of electronic computer. Second, storing W in the computer releases the restriction of finite response time of the PR crystals. The convergence property of the system is thus guaranteed, and therefore, the learning capacity of the network could be improved. Third, since the interconnection weight W is stored in the computer, it is a non-volatile memory and thus there is no need of crystal fixing [13].

6. CONCLUSION

We have presented the growth, characterization, and applications of heavily doped Fe: $LiNbO_3$ photorefractive crystals. Thin plates are prepared from these crystals. It is observed that the temporal behavior of the crystals can be varied by adjusting the dopant concentrations of Fe_2O_3. In general, heavily doped crystals have fast temporal response. Thus, they are suitable for real-time optical processing. We have demonstrated three applications. The first application of the PR plates is a joint-transform correlator. By operating the plates in the Raman-Nath regime, we have demonstrated a real-time, shift-invariant image recognition system. The second application is the optical perceptron. By using the PR plate as the recording medium for dynamic interconnections, we have obtained a distortion-tolerant, shift-invariant image recognition operation. The third application is a hybrid perceptron in which the inner product is performed by optics and the learning algorithm as well as the non-volatile memory are provided by the electronic computer. This system is also successfully trained for a distortion-tolerant image recognition. The experimental results from our study have demonstrated the potentiality of heavily doped photorefractive crystals for real-time image processing.

7. ACKNOWLEDGMENTS

This research is supported by the National Science Council of Taiwan, the Republic of China under the contract NSC 82-0416-E-009-195 and also by Chung Shan Institute of Science and Technology under CS 80-0210-D009-23.

8. REFERENCES

1. D. M. Pepper, J. Au Yeng, D. Fekete and A. Yariv, "Spatial convolution and correlation of optical field via degenerate four-wave mixing", Opt. Lett. 3, pp7 (1978).

2. L. Pichon and J. P. Huignard, "Dynamic joint-Fourier-transform correlator by Bragg diffraction in photorefractive BSO crystals," Opt. Comm., Vol. 36, No. 4, pp. 277-280, February (1981).

3. P. Yeh, " Photorefractive nonlinear optics," Ch. 2, pp. 61-65, John Wiley &Sons, Inc., 1993.

4. D. Psaltis, D. Brady and K. Wagner, "Adaptive optical networks using photorefractive crystals", Appl. Opt. 27, 1752-1759 (1988).

5. C.-J. Cheng, S. H. Lin and T.-C. Hsieh, "Learning-based image recognition by a thin photorefractive crystal plate", SPIE Proceeding, Vol. 1812, pp. 114-124, Hsinchu, Taiwan (1992).

6. L. J. Hu, S. P. Lin, T. S. Yeh, S. H. Lin, T. C. Hsieh and K. Y. Hsu, "A shift-invariant real-time joint-transform correlator", SPIE Proceeding, Vol. 1812, pp. 114-124, Hsinchu, Taiwan (1992).

7. Q. B. He, P. Yeh and Ken Hsu, "Fast photorefractive planar gratings in Fe: $LiNbO_3$ crystal", the OSA Topic Meeting of Nonlinear Optics, Hawaii, August (1992).

8. K.Y. Hsu, H.Y. Li and D. Psaltis, 'Holographic implementation of a fully connected neural network', Proc. IEEE, 78, 1637-1645 (1990).

9. E. G. Paek, J. Wullert and J. S. Patel, "Holographic implementation of a learning machine based on a multicategory perceptron algorithm", Opt. Lett. 14, 1303-1305 (1989).

10. J. Hong, S. Campbell and P. Yeh, "Optical pattern classifier with perceptron learning", Appl. Opt. 29, 3019-3025 (1990).

11. F. Rosenblatt, Principles of Neurodynamics: Perceptron and the Theory of Brain Mechanisms, Spartan Books, Washington DC (1961).

12. K. Hsu, S. H. Lin, C.-J. Cheng, T.-C. Hsieh and P. Yeh, "An optical neural network for pattern recognition", submitted to Journal of Optical Computing (1993).

13. J. J. Amodei and D. L. Staebler, 'Holographic pattern fixing in electro-optic crystals', Appl. Phys. Lett. 18, 540-542 (1971).

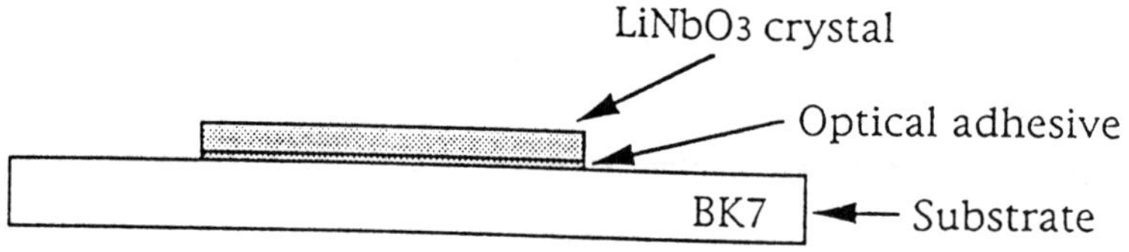

Fig. 1 The structure of the thin Fe-$LiNbO_3$ crystal plate

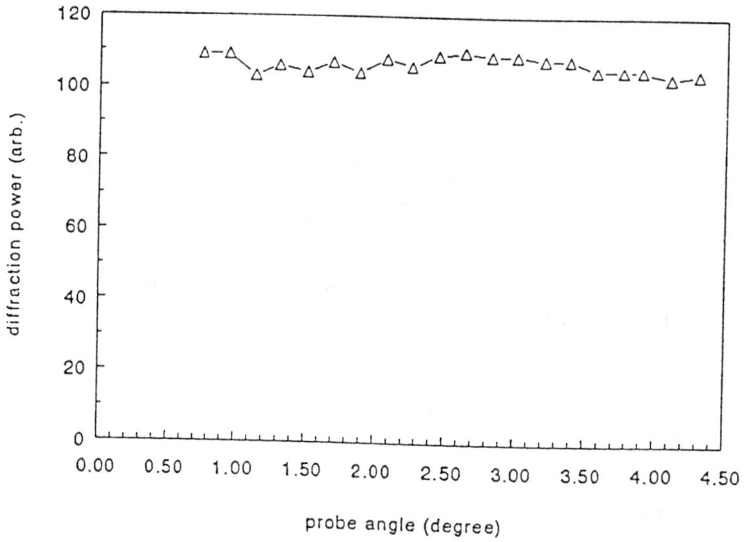

Fig. 2 The diffraction power as a function of the different probe angles at the total write intensity = 2.54 W/cm^2 and read intensity = 0.3 W/cm^2

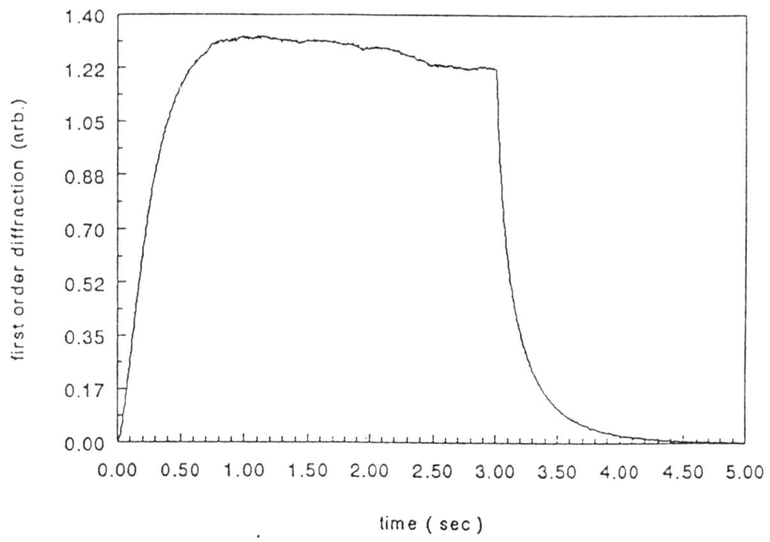

Fig. 3 The temprol response of the diffraction at the total write intensity = 3.48 W/sm^2 and the probe intensity = 0.3 W/cm^2

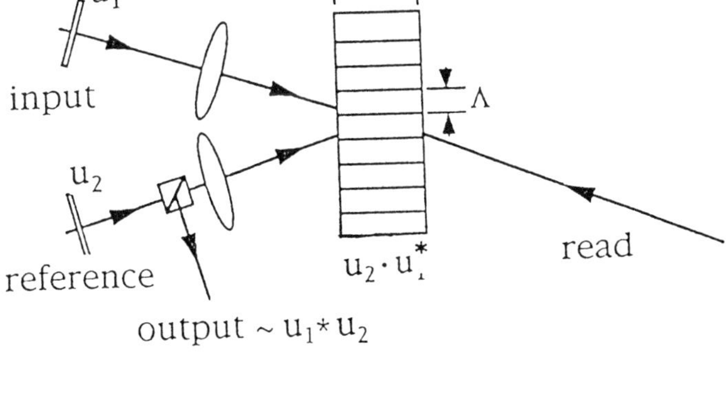

Fig. 4 Joint transform correlator

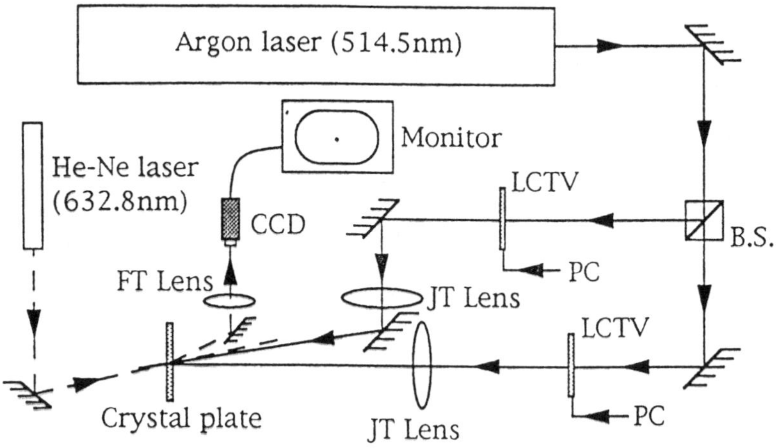

Fig. 5 Schematic diagram of the joint transform correlator

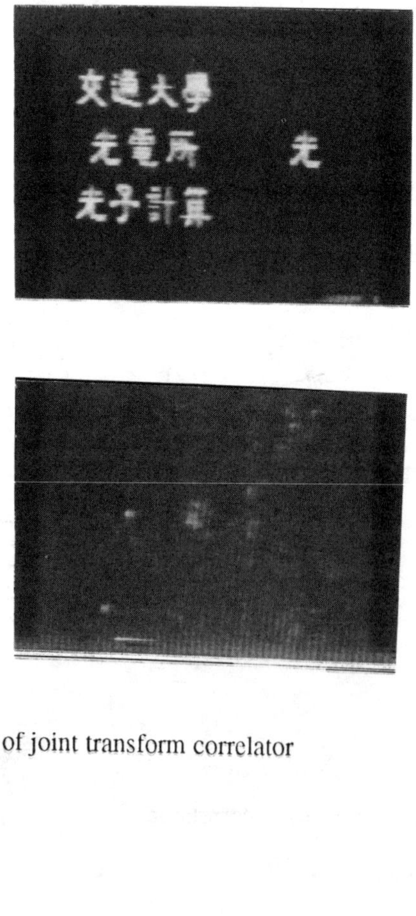

(a) Input images to JTC.

(b) Correlation output.

Fig. 6 Experimental result of joint transform correlator

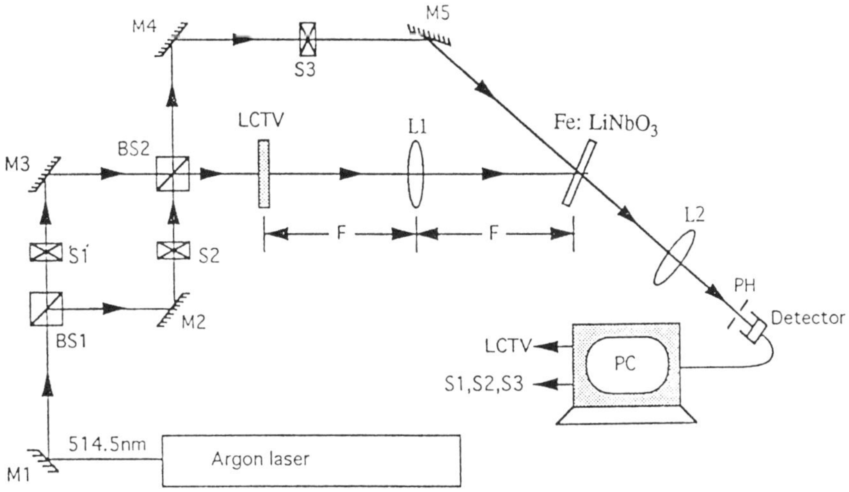

Fig. 7 Optical perceptron

Fig. 8 Training patterns

Fig. 9 Correlation output

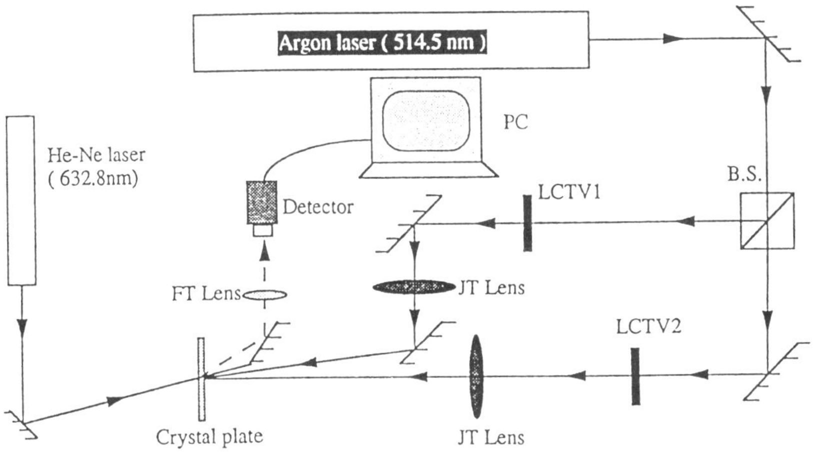

Fig. 10 Schematic diagram of the hybrid perceptron

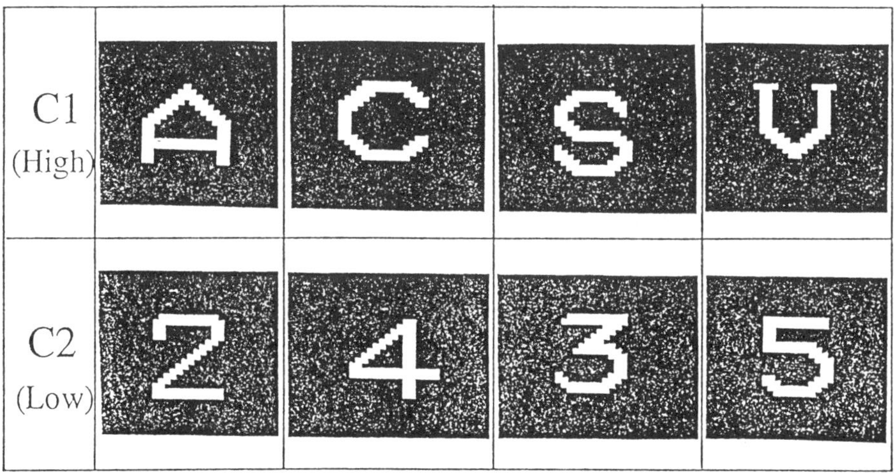

Fig. 11 Training patterns for the hybrid perceptron

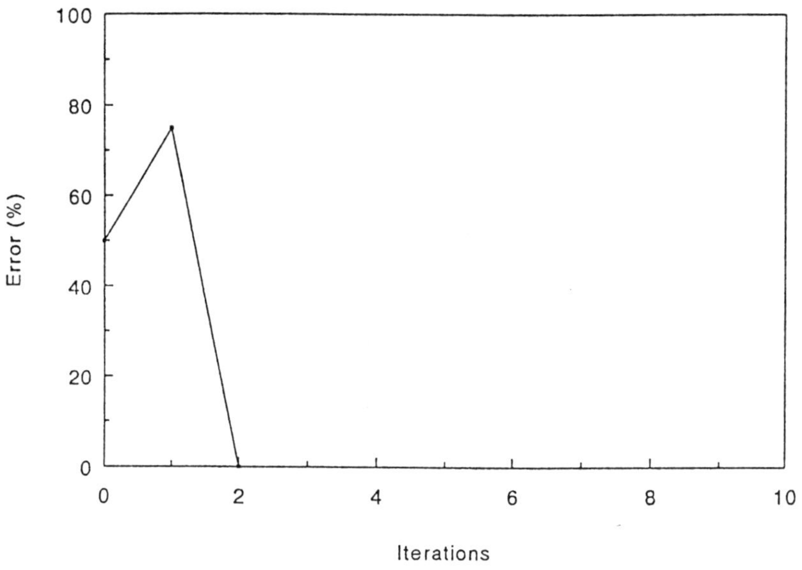

Fig. 12 Learning curve of the hybrid perceptron

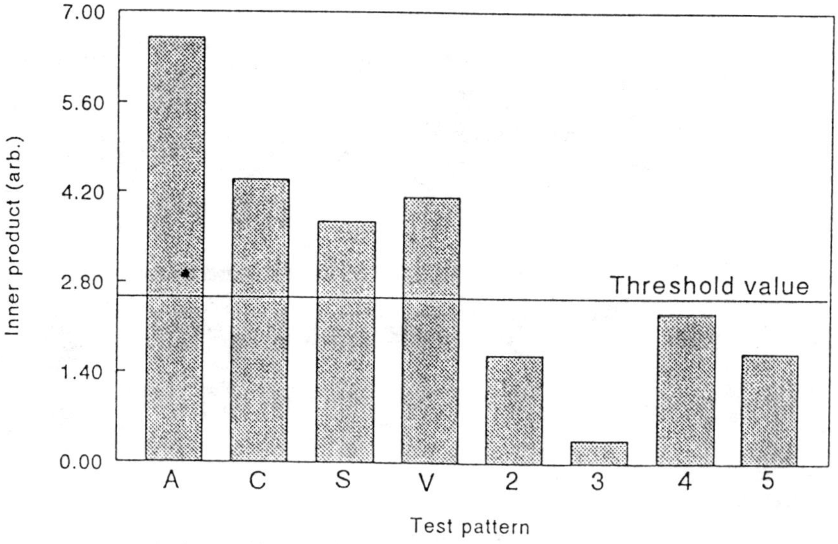

Fig. 13 Pattern recognition by the hybrid perceptron

Photorefractive Memories for Optical Processing and Computing

C. Alves, P. Aing, G. Pauliat, G. Roosen.
Institut d'Optique Théorique et Appliquée, Unité Associée 14 au Centre National de la Recherche Scientifique, Centre Scientifique d'Orsay,
B.P. 147, 91403 Orsay Cedex, France.
Tel: 1 69 41 68 55, Fax: 1 69 41 31 92, Email: roosen at fropt11 (bitnet)

ABSTRACT

After a rapid introduction that indicates the main reasons for the renewal of interest for photorefractive holographic memories, we will first present and compare both the coding techniques and recording procedures used for storage of superimposed images. Potentialities and limitations that are relevant to existing photorefractive crystals will be then discussed. Refreshing procedures that allow to operate the dynamic memory without loss of information will be also described. We will then devote the major part of the article to discussing applications we envision for photorefractive holographic memories.

1. INTRODUCTION.

The possibility of implementing holographic memories taking advantage of the photorefractive effect was proposed about 20 years ago and, since then, has been often demonstrated(1-3). The photorefractive effect is indeed well suited for these implementations. It relies on the existence of traps partly occupied by electric charges. These traps result from either growth defects or crystal doping. In the dark, the charges are frozen in place because of the low dark conductivity of the material. Under illumination, they are excited and can migrate in the crystal. Therefore a spatially periodic illumination, such as the one resulting from the interference of two light beams, produces a charge grating and thus a space charge field of same period. Through the electrooptic effect, an index grating is finally formed in the crystal.

Basic characteristics of the photorefractive effect are now well established(4). The properties of interest for memories are :

* A real time recording of the hologram during illumination without any bleaching.
* A dark memory permitting image storage as the charge distribution remains frozen when the recording beams are turned off.
* A fully reversible behavior. Indeed, illuminating the crystal with a new interference pattern erases the previously recorded grating while the new one is stored.

As the magnitude of the index variation is relatively small, one generally uses thick materials (from a few hundred micrometers up to 1 centimeter) in order to get sizable diffraction efficiencies. As a consequence, the recorded grating exhibits a strong angular selectivity[5] related to the Bragg effect. This is used to increase the number of stored images by superposition of multiple holograms in the same volume of the material[6], forming a stack of images (shared volume). This storage capacity can even be increased by juxtaposition of several stacks in the same crystal[1] (allocated volume). However, the material thickness leads to crosstalk that may limit the memory capacity[7,8]. These limitations are inherent to the use of any kind of thick medium. When working with photorefractive materials, their nonlinear characteristic responses bring requirements that have to be also considered[8].

The recovery of interest for holographic memories might appear surprising. Indeed, between first demonstrations conducted in the late 70's and current results, the overall capacity has not been noticeably increased. This constancy in storage capacity is mostly because nearly no improvement in material performances was obtained during those years despite the great deal of work successfully conducted to get a good understanding of the physics of the photorefractive effect. In fact, the renewal of interest is caused by the disappearance of limitations encountered in the initial development of these memories. Indeed :

* Spatial light modulators (SLM) now become more and more available permitting a dynamic presentation of the images to the memory.
* The development of new addressing techniques such as phase encoding[9,10] of the reference beam allows fast access to the memory.
* The recording of multiple holograms with identical efficiencies is easily achievable using an incremental recording procedure[10-12] that cancels out difficulties previously encountered with the sequential procedure[2].
* Solutions have been proposed and implemented to avoid erasure when using the memory while maintaining its dynamic character[13-16].

2. STORAGE PROCEDURES.

In this paragraph we rapidly compare the main characteristics of the two recording schemes proposed and implemented for photorefractive crystals and show that the incremental recording procedure offers advantageous peculiarities when compared to the sequential recording scheme. To simplify this comparison, we neglect the optical absorption and we consider the simple case where the photorefractive recording τ_w and erasing τ_e time constants are equals. More detailed analyses are presented elsewhere[2,10,11].

2.1. Sequential recording.

In the sequential recording procedure, all N images are recorded one after each other. Because the recording of new images partly erases the previously stored information, the recording time T_m for the mth new image is shorter than the recording time of previously recorded images so that the diffraction efficiencies are identical for all stored holograms. Therefore, the first hologram must be recorded with an index modulation $\delta\bar{n}_1$ as large as possible. However, a too large index modulation may lead to beam coupling which alters the image quality. This phenomenon has been observed(2) and may force to record the first hologram with an index modulation $\delta\bar{n}_1$ lower than the maximum index modulation δn_{max} achievable in the material. However, for a large number of images, the final diffraction efficiency little depends on the efficiency of the first recorded hologram(17). The (N-1) other holograms are then sequentially recorded.

The total time $T_T = \sum_{m=1}^{N} T_m$ required for the total recording and the index modulation $\delta\bar{n}_N$ for each hologram at the end of the recording procedure are:

$$\begin{cases} T_T = \sum_{m=1}^{N} T_m = T_1 + \tau_w \ \ln\left(1 + (N-1)(1 - e^{-T_1/\tau_w})\right) \\ \delta\bar{n}_N = \dfrac{\delta n_{max}(1 - e^{-T_1/\tau_w})}{1 + (N-1)(1 - e^{-T_1/\tau_w})} \end{cases} \quad (1)$$

with T_1 the recording time of the first image and τ_w the photorefractive time constant. Because the time constant τ_w usually depends on the fringe spacing of the interference pattern, these equations can not be fulfilled for complex images and for all the reference beams. Furthermore, because τ_w depends on the intensity, optical absorption decreases the index modulation in the depth of the material. As a result, the diffraction efficiency and the Bragg selectivity decrease(2,18) so that the image capacity of the memory is reduced. In the case where $T_1 >> \tau_w$, (i.e. $\delta\bar{n}_1 \approx \delta n_{max}$), the time T_T required to write the N images is $T_T = T_1 + \tau_w \ \ln(N)$, and the final index modulation $\delta\bar{n}_N$ for each hologram is $\delta\bar{n} \approx \delta n_{max}/N$.

2.2. Incremental recording.

In the incremental recording procedure(10,11), all N images are successively stored in the crystal with identical exposure times T_0 that are much shorter than the photorefractive writing τ_w time constant and the recording sequence of the N images is repeated until the steady state is reached. This procedure completely overcomes the above mentioned problems.

The condition for T_0 is to be much shorter than $(\tau_w/(N-1))$. It leads to the following expressions for the modulation index of the mth hologram at the end of the Qth recording cycle:

$$\begin{cases} \delta\overline{n}_m = \delta n_\infty\left(1-e^{-NQT_0/\tau_w}\right) \\ \text{with } \delta n_\infty = \delta n_{\max}\; e^{-T_0(N-1)/\tau_w}\,\dfrac{1-e^{-T_0/\tau_w}}{1-e^{-NT_0/\tau_w}} \end{cases} \tag{2}$$

One derives for the steady state efficiency of each hologram(8) $\delta n_\infty = \delta n_{\max}/N$ which is reached after an infinite total recording time with an exponential time constant τ_w. This value is the same as the one got for the sequential recording at saturation. It can be easily shown from equations (1) and (2) that the total recording time needed to reach a given modulation index for each hologram is exactly the same for both procedures.

This incremental recording procedure presents the two following important advantages over the sequential procedure:

* For large capacity N, δn_∞ always remains much smaller than $\delta n_{\max}$ so that the beam coupling is kept small and does not alter the image quality and does not introduce crosstalk as pointed out in Ref. 10.
* The same saturation value δn_∞ is reached simultaneously for all holograms so that, at the end of the recording, these holograms are uniform in the crystal thickness even in presence of optical absorption. Thus, this procedure makes full use of the maximum index modulation and of the Bragg selectivity.

Of course, similarly to the sequential recording procedure, the dependence of τ_w on the fringe spacing may lead to some differences between the index modulations.

3. MEMORY ACCESS PROCEDURE.

To selectively record and retrieve holograms superimposed in the same volume of a material, each must have its own address. This address is represented by the reference beam used for the recording. Image retrieval is performed by reading out the hologram with the very same reference beam that was previously employed for recording.

3.1. Reference beam arrangement: sampling grids and square matrices.

To avoid crosstalk between reconstructed images, one may do so that each reference beam only reconstructs the image it has recorded. This is achieved by taking advantage of the Bragg selectivity. As shown in figure 1.a this is obtained by aligning the tips of reference beam wave vectors along a circle (in the θ direction).

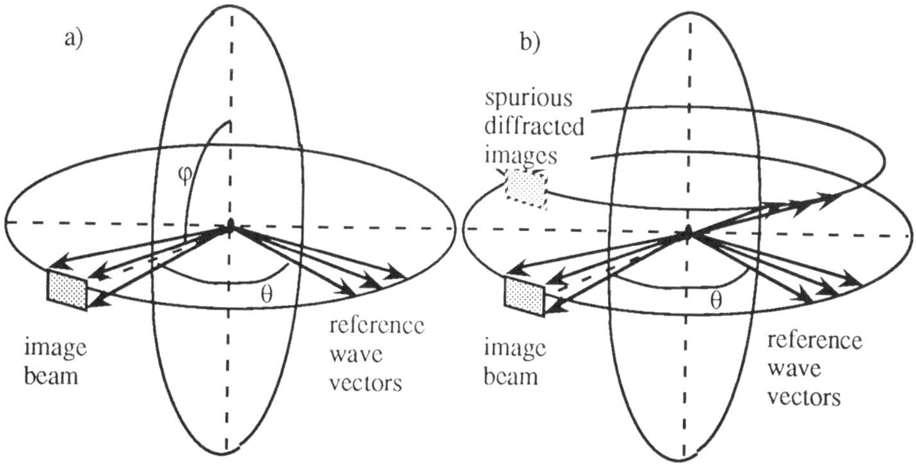

Figure 1: a) arrangement of the reference beam wave vectors in the same incidence plane to use the Bragg selectivity; b) increase of capacity by adding another line of wave vectors. The spurious diffracted images represent the light diffracted by these new reference beams onto the previously recorded holograms.

In this figure each one of the stored holograms fully occupies the same solid angle $\Delta\theta.\Delta\varphi$. The angular spacing A_θ between these reference beams is chosen to be larger than the Bragg angular selectivity $\Delta\theta_B$. If the allowable angular aperture for the reference beams (limited by the photorefractive parameters as discussed in a following chapter) is $\Delta\theta_R$, then the maximum number of images N_θ that can be stored in the crystal is about:

$$N_\theta \approx \frac{\Delta\theta_R}{A_\theta} \leq \frac{\Delta\theta_R}{\Delta\theta_B} \approx \Delta\theta_R \frac{crystal\ thickness}{fringe\ spacing} \tag{3}$$

For example, with a crystal thickness of 5mm, an average fringe spacing of 2μm and an allowable angular aperture $\Delta\theta_R$ of 5° (taken inside the crystal), then we find $N_\theta \leq 200$ superposed holograms.

In order to increase the number of stored holograms, we can add other reference beams whose wave vector tips are aligned along an other circular arc[19] (see figure 1.b). Due to conical Bragg degeneracy, these new wave vectors can read out holograms written by the previous ones. This phenomenon produces spurious diffracted images. If the angular spacing between these two lines A_φ (in the φ direction) is larger than the angular radius $\Delta\varphi$ of the images then these undesired diffracted images can be easily spatially filtered. If $\Delta\phi_R$ is the allowable angular aperture for the reference beams in the φ direction, then the number of identical circles that can be added is:

$$N_\varphi \approx \frac{\Delta\phi_R}{A_\varphi} \leq \frac{\Delta\phi_R}{\Delta\varphi} \tag{4}$$

We see that the capacity $\approx N_\varphi . N_\theta$ now depends on the image size. For example if each image has $\sqrt{M} . \sqrt{M}$ pixels and if it covers a crystal surface of $a.a$ then the diffraction limit gives the minimum angular image size:

$$\Delta\theta = \Delta\varphi \approx 2\sqrt{M}\frac{\lambda}{n\,a} \tag{5}$$

with λ the vacuum wavelength and n the refractive index.

For example, with a crystal area $a.a \approx 5x5mm^2$, at λ = 532 nm (n = 2.4), an average fringe spacing of 2 μm, an allowable angular aperture $\Delta\phi_R$ of 5° (taken inside the crystal) and a pixel number per image $M = 128x128$ then we find $N_\varphi \approx 8$ different circles. The tips of reference beams are now arranged along a rectangular grid which is a particular example of the sampling grids described in Ref 19. These non uniform arrangements may be a nuisance in some applications. It seems thus preferable to dispose these reference beams along a square grid.

In order to describe how to build such a compact square arrangement, we have represented the previous rectangular sampling grid in the θ,φ diagram in figure 2.a. Because we assume that both $\Delta\phi_R$ and $\Delta\theta_R$ are small, the projections of the circular arcs are straight lines in this diagram. We now set the tips of the reference wave vectors along a square matrix, the spacing of which is δ. To achieve that goal, this matrix is rotated by an angle ω in the θ, φ plane. This angle ω is small so that all calculations are conducted assuming that $\omega \ll 1$.

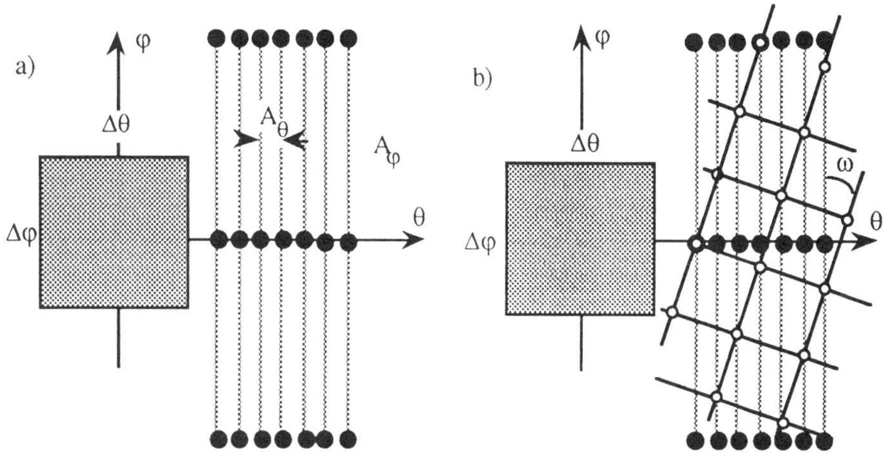

Figure 2: reference (black or white dots) and image (gray square) wave vectors arrangement in the θ,φ diagram; a)with the rectangular sampling grid already depicted in figure 1b; b) with a square sampling grid (white dots).

The reference beam wave vectors are labeled by the indices i, j. The positions of their tips are thus expressed by:

$$\begin{cases} \theta_{i,j} = \overline{\theta} + i\,\delta - j\,\delta\,\omega \\ \varphi_{i,j} = \overline{\varphi} + i\,\delta\,\omega + j\,\delta \end{cases} \tag{6}$$

From these equations we conclude that to avoid image crosstalk we must have:

$$\left|\theta_{i,j+1} - \theta_{i,j}\right| \geq A_\theta \quad \text{i.e.} \quad \delta\,\omega \geq A_\theta \tag{7}$$

Similarly, along the perpendicular direction, two beams aligned along the same arc must be angularly spaced by more than A_φ. Thus, we must have:

$$\left|\varphi_{i+1,j} - \varphi_{i,j+k}\right| \geq A_\varphi \quad \text{i.e.} \quad k\,\delta \geq A_\varphi \tag{8}$$

These two pixels are adjacent along the same arc when integer k verifies:

$$1 = k\,\omega \tag{9}$$

Combining equations (7), (8), (9), leads to the expressions for ω and δ:

$$\begin{cases} \delta = \sqrt{A_\varphi \, A_\theta} \\ \omega = \sqrt{A_\theta / A_\varphi} \end{cases} \tag{10}$$

Equations (9) and (10) demonstrate that this most compact square arrangement is possible if integer k exists, i.e.: if we have correctly chosen A_φ and A_θ.

In that case, the first equation of this set shows that the density of reference beams is the same whatever the chosen arrangement. However the arrangement of the reference beam wave vectors on a square grid presents several advantages, among which a faster addressing speed. For instance if the input reference beam is addressed by two cross acousto optic devices, then the steering speed decays with increasing the number of addresses. In case of rectangular sampling grids, the steering speed is thus limited by the acousto optic device addressing the beam in the θ direction which corresponds to the maximum number of positions ($N_\theta \approx 200$ in our example). With the square grid, the number of positions which are addressed by each acousto optic cell is the same and is thus much smaller ($\sqrt{N_\theta \, N_\varphi} \approx 40$ with our previous example).

3.2. Encoding techniques.

One solution to address the different reference beams consists in rotating the crystal after each image recording(2,20). The angular position of the crystal thus represents the address of the recorded image. The main impediment of this method is its slowness. An alternative is to change the reference beam incident direction. This technique, most often used(1,3), is named angular multiplexing. Recently, up to 5000 images have been recorded in this way(21). The incidence angle of this reference beam represents the address of the stored image. Various methods have been proposed to change this angle(3,21-23). An intensity spatial light modulator may be used to select 1 among N beams(23). However, that technique leads to severe losses of energy. Galvanometric deflectors are slow (a few milliseconds) because the high required accuracy in positioning. Acousto-optic deflectors allow submillisecond deflexions but one needs two devices in order to compensate for the Doppler shift(21). This leads to more complicated implementations.

Another solution is to encode by changing the wavelength of the recording beam(24,25). This method potentially leads to a larger storage capacity provided we record reflection holograms. Indeed, reflection volume holograms have a much higher wavelength selectivity than transmission holograms while the angular sensitivity remains similar(5). However, as the single diffraction efficiency decreases as the square of the number of stored holograms, the storage capacity will be mostly limited by the weakness of the retrieved signal. Wavelength tuning in semiconductor lasers (for instance piezoelectric drive of external cavity) is achieved in the submillisecond range. The access time is thus similar to what is obtained in angular multiplexing with acousto optic deflectors.

A novel technique, based on the phase encoding of the reference beam has been recently proposed and demonstrated(9,10,12). Each image, say image # m, is stored by interfering the image beam with N reference beams whose set of relative phases, $\left(\varphi_1^m, \varphi_2^m, \ldots \varphi_N^m\right)$, represents the address of that image. All reference beams overlap the image in the recording volume. The angular spacing between adjacent reference beams is chosen to be large enough to avoid image crosstalk. As a result, N independent holograms of each image are recorded. During the readout, each reference beam reconstructs the hologram it has recorded only. Image #m is retrieved by reading out with all the N reference beams having the very same phase code than the one used during the recording of that image. Noiseless reconstruction appears because the sets of phase codes are chosen so that, in the reconstruction, unwanted images interfere destructively to produce zero intensity. The set of orthogonal phase addresses required to get noiseless reconstruction can be easily calculated(8,26). For the binary phase codes we are interested in ($\varphi_n^m = 0$ or π), we may use the Walsh Hadamard(27) transform to construct the orthogonal set of addresses. In case where N is a power of 2, this transform allows the construction of N orthogonal phase codes. Thus, with N reference beams, N images can be stored in the material.

Phase only addressing is naturally light efficient and provides the same theoretical storage capacity as angular multiplexing. In real systems, this image capacity is in fact limited by the imperfections of available optical components(26). Limiting factors of realistic systems arise from the recording procedure used to store the images, the imperfections of the recording medium and the imperfections of the phase coding element. This last point will be illustrated later when presenting the system we are implementing. Access time as short as a few microseconds can be achieved using ferroelectric liquid crystals to build the phase only modulator (ΦSLM). Indeed, for the application we consider and that will be described below, the number N of stored images will remain limited (a few hundreds). As seen above, N is also the number of reference beams and thus of pixels required in the phase only modulator. This corresponds to a low resolution ΦSLM where each pixel can be driven by its own electrical line. All pixels may thus be addressed in parallel and the addressing time is the ferroelectric liquid crystal switching time (a few microseconds).

4. THEORETICAL AND PRACTICAL LIMITS OF PHOTOREFRACTIVE STORAGE.

There is an important limitation to capacity due to a crosstalk effect that is inherent to all memories using thick materials whether they are photorefractive or not(7). This source of noise results from successive diffraction processes of the input light beams on the different superposed holograms. In the following we will only pay attention to the limitations relative to photorefractive materials. We will show that the maximum capacity and data rate are determined by both the performances of the storage material and the recording geometry. In particular, the maximum capacity is much lower than the theoretical limit Vn^3/λ^3 (6). As an example, for a crystal of volume 5x5x5 mm^3, a wavelength $\lambda = 532$ nm, the theoretical limit gives a capacity $C \approx 10^{13}$ bits. We will see that the available maximum capacity is much lower. Moreover the effective capacity will strongly depend on the requirements on data rate.

4.1 Capacity limitation due to the finite number of photorefractive centers.

We here present a simplified approach for the determination of the signal to noise ratio limitation due to the available number N_A of charges that can be redistributed in photorefractive crystals. A more detailed analysis is given in ref. 28.

In photorefractive crystals, this is the spatial charge redistribution in deep levels that gives the index modulation. These levels are randomly distributed in the dark. From Poisson's equation, the spatially modulated deep level concentration $m\, N_d^{+1}$ resulting from the interference of two planes waves is(4):

$$m\, N_d^{+1} = m\,\frac{\varepsilon k_g E_1}{e}\sin\left(k_g x\right) \tag{11}$$

with m the modulation ratio of the interference pattern, k_g the grating wave vector, ε the dielectric constant, e the charge of the electron and mE_1 the photoinduced space charge field. In absence of external field, one has:

$$E_1 = \left(\frac{E_d\, E_q}{E_d + E_q}\right) \tag{12}$$

where E_d and E_q are the diffusion and maximum fields respectively as defined in ref.4.
In steady state, the charge distribution is written as:

$$N_d^+ = N_A\left(1 + m\frac{N_d^{+1}}{N_A}\sin(k_g x)\right) \tag{13}$$

where N_A represents the ionized trap density in the dark[4].

Following a macroscopic approach, this represents the average distribution of charge density. For large N_d^+, in a microscopic approach, the mathematical expectation equals this mean value. Thus the mathematical expectation of charge number *n(x)* (random variable) distributed in the slice of volume *h.l.dx* (thickness *l*, width *dx*) is:

$$E[n(x)] = N_A\left(1 + m\frac{N_d^{+1}}{N_A}\sin(k_g x)\right) h\, l\, dx \tag{14}$$

Assuming that all charges are independent, i.e. fringe spacing $\Lambda >> l_d$ the Debye screening length, we have a normal law and the probability law for *n* charges follows a Poisson's law, giving a variance:

$$\sigma^2[n(x)] = E[n(x)] \tag{15}$$

The Coulomb interaction is taken into account in ref.28.

The diffracted amplitude is proportional to the Fourier transform of *n*. Again, for independent charges, one gets the mathematical expectation of the Fourier transform of *n(x)*:

$$E[\mathrm{FT}(n(x))] = h\, l\, m\, N_d^{+1}\, \mathrm{sinc}\left[\left(k_g - k_g'\right)a\right] \tag{16}$$

a being the crystal size along the *x* direction and k_g' the wave vector of the observed Fourier component.

Its variance is:

$$\sigma^2 = 2\, N_A\, h\, l \,/\, a \tag{17}$$

The signal to noise ratio in amplitude corresponds to the ratio of the amplitude received at $k_g = k_g^{'}$ by the noise amplitude. Note that the noise is independent of the grating wave vector, i.e. of the fringe spacing. One thus gets:

$$SNR = m\frac{N_d^{+1}}{N_A}\sqrt{\frac{N_A h a l}{2}} \leq m\sqrt{\frac{N_A V}{2}} \text{ as at maximum, } N_d^{+1} = N_A \tag{18}$$

where V is the crystal volume.

For the storage of 1 hologram of M pixels, considered as multiple gratings formed by plane waves, the modulation for each pixel is at maximum $m_1 = \frac{1}{\sqrt{M}}$. For N superposed stored images, we will thus have a pixel modulation $m_1 = \frac{1}{N\sqrt{M}}$. The amplitude signal to noise ratio is consequently given by:

$$SNR \leq \frac{1}{N\sqrt{M}}\sqrt{\frac{N_A V}{2}} \tag{19}$$

This equation defines the ultimate capacity of the material and shows that it is worth having few images with a large number of pixels rather than the opposite.

The intensity signal to noise ratio SNR_{int} is about the square of the amplitude signal to noise ratio SNR. For a crystal of volume 5x5x5 mm^3, taking the generally admitted value of $N_A \approx 10^{22}\,\text{m}^{-3}$, $M = 256x256$ and an intensity $SNR_{\text{int}} = 4$ the image number will be $N \approx 50000$. The effective capacity is $C_{eff} \approx 3\ 10^9$ which is a few orders of magnitude smaller than the theoretical limit(6).

We must point out that this relatively large number of images, $N \approx 50000$, can be stored in the crystal at the expense of the amplitude signal to noise ratio. Such a low signal to noise ratio $SNR_{\text{int}} = 4$ only allows to store binary images. If one wants to record gray level images with g different levels, then the signal to noise ratio must be larger than $SNR_{\text{int}} \geq p(g-1)$ with $p>1$ defining the error rate. For example with $p=4$ and 128 gray levels then the image number reduces to $N \approx 4000$.

This short calculation shows that to store many images in a small volume, one has to choose a photorefractive sample with a high ionized trap density N_A. For instance with iron doped $LiNbO_3$ samples the trap density N_A can be as large as 10^{24}m^{-3}. The image capacity is thus increased by a factor 10.

4.2. Data rate-Capacity trade-off.

We do not only look for capacity but we also desire high data rates. This is particularly true for pattern recognition applications. Of course, the power P of the read-out beam is limited. We thus must integrate the diffracted signal during a given time τ for detecting a sufficient enough photon number per pixel of the reconstructed image[29]. If η^H is the diffraction efficiency per hologram, M is the number of pixels in the image, n_b is the photon number required for deciding the pixel value, the data rate is thus expressed as:

$$\frac{M}{\tau} = \frac{P\,\eta^H\, n_0}{n_b} \tag{20}$$

n_0 being the photon number in 1 Joule ($n_0 = \lambda / hc$, h Planck constant and c light velocity in vacuum).

The detection error rate depends on signal n_b and on detector noise. If we deal with binary images, we may choose a reasonable value $n_b = 10^3$ photons. Thus for a reading power of 10 mW at 532 nm, a diffraction efficiency per hologram of 0.1% (which is readily achievable with N = 256 stored images) we get a maximum information rate of $M/\tau \approx 3.10^{10}\ s^{-1}$. Consequently, for images of M = 256 x 256 pixels, i.e. a total capacity of $\approx 2.10^7$ bits, the minimum image reading time becomes $\tau \approx 2\mu s$.

However, if we want to store gray level images, because of the Poissonnian distribution of photons, then there is the insurmountable quantum noise: $noise = \sqrt{n_b}$. Even if we neglect all other noise sources, if we want to record images with g gray levels then we must have a signal to noise ratio: $SNR_{int} = \sqrt{n_b}/(g-1) \geq p$. For example, with the same parameters as previously, but with $p = 4$ and $g = 128$ gray levels, we now find $M/\tau \approx 10^8\ s^{-1}$. The minimum image reading time now becomes $\tau \approx 600\mu s$.

Note that the maximum value for M is defined by the crystal entrance surface and angular bandwidth[8].

5. REFRESHING PROCEDURES.

We here consider the problem of erasure of the holographic memory when it is used. Fixing techniques have been proposed[30-32] but they are not adequate as they generally hinder further updating of the memory. Indeed fixing being globally achieved for all holograms, it is necessary to erase them all before recording any new information. Moreover, for temperature fixing(30), the holograms are recorded at high temperature and then read-out at room temperature. This temperature difference gives rise to a crystal expansion that modifies the value of the Bragg angle. This can make the read-out impossible. This impediment does not exist in electrical fixing(31,32) but this technique has been demonstrated for high modulation gratings only what is not the case for multiple superimposed holograms.

In order to overcome the problem of erasure of dynamic memories, new architectures have been proposed. For instance, it is possible to use two holographic media(13) that can be two photorefractive crystals(14). All the data stored in one of them are saved in the second crystal before the first crystal is read until complete erasure. An optical amplifier is used to intensify the useful part of information that is transferred. The process is then repeated by exchanging the role of the two crystals.

The use of two phase conjugate mirrors associated to the memory is another approach(15). The transmitted part of the reference beam and the diffracted one (the retrieved image) are reflected back by two phase conjugate mirrors leading to a new recording of this image. However, up to now, this technique has been demonstrated for one image only.

Adding a feedback loop, either opto-electronical or purely optical, to the memory permits a more flexible implementation of a "non destructive read-out-refreshing" scheme(16). The images retrieved from the memory are fed back and rewritten inside the crystal after being processed, thresholded and amplified. Experimental demonstration has shown evidence of a memory operation for days without loss of information and with the use of one dynamic (write/erase) holographic medium only, here a single photorefractive crystal(8,16). In the opto-electronic approach, the images retrieved from the memory are detected by a CCD camera and processed electronically. An all optical feedback loop, using an optically addressed spatial light modulator (OASLM) as thresholding and amplifying element, has the great asset of allowing parallel processing of the pixels of the whole image. This opto-optical refreshing scheme has been recently implemented using a sampling technique for image stabilization(33).

The optoelectronic feedback loop demonstrated in ref. 16 was lately used for analyzing the dynamics of selective erasure in dynamic photorefractive memories(34).

6. LOW CAPACITY AND FAST ACCESS TIME PHOTOREFRACTIVE MEMORIES

High capacity photorefractive memories are currently studied by various groups. Recently, 5000 transmission holograms have been stored in a 2x1.5x1 cm^3 $LiNbO_3$

crystal[21] using angular multiplexing and a so-called fractal-space multiplexing[19,35]. In other approaches, different stacks of images are stored. This corresponds to either SBN fiber microcrystals[36] or different locations on a rotating $LiNbO_3$ disk[37]. The latter solution appears to be the most attractive as it will not encounter the aperture limitations caused by optical systems[38].

Another exciting approach is to use the photorefractive memory as a buffer memory for optical images[8]. A photorefractive memory can easily provide high resolution images with a flow rate of more than 10^4 images per second (see before). This appears to be a very promising solution in the scope of both digital computing and optical pattern recognition using correlators.

6.1 Photorefractive memories for optical computing

We may take advantage of the fast access time of memories to implement optical reconfigurable interconnects[39,40] for digital computing. Most of these systems are based on multiplications between an input data vector and a control matrix. The input vector is formed by a spatial light modulator linear array and the matrix is a 2 dimensional spatial light modulator (2DSLM) (see figure 3a). For instance, in the scheme proposed by P.S. Guilfoyle[41] both input vector and control matrix must be changed at each computation cycle.

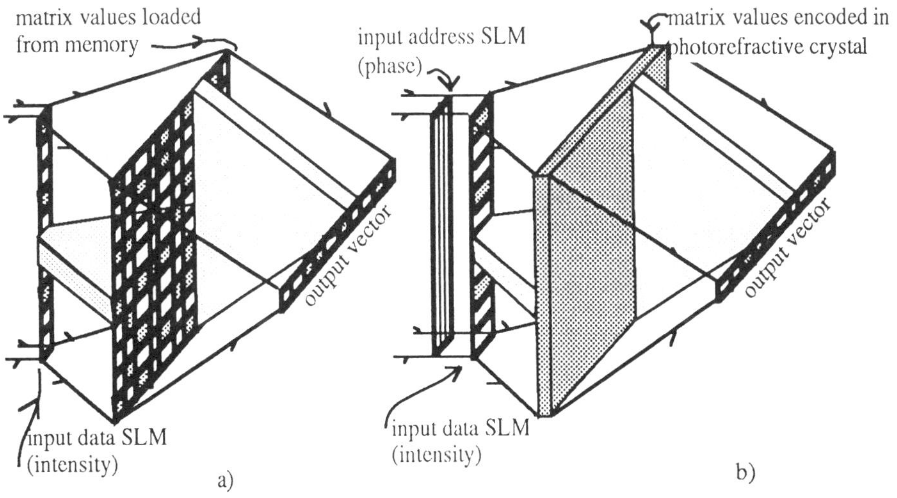

Figure 3.: a)reconfigurable optical interconnections with a spatial light modulator; b) with volume holograms. The relay optics are not represented.

The time required to load the matrix from the computer memory onto the 2DSLM thus hinders the system performance. Photorefractive memories can easily overcome this limitation if each pixel of the input optical vector simultaneously carries the input numerical data encoded on the pixel intensity (0 or 1) and an address encoded, either on the optical frequency, on the spatial phase φ(x,y) of the pixel wavefront, or on the incident angle. According to that address, the pixel reads out the volume hologram corresponding to the desired interconnection matrix only. In the system we propose, all interconnection matrices are stored in the photorefractive crystal and are thus readily attainable.

Such a possible interconnection system is depicted in figure 3b for phase encoding.

6.2 Photorefractive buffer memories for optical correlators

Optical correlation seems to be quite promising for pattern recognition. However, the correlation of a single input image to a single filter at the video rate (imposed by the existing spatial light modulators) limits its interest. One previously proposed solution to overcome that bottleneck is to use spatially or temporally multiplexed filters(42). Such systems have been implemented with the filters stored on optical disks. Spinning the disk allows to rapidely present all the filters to the correlator and thus to compare each image to all filters at the video rate(37).

Using a photorefractive crystal for the memory is an intersting alternative to optical disks. The correlator itself (figure 4) encompasses a nonlinear material, i.e. an opto-optical SLM or a photorefractive crystal, in which is holographically recorded the Fourier transform of the image to be analyzed.

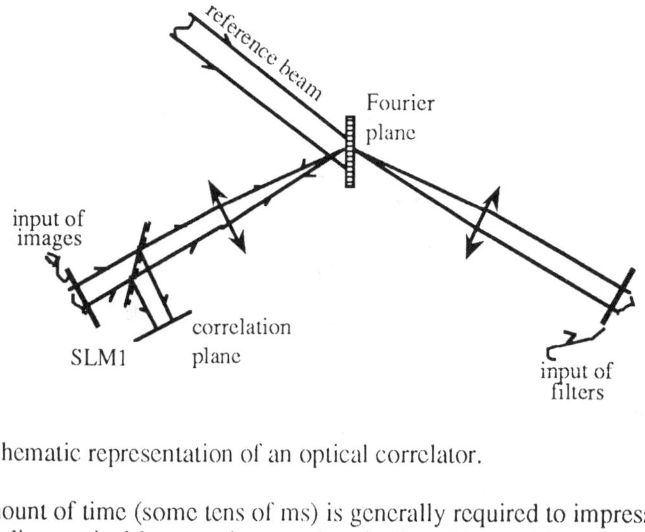

Figure 4: Schematic representation of an optical correlator.

A given amount of time (some tens of ms) is generally required to impress that image on the recording optical beam using an electrically addressed SLM (SLM1) and to record its hologram in the nonlinear medium. Subsequent read-out by a beam carrying the Fourier transform of a filter gives the correlation product between the image and that filter. To realize a fast image recognition process, several filters have to be rapidly presented. Thus the correlation speed is determined by the time required for the presentation of a new filter. If these filters have to be set on the optical reading beam by using an electrically addressed SLM (SLM2) with a large enough number of pixels ($>10^6$ for example), it will take tens of milliseconds.

But if these filters were previously stored in a photorefractive memory, each can be retrieved in few microseconds only using the phase encoding technique. A comparison of the unknown image with the whole set of filters previously stored in the photorefractive memory (say $N \approx 1000$) can be realized at video rate, that corresponds to the presentation time of the unknown image with SLM1 (figure 5).

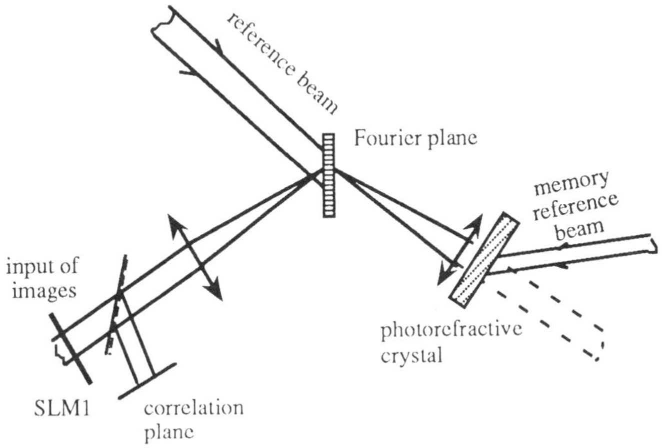

Figure 5: Optical correlator with a photorefractive memory for fast comparison to multiple filters.

Note that for this application, the refreshing of the memory can be easily accomplished when a new image is presented to the correlator. Indeed, little erasure will take place during each comparison with the whole set of filters as the erasure time of the photorefractive crystal used for the memory is much longer than the video rate. It will be thus sufficient to refresh 1 among the total number of N stored filters each time a new image to analyze is introduced on SLM1 to be presented to the correlator. The required time ($\approx$ 40ms) for setting this new image is long enough for permitting the memory refreshing. Consequently, for this application, the previously presented feedback loops are not required. That considerably alleviates the system implementation.

7. COMPONENTS FOR MEMORY IMPLEMENTATION

We are currently assembling a prototype of such a buffer image memory with currently available components[43]. Our goal is to analyze the potentialities and limitations in the "real world".

Figure 6 shows a diagram of the memory implementation during the phase of filter storage.

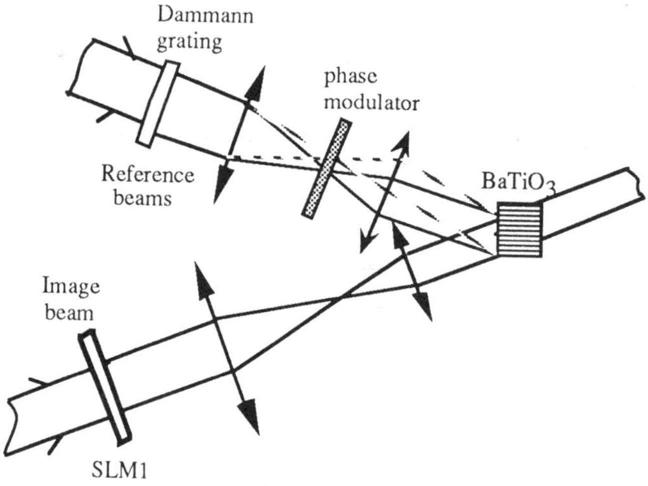

Figure 6: Layout of the memory storage scheme

In the reference arm, a Dammann grating diffracts the incident beam in $\sqrt{N} \times \sqrt{N}$ plane waves of equal intensities (limited to 8x8 in a first implementation because of component availability). The Dammann grating is imaged onto the photorefractive crystal by means of a refractive telescope (4-f system). All the $\sqrt{N} \times \sqrt{N}$ reference beams thus overlap the image beam in the crystal.

The Dammann grating is also Fourier transformed and produces a square array of $\sqrt{N} \times \sqrt{N}$ spots. These spots coincide with the pixels of the phase only modulator so that each reference beam can be phase modulated separately.

In the other arm (image arm), the filter to be holographically stored in the memory is introduced by an intensity modulator that is imaged onto the photorefractive crystal with again a refractive telescope used in a 4-f configuration.

7.1. Limitations due to imperfections of the phase modulator

A. Phase noise

The phase modulator may suffer from imperfections so that the phase modulation is not exactly equal to π: $\varphi_n^p - \varphi_n^m = \pi + \varepsilon$ instead of π with ε the phase noise. Thus, during reconstruction, the undesired images do not interfere destructively any more. This phase noise ε consequently leads to a severe capacity reduction. The main capacity reduction is induced if the mathematical expectation $\langle\varepsilon\rangle$ ($\approx$mean value of variable ε for different reference beams) is non zero. In that case, we showed(26) that the signal to noise ratio in intensity is given by:

$$SNR_{Int} \approx \left[\frac{2}{(N-1)\ \langle \varepsilon \rangle} \right]^2 \tag{21}$$

For a given image capacity and a desired signal to noise ratio SNR_{Int}, this equation sets an upper limit on $\langle \varepsilon \rangle$. For example, for our first demonstrator, we require N=64 images and with a $SNR_{\text{int}} \approx 100$, these conditions lead to:

$$\langle \varepsilon \rangle \leq 3\ 10^{-3}\ \text{radian} \tag{22}$$

B. Amplitude noise

We also demonstrate(26) that the non uniformity in the diffracted amplitude gives rise to crosstalk between images. The diffracted amplitude is proportional to:

$$u_{nm} = \frac{P_n^r\ \left(P_n^w\right)^*}{I_m + I_R}\ \delta n_n \tag{23}$$

where:

- δn_n is the index modulation of the grating induced by interfering the image beam with the nth reference beam;
- P_n^w is the amplitude inside the crystal of the nth reference beam during the recording of the hologram;
- P_n^r is the amplitude inside the crystal of the nth reference beam during the reading out of the hologram;
- I_m and I_R are respectively the intensity of the image beam and the sum of intensities of the reference beams.

This equation with the implementation of figure 6 shows that the diffracted amplitude depends on components performances:

$$u_{nm}\ \alpha\ T_n^{Dammann} \sqrt{T_n^{phase\ mod.,r}} \sqrt{T_n^{phase\ mod.,w}}\ \delta n_n \tag{24}$$

with $T_n^{Dammann}$ the transmission of the Dammann grating, $T_n^{phase\ mod.,r}$ and $T_n^{phase\ mod.,w}$ the transmission of the phase modulator during the readout and the writing processes respectively. Deriving this equation leads to:

$$\frac{\Delta u_{nm}}{u_{nm}} = \frac{\Delta T_n^{Dammann}}{T_n^{Dammann}} + \frac{\Delta T_n^{phase\ mod.}}{T_n^{phase\ mod.}} + \frac{\Delta \delta n_n}{\delta n_n} \tag{25}$$

This indicates that the fluctuations Δu_{nm} of u_{nm} have 3 different origins. $\Delta\delta n_n$ originates from the dependence of the photorefractive effect on the reference beam angle. We will examine this effect in a next section. For an idealistic phase modulator, $\Delta T_n^{phase\ mod.}$ should equal zero. However, for the phase modulator we will use, the phase modulation induces a small amplitude modulation so that $\Delta T_n^{phase\ mod.} \neq 0$ between the two different states corresponding to the two phase shifts 0 and π. $\Delta T_n^{Dammann}$ represents the spatial variation of the intensity between the different beamlets produced by the Dammann grating.

If all these variables are random with a gaussian distribution, the probability for the amplitude signal to noise ratio r to be larger than r_0 is given by(26):

$$\text{Prob}(|r| > r_0) = \frac{2}{\sqrt{2\pi}} \int_0^{1/r_0\sigma'} e^{-t^2/2} dt \quad \text{with} \quad \sigma'^2 = \frac{(N-1)}{2N} \frac{\sigma_u^2}{\langle u \rangle^2} \tag{26}$$

with $\langle u \rangle$ and σ_u^2 the mathematical expectation and variance of variable u_{nm}. With obvious notations, the fluctuations in the diffracted amplitude are thus lower than:

$$\frac{\sigma_u}{\langle u \rangle} \leq \frac{\sigma_{T^{Dammann}}}{\langle T_n^{Dammann} \rangle} + \frac{\sigma_{T^{phase\ mod.}}}{\langle T^{phase\ mod.} \rangle} + \frac{\sigma_{\delta n}}{\langle \delta n \rangle} \tag{27}$$

These two last equations will permit to estimate an upper limit of acceptable defects for the optical components. For instance, if the fluctuations in the diffracted amplitude are about 20% ($\sigma_u / \langle u \rangle = 0.2$) we find $\text{Prob}(|r| > 3) = 0.98$.

C. Phase only modulator realization.

The phase modulator must fulfill the two following conditions in order to fit our system:

*the binary phase modulation must exactly equal π with a systematic error lower than $<\varepsilon> \approx 3.10^{-3}$;

*the access time must be in the submillisecond range.

Pure phase modulation may be achieved with a nematic liquid crystals display[44]. However, these systems do not allow to set the phase shift with a high accuracy. A much better accuracy can be achieved by using a binary polarization spatial light modulator sandwiched between two polarizers, for instance a magneto-optic spatial light modulator[45].

For our system, we use an electrically addressed ferroelectric liquid crystal spatial light modulator sandwiched between two polarizers properly oriented. The ferroelectric liquid crystal thin layer used in these modulators is a uniaxe birefringent medium whose fast axis can be switched between two stable states by applying a control voltage $\pm V$ to the layer. This fast axis corresponds to the molecule director and is constrained to stay in the layer plane. If the medium thickness is so that $(n_+ - n_-)e = \lambda / 2$ (with n_+ and n_- the refractive index for light propagating along z direction and polarized along the x and y axes and e the layer thickness), then the liquid crystal layer is a controllable half wave plate.

Phase modulation is achieved by adding two polarizers to the ferroelectric liquid crystal modulator. The first polarizer axis is approximately aligned along to the bisector of the two stable states of the optical axis. The second polarizer is crossed to the first one.

An exact π phase shift is obtained whatever the birefringence or total phase shift $\Psi = 2\pi(n_+ - n_-)e/\lambda$ and the absolute values α_1 and α_2 between the incoming light polarization and the liquid crystal axis as long as the two polarizers are perfectly crossed and the signs of α_1 and α_2 are opposite. Such a phase modulator is thus very robust and for example insensitive to any temperature variation which affects the values for α_1, α_2 and Ψ.

However, when the two polarizers are not perfectly crossed, the phase modulation is not exactly equal to π so that the memory capacity or signal to noise ratio is reduced as presented in 7.1.A. To compute the value for ε, we consider that the angle between the two polarizers is $\delta+\pi/2$. Then we derive the electric fields of the optical waves transmitted by the second polarizer and find that, for $\alpha_1 = -\alpha_2 = 22.5°$ (as in our modulator), the phase modulation becomes $\pi + \varepsilon$ with:

$$\varepsilon \approx \sqrt{2}\,\delta \sin(\Psi - \pi) \tag{28}$$

We thus conclude that it is possible to achieve a very accurate π phase shift provided that either δ is close to zero or Ψ close to π. Both conditions are easily fulfilled with good polarizers and with commercially available ferroelectric modulators which are designed to be controllable half wave plates i.e. with $\Psi \approx \pi$.

Another limitation may occur because α_1 may be different from α_2. This will then induce an amplitude modulation. We compute the corresponding modulator transmission variation (with $\alpha_1 \approx -\alpha_2 \approx 22{,}5°$) and find:

$$\frac{\Delta T_n^{phase\ mod.}}{T_n^{phase\ mod.}} \approx 4\left(\alpha_1 + \alpha_2\right) \qquad (29)$$

Our experiments have shown that this intensity modulation is indeed negligible with our phase modulator.

D. Phase only modulator characterization.

We chose to make our first implementation with the 10x10 ferroelectric liquid crystal spatial light modulator from Displaytech (model 10x10PM). This square array allows independent phase modulation of all 10x10 pixels with a reconfiguration rate smaller than 150µs. We will thus be able to phase encode $N=2^6=64$ reference beams and store 64 images. All images will be retrieved during one video frame.

We fully characterized this phase modulator in order to determine, for each of its pixel (i.e. for each reference beam) the exact value of the phase noise ε and of the transmission. These parameters are computed from the measured values for δ, α_1, α_2 and Ψ.

The spatial light modulator being placed between two crossed polarizers the pixels are then periodically switched between the two states leading to an intensity modulation detected by a photodiode. The two polarizers are then simultaneously rotated (keeping a 90° angle between them). From this, we measured $\alpha_1 \approx -\alpha_2 \approx 26°\,50'$. We checked that same values are found for all pixels within an uncertainty of 10' showing nearly no variation in liquid crystal alignment. Therefore, the induced modulator transmission variations (see above) are lower than $\Delta T_n^{phase\ mod.} / T_n^{phase\ mod.} < 2\,\%$.

The phase difference Ψ is determined by measuring the intensity modulation of the output wave at λ=532nm. From this, we derive the value for the phase difference Ψ, i.e. $|\Psi - \pi| \approx 8.5°$. Combining the measured value for $|\Psi - \pi|$ and for δ, we compute an upper limit of ε, $\varepsilon \le 6.10^{-4}$ radian. Thus, the mean value $<\varepsilon>$ of ε is lower than 5.10^{-4}. We thus conclude that, for our current implementation (N=64 images), the intensity signal to noise ratio limited by the systematic phase noise is larger than $SNR_{Int} \geq 3000$. The memory capacity will thus not be limited by this phase noise.

E. Geometrical positioning of the phase encoder

Due to the phase modulator presently used (other designs are currently being manufactured) the reference beam wave vectors are arranged on a square grid. The way to use the Bragg selectivity is thus to rotate this matrix (see paragraph 3.1). Spurious diffraction is avoided if the rotation angle ω is so that the forbidden area for the different reference beams do not overlap.

7.2. Photorefractive crystal characteristics

To implement a photorefractive memory, a sample with a high diffraction efficiency and a long dark storage time is required. Only few photorefractive crystals fulfill these requirements: $LiNbO_3$, $Sr_{1x}Ba_xNb_2O_2$ and $BaTiO_3$. $LiNbO_3$ is the most commonly used crystal. However, large index modulation can be induced in these samples by using the photovoltaic effect only. This implies working in the blue range of the spectrum while we are using a 532nm laser source. $Sr_{1-x}Ba_xNb_2O_2$ is not easily available.

For these reasons, we employ $BaTiO_3$ samples. In order to increase their diffraction efficiencies, these samples are cobalt doped. As shown below, this doping also presents the following advantages. First it increases the acceptable angular range for the image and reference beam incident angles. Second, it reduces the average fringe spacing of the induced grating and thus increases the Bragg selectivity.

A. Recording geometry

The recording geometry is imposed by the two following requirements:
i) image distortion must be minimized;
ii) diffraction efficiency must be equal for all pixels.

Image distortion comes from the thickness of the photorefractive medium and may also be increased by index inhomogeneities present in most photorefractive samples. To minimize this distortion we must first form the image of the input modulator in the photorefractive medium and second, all the pixel beamlets must be incident at zero angle on the crystal input face.

To insure equal diffraction efficiency for all these parallel pixel beamlets, the angle between them and a given reference beam must not depends on the pixel we consider (i.e. it must be independent on the location in the crystal). The reference beams must thus be plane waves.

B. Crystal cut

The cobalt doped sample we use(46) in this first realization has a doping of 100ppm leading to an absorption coefficient of α =3 cm^{-1} at 532 nm. Because of this absorption, there is an optimum thickness l ($\approx$ interaction length) which optimizes the diffraction efficiency. It corresponds to $\alpha l = 2$.

Using the parameters for this $BaTiO_3$ sample ($N_A \approx 5.2\,10^{22} m^{-3}$) and other parameters given in Ref. 46, we plotted figure 7. It represents the refractive index modulation δn of the grating induced by interfering two extraordinary polarized plane waves of equal intensity. θ_1 and θ_2 are the angles between their wave vectors and the $\hat{a}$ crystallographic axis (see insert).

As shown by figure 7, the index modulation is symmetrical about $\theta_1 = \theta_2$ for which $\delta n = 0$. One sees that the index modulation strongly depends on the difference $\theta_1 - \theta_2$ that corresponds to the grating spacing. It also varies, but more smoothly, with $\theta_1 + \theta_2$, that is with the orientation of the grating planes. These dependencies are valid for all photorefractive materials. It demonstrates that only a small part of the (θ_1,θ_2) plane is permitted if one wants a sizeable refractive index modulation. This does not depend on the multiplexing technique used.

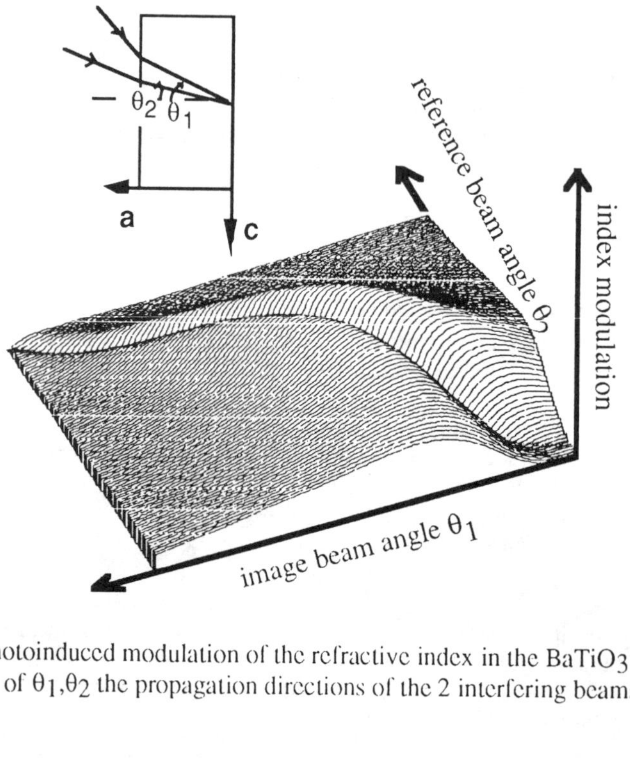

Figure 7: Photoinduced modulation of the refractive index in the $BaTiO_3$ as a function of θ_1,θ_2 the propagation directions of the 2 interfering beams

The pixel beam is a plane wave limited by diffraction so that the useful area reduces to a small strip parallel to the reference beam axis θ_2. Its length $\Delta\theta_R$ is the angle variation for the reference beams. It is limited by the acceptable variation of the index modulation which must be kept low enough to maintain the noise at a reasonable level. We choose a reasonable value of $\frac{\sigma_{\delta n}}{\delta n} < 5\%$ that yields $\Delta\theta_R \approx 5°$ around the maximum of the δn surface. This top is located at $\theta_{pixel} \approx 30°$ and $\overline{\theta} \approx 40°$. It corresponds to the following maximum index modulation and average fringe spacing:

$$\delta n_{\max} \approx 3.5\,10^{-4} \quad \text{and} \quad \overline{\Lambda} \approx 3\,\mu\text{m} \tag{30}$$

To avoid image distortion, the sample will thus be cut at 30° of the crystallographic axis so that the pixel beams will be at normal incidence and the reference beam will be incident around 25° (refractive index $n \approx 2.4$).

It is interesting to compare these results to those obtained for crystals without Cobalt doping. In that case, the photorefractive trap density is lower and about $N_A \approx 10^{22}\,m^{-3}$. The maximum index modulation, the average fringe spacing and the allowable angle range for the reference beams become $\delta n_{\max} \approx 1.8\,10^{-4}$, $\overline{\Lambda} \approx 7.2\,\mu\text{m}$ and $\Delta\theta_R \approx 3.2°$. These values clearly demonstrate that the Cobalt doping increases the memory capacity.

8. CONCLUSION.

In previous sections, we have stressed some important features of holographic photorefractive memories. Taking the example of low capacity and fast access time memories, we have shown that it is now possible to implement prototypes that will allow to effectively conduct experimental investigations of performances and limitations.

Indeed, current photorefractive crystals exhibit a sufficient refractive index modulation for permitting the storage of multiple images with low variations in their efficiency giving a little contribution to the signal to noise ratio.

Several convenient solutions exist for both recording processes and fast addressing procedures and SLMs becomes commercially available allowing, for example, an excellent phase only modulation.

However, a problem inherent to all thick memory materials remains. This is the amount of crosstalk due to multiple order Bragg diffraction that is not fully studied yet.

9. REFERENCES.

1. L. d'Auria, J.P. Huignard, C. Slezak, E. Spitz, "Experimental holographic read-write memory using 3-D storage", Appl. Opt., **13**, 808-818, (1974).
2. D.L. Staebler, W.J. Burke, W. Phillips, J.J. Amodei, "Multiple storage and erasure of fixed holograms in Fe-doped $LiNbO_3$", Appl. Phys. Lett., **26**, 182-184, (1975).
3. F.H. Mok, M.C. Tackitt, H.M. Stoll, "Storage of 500 high-resolution holograms in a $LiNbO_3$ crystal", Opt. Lett., **16**, 605-607, (1991).
4. "Photorefractive Materials and Their Applications I", P. Günter, J. P. Huignard Ed., Springer Verlag **61**, Berlin 1988.
5. H. Kogelnik, "Coupled wave theory for thick hologram gratings", Bell Syst. Tech. Journ., **48**, 2909-2947, (1969).
6. P.J. Van Heerden, "Theory of optical information storage in solids", Appl. Opt., **2**, 393-400 (1963).
7. H. Lee, X. Gu, D. Psaltis, "Volume holographic interconnections with maximal capacity and minimal crosstalk", J. Appl. Phys, **65**, 2191-2194, (1989).
8. G. Pauliat, G. Roosen, "New advances in photorefractive holographic memories", Int. Journ. Opt. Computing, Special issue on "Photorefractive materials", Vol.2, Issue 3, (1993).
9. C. Denz, G. Pauliat, G. Roosen, T. Tschudi, "Volume hologram multiplexing using a deterministic phase encoding method", Opt. Comm., **85**, 171-176, (1991).
10. Y. Taketomi., J. E. Ford, H. Sasaki, J. Ma, Y. Fainman, S.H. Lee, "Incremental recording for photorefractive hologram multiplexing", Opt. Lett., **16**, 1774-1776, (1991).
11. D. Psaltis, D. Brady, K. Wagner, "Adaptative optical networks using photorefractive crystals", Appl. Opt., **27**, 1752-1759, (1988).
12. E. S. Maniloff, K. Johnson, "Maximized photorefractive holographic storage", J. Appl. Phys. **70**, 4702-4707, (1991).
13 D. Brady, K. Hsu, D. Psaltis, "Periodically refreshed multiply exposed photorefractive holograms", Opt. Lett., **15**, 817-819, (1990).
14. H. Sasaki, Y. Fainman, J. E. Ford, Y. Taketomi, S. H. Lee, "Dynamic photorefractive optical memory", Opt. Lett., **16**, 1874-1876, (1991).
15. Y. Qiao, D. Psaltis, C. Gu, J. Hong, P.Yeh, R.R. Neurgaonkar, "Phase-locked sustainment of photorefractive holograms using phase conjugation", J. Appl. Phys., **70**, 4646-4648, (1991).
16. S. Boj, G. Pauliat, G. Roosen, "Dynamic holographic memory showing readout refreshing, and updating capabilities", Opt. Lett., **17**, 438-440, (1992).
17. A.C. Strasser, E.S. Maniloff, K.M. Johnson, S.D.D. Goggin, "Procedure for recording multiple-exposure holograms with equal diffraction efficiency in photorefractive media", Opt. Lett., **14**, 6-8, (1989).
18. N. Uchida, "Calculation of diffraction efficiency in hologram gratings attenuated along the direction perpendicular to the grating vector", J. Opt. Soc. Am., **63**, 280-287, (1973).

19. D. Psaltis, X. G. Gu, D. Brady "Fractal sampling grids for holographic interconnections. SPIE proceeding **963**, 468-474 (1988).
20. H. Kurz, "Photorefractive recording dynamics and multiple storage of volume holograms in photorefractive $LiNbO_3$", Optica Acta, **24**, 463-473, (1977).
21. F.H. Mok, "Angle multiplexed storage of 5000 holograms in lithium niobate" Opt. Lett., **18**, 915-917, (1993).
22. L. d'Auria, J.P. Huignard, E. Spitz, "Holographic read-write memory and capacity enhancement by 3-D storage", IEEE Transactions on Magnetics, **Mag-9**, 83-94, (1973).
23. E. S. Maniloff, K. M. Johnson, "Dynamic holographic interconnects using static holograms", Opt. Eng., **29**, 225-229, (1990).
24. F. T.S. Yu, S. Wu, A.W. Mayers, S. Rajan, "Wavelength multiplexed reflection matched spatial filters using $LiNbO_3$", Opt. Comm., **81**, 343-347, (1991).
25. G.A. Rakuljic, V. Leyva, A. Yariv, "Optical data storage by using orthogonal wavelength-multiplexed volume holograms", Opt. Lett., **17**, 1471-1473, (1992).
26. C. Denz, G. Pauliat, G. Roosen, T. Tschudi, "Potentialities and limitations of hologram multiplexing by using the phase encoding technique", Appl. Opt., **31**, 5700-5705, (1992).
27. A.K. Jain, "Fundamentals of Digital Image Processing", Prentice Hall, N.J., 155, (1988).
28. C. Gu, P. Yeh, "Scattering due to randomly distributed charge particles in photorefractive crystals", Opt. Lett., **16**, 1572-1574, (1991).
29. K. Bløtekjaer, "Limitations on holographic storage capacity of photochromic and photorefractive media", Appl. Opt., **18**, 57-67, (1979).
30. J. J. Amodei, D. L. Staebler, "Holographic pattern fixing in electro-optic crystals", Appl. Phys. Lett., **18**, 540-542, (1971).
31. F. Micheron, G. Bismuth, "Electrical control of fixation and erasure of holographic patterns in ferroelectric materials", Appl. Phys. Lett., **20**, 79-81, (1972).
32. F. Micheron, G. Bismuth, "Field and time thresholds for the electrical fixation of holograms recorded in $(Sr_{0.75}Ba_{0.25})Nb_2O_6$ crystals", Appl. Phys. Lett., **23**, 71-72, (1973).
33. Y. Qiao, D. Psaltis, "Sampled dynamic holographic memory" Opt. Lett., **17**, 1376-1378, (1992).
34. H. Sasaki, J. Ma, Y. Fainman, S.H. Lee, Y. Taketomi, "Fast update of dynamic photorefractive optical memory" Opt. Lett., **17**, 1468-1470, (1992).
35. G. Pauliat, G. Roosen, "Interconnects with optically thick elements" in "Perspectives for parallel optical interconnects", pp. 297-331, Ph. Lalanne and P. Chavel eds., Springer Verlag, Heidelberg, (1993).
36. S. Redfield, L. Hesselink, "Data storage in photorefractives revisited" SPIE, **963**, 35-45, (1989).
37. A. A. Yamamura, M.A. Neifeld, S. Kobayashi, D. Psaltis, "Optical disk based artificial neural systems" Opt. Computing and Processing, **1**, 3-12, (1991).
38. A. Vander Lugt, "Design relationships for holographic memories" Appl. Optics, **12**, 1675-1685, (1973).
39. G. Pauliat, G. Roosen, OCCC'92, Hsinchu, Taiwan (1992), SPIE Proceedings "Optical Computing and Neural Networks", Vol. 1812, in press.
40. G. Pauliat, G. Roosen, Technical digest of the Indo-French Workshop on Optoelectronics, Goa, Dec.10-12, (1992).
41. P.S. Guilfoyle, Critical Reviews, Vol. CR35, p. 288, (1990).

42. D. Psaltis, M. A. Neifeld, A. Yamamura, “Image correlators using optical memory disks”, Opt. Lett. **14**, 429-431, (1989).
43. Project POPAM (Parallel Optical Processors and Memories) partly funded by the Commission of the European Community under the European Strategic Program for Research and Development.
44. B.S. Lowans, B. Bates, R.G. H. Greer, J. Aiken, "Binary phase modulation properties of a programmable liquid crystal television display", Appl. Opt. **31**, 7393-7395, (1992).
45. D. Psaltis, E. G. Paek, S. S. Venkatesh, “Optical image correlation with a binary spatial light modulator”, Opt. Eng. **23**, 698-704, (1984).
46. D. Rytz, B.A. Wechsler, M.H. Garrett, C.C. Nelson, R.N. Schwartz, "Photorefractive properties of $BaTiO_3$:Co", JOSA B, **7**, 2245-2254, (1993).

Applications II

Photorefractive semiconductors and applications

Li-Jen Cheng

Center for Space Microelectronics Technology
Jet Propulsion Laboratory
California Institute of Technology
Pasadena, California 91109

and

Keung L. Luke

Department of Physics and Astronomy
California State University
Long Beach, California 90840

ABSTRACT

Photorefractive semiconductors are attractive for information processing, because of fast material response, compatibility with semiconductor lasers, and availability of cross polarization diffraction for enhancing signal-to-noise ratio. This paper presents recent experimental results on information processing using photorefractive GaAs, InP and CdTe, including image processing with semiconductor lasers. The results demonstrate the feasibility of using photorefractive compound semiconductors as dynamic holographic interaction media for information-processing applications in low-power, compact configurations.

1. INTRODUCTION

Use of photorefractive semiconductors provides a promising fast, parallel approach to information processing. Based on the principles of nonlinear optics and holography, it owes its unique capability to the photorefractive effect observed in photoconductive electro-optic materials.

Several semi-insulating compound semiconductors have been demonstrated to be photorefractive. They include undoped and chromium-doped gallium arsenide (GaAs, GaAs:Cr),[1,2] iron-doped and titanium-doped indium phosphide (InP:Fe, InP:Ti),[1,3] undoped gallium phosphite (GaP)[4] and vanadium- and titanium-doped cadmium telluride (CdTe:V, CdTe:Ti).[5] These photorefractive semiconductors provide several attractive features for information-processing applications and could lead to a new generation of integrated optical information processors.

The purpose of this paper is to review the progress of this technologically important area, based mainly on the work carried forward at the Jet Propulsion Laboratory.

This paper is divided into eleven sections. Following the introduction, we describe five important photorefractive properties of GaAs in section 2. We then describe two generic wave mixing configurations for image processing and spatial resolution of GaAs four-wave mixing systems in section 3. Phase conjugate interferometer and its applications are covered in section 4. Image correlation with photorefractive GaAs is treated in section 5. A novelty-filtered optical correlator using photorefractive CdTe is described in section 6. Image processing with photorefractive InP and with semiconductor lasers are covered in section 7 and 8, respectively. Time correlation is discussed in section 9. Image correlation using a new class of thin photorefractive devices is described in section 10. The review is summarized in the concluding section 11.

2. PHOTOREFRACTIVE PROPERTIES OF GaAs

Because the demonstration experiments were mainly on information processing using photorefractive GaAs, this section presents the photorefractive properties of GaAs. Some photorefractive properties of InP and CdTe are stated for comparison purpose.

2.1. Basic mechanism

The photorefractive effect refers to the phenomenon of the light-induced refractive index change of an electro-optic material, analogous to the light-induced absorption change known as the photochromic effect. However, the physical mechanism is different.

The light-induced refractive index change in a material is based on the spatial modulation of electrical carrier generation by periodic illumination. When two coherent laser beams interfere in the material to form a fringe pattern, the photoexcited carriers generated in the bright regions diffuse or drift to the dark regions in which they are trapped at localized energy levels in the bandgap, leaving behind in the bright regions fixed charges of opposite sign. This periodic charge redistribution creates a periodic local space-charge field pattern, giving rise to a refractive index grating via the linear electro-optic effect with the same spatial period as the fringe pattern, but shifted one quarter of a period in space from it.[6] This grating acts as a volume hologram and can be used for optical information processing and storage. Because of the charge trapping, the photorefractive effect can be observed with light beams of moderate intensities. For example, photorefractive effects in GaAs can be observed with a total beam intensity less than 1 mW/cm^2.[7]

In addition to possessing a non-zero electro-optic coefficient, a photorefractive material must have adequate densities of localized energy levels which can act as the donors and the acceptors for supplying and receiving the transferred charges, respectively. For example, the existence of both neutral and ionized states of the EL-2 center in LEC-grown, undoped, semi-insulating GaAs makes this material photorefractive. Typical concentration of the EL-2 centers in LEC-grown, undoped GaAs is about 1×10^{16} /cm^3, among which about 1×10^{15} /cm^3 are positively charged acceptors.

The energy band gap of the crystal and the position of the level in the bandgap determine respectively the short and long limits of the wavelength range within which the photorefractive effect exists. For example, the band gap of GaAs is about 1.43 eV and the responsible EL-2 level in LEC-grown, undoped, semi-insulating GaAs is about 0.75 eV below the edge of the conduction band. Thus, the GaAs crystal can have the photorefractive effect in the wavelength range of about 0.8-1.6 μm. The effect in CdTe exists approximately in the same wavelength range. The short wavelength limit of InP is about 0.85 μm, because of smaller energy bandgap.

Furthermore, the photorefractive material has to be insulating or semi-insulating in order to avoid the Coulomb screening around the charged centers. Typical resistivity of semi-insulating GaAs is in the order of 5×10^7-2×10^8 ohm-cm whereas typical resistivity of InP is about 10^7 ohm-cm. This difference is due to the fact that the energy bandgap of InP (1.27 eV) is lower than that of GaAs (1.43 eV). Consequently, photorefractive effect of InP requires higher light intensity or at lower temperatures as illustrated experimentally.

2.2. Material response times

The photorefractive response time of a material is the time required for writing an index grating in this material. It depends on the beam intensities, grating periodicity, as well as material properties such as carrier mobility, donor concentration, photoionization cross section of the donor, and carrier capture rate of the acceptor. It has been reported that the response time in LEC-grown, undoped GaAs is about 1 ms[8] and 20μs[2] under a total intensity of 0.1 and 4 W/cm^2, respectively. These values are, at least, about two orders of magnitude smaller than those of the oxides under the same intensities. This provides semiconductors with a clear advantage in the image processing speed. The response time of GaAs can be in the picosecond time scale, if intense light pulses are used. For example, under a total intensity of 5×10^7 W/cm^2, the response time of GaAs was reported to be 43 ps.[9] However, this picosecond-response photorefractive effect uses two-photon absorption, different from that of the conventional photorefractive effect.

The volume holographic grating written in a photorefractive material has a finite lifetime. This lifetime determines the image storage time in the crystal. It is an important parameter for evaluating the image processing capability. For a given material, the read beam intensity and the grating periodicity are two parameters determining the lifetime. Recently, a beam coupling technique using a 1.15-μm He-Ne laser was employed to measure the grating lifetime in GaAs.[10] The largest lifetime measured is about 8μs under a read beam intensity of 0.7 mW/cm2 with the grating periodicity being 0.63 μm. The measured value decreases to milliseconds as the read beam intensity and the grating periodicity increase to about 10 mW/cm2 and 4 μm, respectively. In addition, the results suggest that lifetime is sensitive to residual imperfections in the crystal.

2.3. Figure of merit

It is known that the figures of merit, n^3r/ε, of these semiconductors are similar to those of the oxides,[1] where n is the refraction index, r is the electro-optic coefficient, and ε is the relative dielectric constant. n^3r/ε is for maximum photorefractive sensitivity of the material. The figures of merit of GaAs, GaP, InP, and CdTe are 3.3, 3.7, 4.1, and 16, respectively, whereas those of BSO, SBN, $BaTiO_3$, $LiNO_3$ and $KNbO_3$ are 1.8, 4.8, 4.9,

11, and 14, respectively.[11] CdTe has the highest figure of merit among all the materials stated, slightly larger than $KNbO_3$, and is a potentially superior image processing medium in the near infrared range. Among the semiconductors, the figure of merit of CdTe is about four times larger than those of GaAs, GaP, and InP. This is consistent with experimental results. For example, it was observed that vanadium and titanium doped CdTe crystals had larger photorefractive effects than GaAs and InP without an applied electric field.[5]

A detailed discussion of photorefractive sensitivity is available in the literature.[12]

2.4. Material properties

The photorefractive effect is extremely sensitive to the existence of crystal imperfections, because these defects can drastically reduce the net charge separation efficiency. For example, the concentration of the ionized EL-2 center in LEC-grown, undoped GaAs which acts as the acceptor in the photorefractive process can be increased using an inverted thermal treatment, but the photorefractive gain in the sample was poor. This is due to the fact that the treatment creates a new defect which reduces the net charge separation.[13] Although the technology of GaAs crystal growth is the most advanced one among those of all the compound semiconductors available, it is evident that the photorefractive effect of GaAs crystals (available commercially and from research laboratories) varies considerably among suppliers as well as from one ingot to another.[13] The best observed gain coefficients in GaAs without and with an applied DC electric field (with and without moving grating technique) are reasonably close to those of theoretical prediction based on known concentration of the EL-2 center in comparison with those of photorefractive oxides.[14] This indicates that the state-of-the-art technology of GaAs can produce good photorefractive crystal. However, given a crystal's growth condition, there is no way to know whether or not it is photorefractive crystal, unless photorefractive measurement is done on it. More research is needed to understand the origin of the detrimental effects and to develop a growth condition of optimized crystals for the photorefractive effect. This argument can be applied to InP as well. However, the CdTe material technology is still not mature in comparison with those of GaAs and InP. More material research effects on CdTe are needed.

2.5. Electric field Effects

It is known that several techniques can be used to enhance photorefractive effects in semiconductors, for example, applications of a high DC electric field without and with moving grating.[14,15] In addition to the enhancement, the application of an electric field also causes two effects, namely, the increase of material response time and the creation of signal fluctuation. The latter is due to moving charge domains created by the high electric field with speed which increases the field strength and light intensity. The increase of material response time will make the use of photorefractive semiconductors for some applications less attractive. Nevertheless, good correlation images with enhancement of 100 in diffraction efficiency were obtained as illustrated by the results to be presented later. In addition to the conventional photorefractive effect using the linear electro-optic property, the large photorefractive effect based on the Franz-Keldysh effect under a DC electric field with and without moving grating technique was recently observed in GaAs[16] and InP.[17] This enhanced effect has not yet been investigated for image processing applications.

3. WAVE MIXING CONFIGURATIONS FOR IMAGE PROCESSING

There are two generic configurations for optical information processing using the photorefractive effect, namely, two-wave mixing (beam coupling) and four-wave mixing. The latter includes degenerate and non-degenerate four-wave mixing configurations with and without polarization switching. A brief discussion on these configurations is given next.

3.1. Two-wave mixing

In the two-wave mixing configurations, two coherent beams intersect in a crystal and create an index grating via the photorefractive effect. In general, the interference pattern of these beams and the induced index grating are 90 degrees out of phase. This leads to an unsymmetrical diffraction of the incident beams by this index grating and results in a net intensity transfer from one beam to another. If the intensity gain of one beam is larger than the absorption in the material, the intensity of this beam is amplified. This effect, initially observed in photorefractive $LiNbO_3$ crystals,[18] can be used for amplifying the intensity of an information-bearing optical beam as initially demonstrated by Kukhtarev et. al in $LiNbO_3$.[19] Two-beam coupling together with the polarization switching property of GaAs crystals can also be used to implement spatial light modulation as recently demonstrated.[20]

3.2. Degenerate four-wave mixing

In the four-wave mixing configurations, two coherent beams write an index grating and a third beam reads the grating, creating a fourth (i.e. output) beam by diffraction. In the degenerate four-wave mixing, the diffracted beam travels against one of the write beam and is phase conjugate of the write beam. The degenerate four-wave mixing configuration is more useful and versatile. A number of basic image processing operations, including phase conjugate imaging,[8] edge-enhancement,[8] correlation[21,22] and convolution, vector-vector multiplication,[23] and interferometry,[24,25] were demonstrated using GaAs in the degenerate four-wave mixing configuration. The crystals can be used in both the reflection and transmission modes, depending on the beam geometry and the crystal orientations. For GaAs, the transmission mode has a faster response time but a smaller diffraction efficiency, whereas the reflection mode has a slower response time but a larger diffraction efficiency. This is because these two modes have different grating spacing ranges and the response time and the diffraction efficiency decreases as the grating spacing becomes larger.

3.3. Non-degenerate four-wave mixing

In the non-degenerate four-wave mixing configuration, one can write the information-bearing grating with one wavelength and read it with another. A potential use of non-degenerate four-wave mixing is that if the information-bearing index grating is to be preserved after the readout operation, one can read the grating with a longer wavelength that does not erase the grating. Image transfer from a beam at 1.06 μm to another at 1.3 μm was demonstrated using GaAs.[23]

3.4. Polarization switching

Crystal structures of GaAs, InP, CdTe, and GaP are $\bar{4}3m$ cubic group symmetry. The optical isotropy and tensor nature of the electro-optic coefficients of these crystals permit the possible cross polarization switching in beam coupling and four-wave mixing.[26-28] Such cross polarization switching is possible only along a special direction in $BaTiO_3$ and SBN because of their optical anisotropy. In BSO, the circular birefringence strongly influences the polarization of the diffracted beam. The polarization switching has been very important for image processing using GaAs because its electro-optic efficiencies are small. Use of polarization switching is an effective way to enhance the signal-to-noise ratio. Most experiments on image processing using GaAs were done with a polarization switching configuration.

Figure 1 gives two four-wave mixing configurations with and without polarization switching effect. The two configurations use the crystal in different orientation with respect to the beam arrangement as indicated in the figure. Two write beams propagate codirectionally. In the configuration with polarization switching, the [001] direction of the crystal is perpendicular to the incident plane of the beams. The polarization directions of the write beams and the read beam are in the in the incident plane. Because of polarization switching, the diffracted beam is polarized in the direction perpendicular to the incident plane and can be separated easily by a polarizing beamsplitter. This configuration provides excellent signal-to-noise ratio. In the configuration without

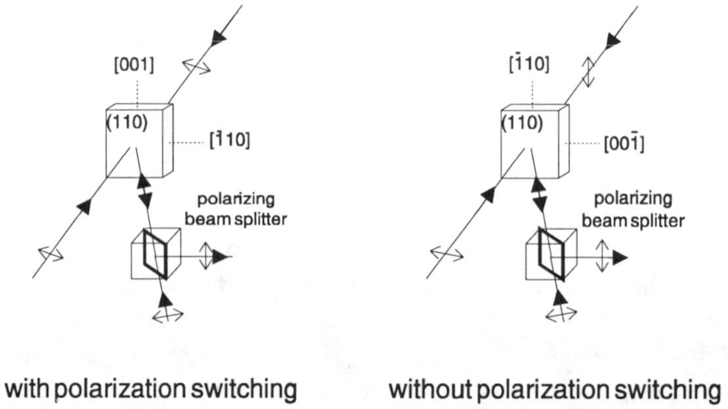

Figure 1. Examples of degenerate four-wave mixing configuration.

polarization switching, the [110] direction is perpendicular to the incident plane. The polarization of the read beam has to be perpendicular to the incident plane in accordance with the electro-optic tensor properties. In this configuration, some scattered light from the read beam can mix with the diffracted beam. Therefore, the signal-to-noise ratio is lower.

3.5. Spatial resolution of GaAs four-wave mixing systems

Four-wave mixing systems can resolve an image with a spatial resolution limited by the index grating periodicity which is a function of the wavelength and the incident angles. It has been demonstrated that degenerate four-wave mixing in a ruby crystal can produce phase conjugate images with a resolution better than 500 lp/mm using 0.514-µm laser light.[29]

To investigate the spatial resolution of photorefractive GaAs image processing systems, a vector-vector outer product experiment and a phase conjugation experiment were performed using Ronchi gratings as the input patterns. Figure 2 shows a schematic of the experimental setup used.[25] Experimental results show that this setup could resolve the rectangular dot array clearly when a pair of 20 lp/mm Ronchi gratings were used, but not with a pair of 100 lp/mm Ronchi gratings. However, if the Ronchi grating in the pump beam 1 was removed, the line array picture of another Ronchi grating (namely the phase conjugate image of the input pattern in the probe beam) could be clearly seen. When a 250 lp/mm Ronchi grating is placed in the probe beam, the phase-conjugation setup also failed to resolve it.

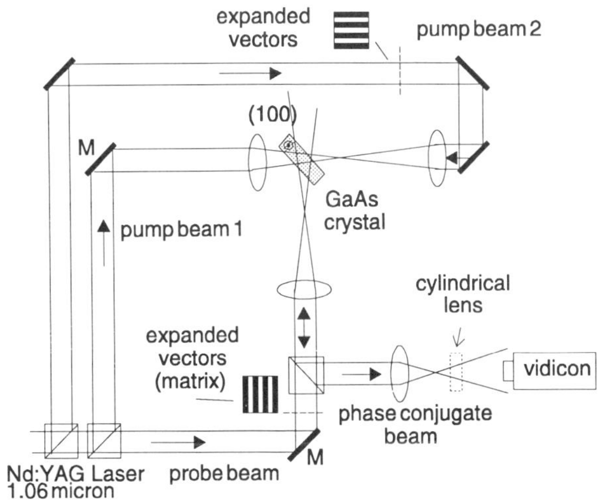

Figure 2. Schematic of experimental set-up used to investigate spatial resolution.

4. PHASE CONJUGATE INTERFEROMETER AND ITS APPLICATIONS

4.1. General

It is well-known that photorefractive crystals are volume holographic media. Because of the nature of volume holograms, in principle, it is possible to read and write multiple volume holograms with slightly different orientations superimposed in the same photorefractive crystal without causing notable cross talk. Figure 3 shows the schematic diagram and results of a multiple hologram writing and reading experiment. The "common" write beam generates a hologram with each of the image-bearing write beams. Each hologram carries a different spatial information and is superimposed on another in the crystal. The read beam is diffracted by each hologram separately. The output beams are then coupled out by a beam splitter and captured by a CCD camera. This experiment has established a base on which a phase conjugate interferometer can be built.

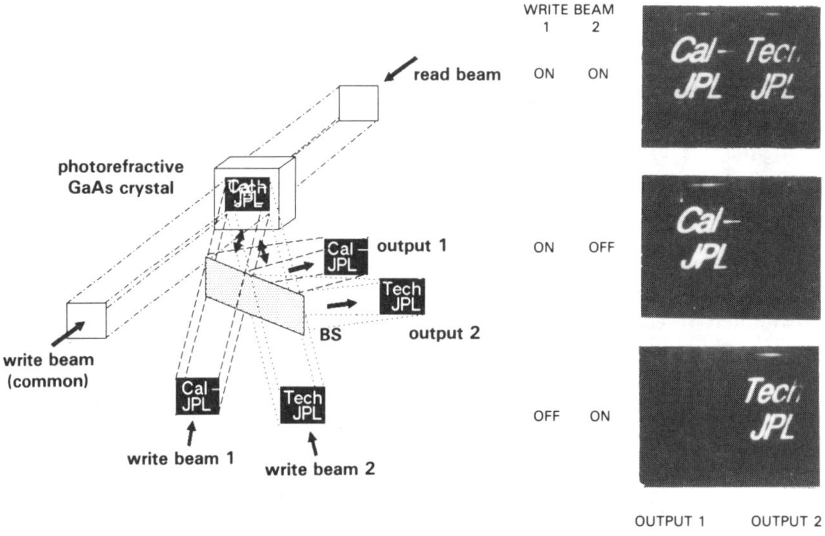

Figure 3. Schematic of a multiple degenerate four-wave mixing configuration with overlapping gratings (left). Photographs (right) illustrate the experimental results.

4.2. Phase-conjugate interferometer

Coherent image substraction and addition can be achieved with an interferometer, such as the Michelson interferometer. In practice, the image intensity drifts between the extremes because of the phase fluctuation due to ambient air currents and/or thermal drift and also is extremely sensitive to misalignment. The use of fast photorefractive phase conjugate mirrors which replace the ordinary mirrors can overcome both problems, as demonstrated by experimental observations.

The use of an interferometric configuration with a phase conjugate mirror of $BaTiO_3$[30-34] and BSO[35] for parallel image operations was previously reported. Because of the nature of phase conjugate mirrors, these phase conjugate interferometers are more stable in operation and easier to align than those using conventional mirrors only. However, operations with these materials are slow and sensitive to environmental fluctuations, such as air turbulence and low-frequency vibrations of optical components.

The fast response of compound semiconductors, such as GaAs, not only provides high speed operation of a phase conjugate interferometer, but also makes the system immune to low-frequency vibration of optical components. The latter is due to the fact that the light-induced grating can follow the slow motion induced by low-frequency vibration.

4.3. Applications

Several basic computing operations using a GaAs phase conjugate interferometer have been demonstrated. These include image subtraction, coherent and incoherent image addition, image inversion, OR and exclusive OR (XOR) logic operations.[25]

Time variation of spatial information can be detected using a phase conjugate interferometer. This property is important for motion detection, novelty filtering and target tracking, which were demonstrated using $BaTiO_3$ by Anderson and coworkers.[36,37] Fast semiconductors can be used to make high-resolution fast novelty filters to track a target in a more precise way.[38]

5. IMAGE CORRELATION USING PHOTOREFRACTIVE GaAs

5.1. General

Optical Image correlation takes full advantages of light, namely, parallel operation and global interconnection in implementing Fourier transforms by a lens. Photorefractive compound semiconductors can provide this type of high-speed implementation.[21,22]

According to the correlation theorem, the correlation contains Fourier transforms of input and reference images, multiplication of transformed images in the Fourier domain, and inverse Fourier transform of the product.[39] This process can be implemented using two lenses and a photorefractive crystal to provide Fourier transform and multiplication, respectively. In addition, the photorefractive crystal also serves as a temporal recording medium.[40-48] There are two configurations of implementation, namely, matched filter[49] and joint transform.[39] In the matched filter mcthod, the hologram of the Fourier transformed incoming scene is recorded in the crystal, whereas, in the joint transform method, the hologram of the product of the two transformed images is written. The matched-filter method is the preferred one, because it does not require rewriting of the hologram whenever any reference image is changed.

It is worthwhile to note that the space bandwidth product of the thick volume hologram will limit the spatial resolution of the photorefractive-crystal-based image correlator.[25] Presumably, the spatial resolution of the GaAs image correlator could be of the same order of magnitude of 20 lp/mm, because the same lenses were used for both experiments. Obviously, more detailed experiments are needed to determine the accurate values.

In the GaAs photorefractive correlator used in this investigation, the incoming scene is stored in the crystal as a hologram in the Fourier domain and then the template matching process occurs at the diffraction of the reference object-bearing read beam at the hologram as illustrated in the upper part of Fig. 4. This configuration provides two distinctive advantages: (1) incoming scene can have a large incident angle, and (2) an extremely fast template matching process.

When the time needed for writing a hologram is about equal to or shorter than the time of inputting the scene, the correlator can track targets in real time. Because of fast material response, the photorefractive semiconductor correlator is suitable for real-time tracking applications. Slow photorefractive crystals, such as $BaTiO_3$, cannot provide real-time target tracking, because the grating formation in the crystal is too slow to follow the change of the incoming scene.

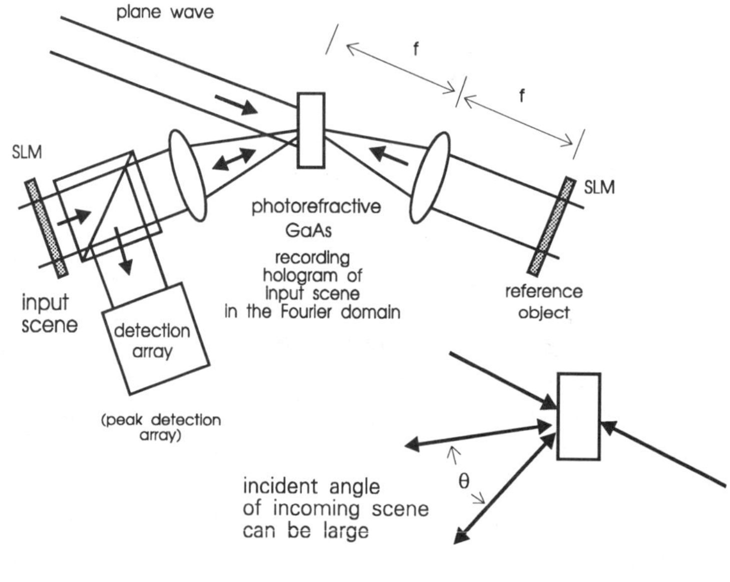

Figure 4. Matched-filter image correlation configuration using degenerate four-wave mixing in photorefractive GaAs which records a hologram of the input scene in the Fourier domain; the template-matching process occurs at the diffraction of the reference object-bearing read beam at the hologram (upper figure). The time required for the template-matching process is the travel time of the light beam from the reference object input device through the diffraction in the crystal to the detector array, typically in the time scale of nanoseconds. The incoming scene can have a large incident angle because of the nature of degenerate four-wave mixing, as illustrated by the schematic at the lower right.

5.2. Experiment

In our previous work,[21] input images were put on photographic films. Recently, experimental data on an optical image correlator using a photorefractive GaAs crystal and two modified liquid crystal television (LCTV) screens as spatial light modulation input devices were reported.[22] The LCTV screens are from a low-cost commercial home grade projection TV (EPSON Crystal Image Video projector). Because of the high contrast ratio (100:1) of these LCTVs, the quality of some output images of the current correlator is very close to those obtained with transparencies. The frame rate of this correlator is limited by the speed of the LCTVs and vidicon camera used, which is about 30 frames/s. Figure 5 shows a sketch of the experimental setup which is in a matched-filter configuration.

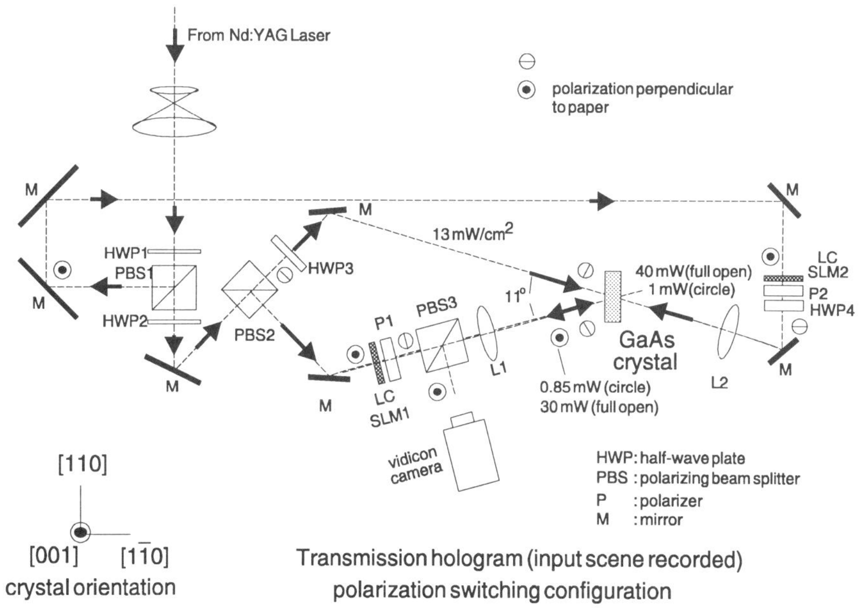

Figure 5. Schematic of the experimental set-up using degenerate four-wave mixing in photorefractive GaAs with two electrically addressed liquid crystal spatial light modulators modified from an EPSON liquid crystal television projector. The experiment used polarization switching to minimize stray light.

5.3. Experimental observations

5.3.1. High-quality correlation

Two pictures on the left side of Fig. 6 give autocorrelation images of a gray-scale car and an edge-enhanced image of the same car (produced with by a digital computer) and signal profiles taken along the path indicated by solid lines. The autocorrelation signals

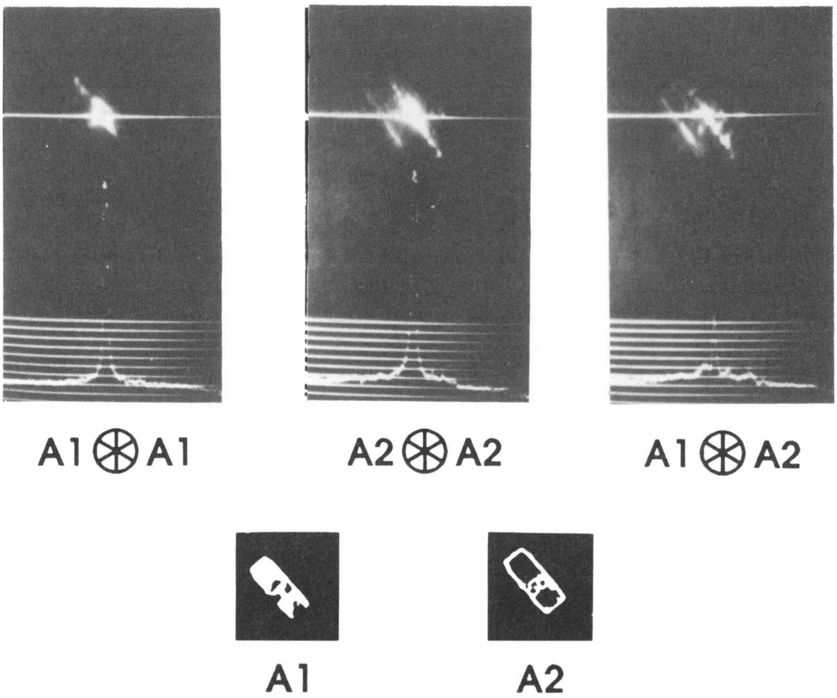

Figure 6. Autocorrelation images of a gray-scale car and an edge-enhanced image of the same car by a digital computer, illustrating that the photorefractive correlator can perform edge enhancement automatically. However, the amplitude of the cross-correlation peak of the grey-scale car and the edge-enhanced image is only one-third of that obtained from the autocorrelation, indicating that the result from optical edge enhancement is slightly different from that made by a digital computer.

are strong and sharp as shown in the pictures. This is expected from a true matched-filter implementation.

5.3.2. Automatic edge enhancement

A comparison of the scan signals in the two pictures reveals that the two correlation peaks have the same size, which is consistent with the demonstrated fact[8] that edge-enhancement process does occur in the photorefractive process because of the saturation of the DC component of the hologram in the Fourier domain. However, the signal of the cross correlation between the gray-scale car and the edge-enhanced car is much smaller as indicated in the picture at the right. This is due to the fact that the edge-enhanced image of the car generated with a digital computer is slightly different from that by the photorefractive process, demonstrated in a separate experiment. The results illustrate that the edge-enhancement is an inherent phenomenon in the correlation operation using a photorefractive material.

5.3.3. Submillisecond response time

Because the frame rate of the LCTV spatial light modulators is about 30 Hz and the material response time is much shorter than 1/30 s, it is necessary to insert a fast chopper device in the path of the probe beam for switching light on and off to measure the response time of the GaAs crystal which is an important parameter limiting the input speed of incoming scene for future GaAs correlator. The signal of the GaAs correlator was measured by a Ge detector with the probe beam being chopped. The rise time of the correlation signal was measured as a function of the total laser intensity for a beam ratio of read beam : probe beam : reference beam = 100 : 70 : 1 (measured before the beams transmitted through LCTVs). Since the maximum transmission of the LCTVs used is about 8%, the total intensity used for the correlation process in the photorefractive crystal is only about 8% of the total laser intensity or less. The measured response time of the correlator is inversely proportional to the light intensity, which basically agrees with the photorefractive theory. The shortest response time measured was 0.8 ms at a total laser intensity of about 1.5 W/cm^2 and a grating spacing of about 6 μm (angle between write beams = 11°).[50] In our experiment, about 92% of the total laser intensity is wasted because the poor transmission of the LCTV. If the transmission can be improved, the correlator can either operate at a much faster rate for the same laser power or at a fixed rate with a much reduced laser power, or a combination. The experimental data obtained so far have demonstrated that the GaAs correlator can operate at a rate of at least 1000 frames/s. Based on the known material response time, the speed of the GaAs correlator can be in the order of several tens of thousand frames/s when the intensity increases.

5.3.4. Enhancement of correlation signal by DC electric field

Application of a DC voltage can substantially enhance photorefractive effects. An experiment was performed to examine the effect of a DC electric field on the signal from the GaAs correlator. The data reveal that a 4-kV voltage can enhance the correlation signal by a factor of about 100 and also increase the response time by a factor of 20.[50] There is a tradeoff for the system operation when an enhancement of the signal by an applied voltage is desired.

5.3.5. Noise effects

The capability of recognizing objects in a cluttered environment is a very important parameter for evaluation of a potential correlation system. An experiment was carried out with the objective to examine the effect of random dot noise added to the input scene. The noise density, defined as the percentage of the total area covered by the random dot noise, increases from 10% to 60%. The correlation peak with 10% is found to be the same as that without noise, showing that noise of 10% or less has no effect on the correlation. However, 20% noise reduces 40% of the peak. When 60% of the area covered by the noise, the peak is reduced to about one twelfth of the original size, but the peak is still observable. The peak is not observable with the noise occupying 70% of the area. The present result reveals clearly that the GaAs correlator can perform well when the noise is less than 60%.[50] The mechanism of the peak reduction due to the noise is not clear at this moment. Further investigation is needed in this important area.

5.3.6. Other characteristics of GaAs correlators

Capability of tracking objects

Because of fast response, the correlation using photorefractive GaAs is capable of tracking objects, as demonstrated experimentally. It is noted that this is not possible if a photorefractive material with a slow response, such as $BaTiO_3$, is used.

No need for preprocessing Fourier transform of reference objects

The photorefractive correlation system uses optical lenses for Fourier transform. Therefore, there is no need for any preprocessing Fourier transform of reference objects and the system is suitable for applications where real-time selection of reference objects are needed.

Easy alignment

It is worthwhile to mention that the correlation process using photorefractive effect requires only the three light beams intersecting in the crystal. There is no need for alignment in pixel accuracy as required by some correlation architectures, such as a binary phase-only filter correlator.

6. NOVELTY-FILTERED OPTICAL CORRELATOR

Recently, several authors[21,40-48] have demonstrated real-time optical correlator using photorefractive crystals. In Ref. 48, a video frame rate operation was achieved using a photorefractive GaAs crystal and two Liquid Crystal TV (LCTV) based spatial light modulators. This optical correlator could potentially operate at a frame rate of > 1000 frames/s, provided other parts of the system have comparable speeds. Furthermore, because both the input and reference images are automatically edge enhanced in the correlator, the profile of the autocorrelation peak is sharper and most background caused by the clutters are reduced.

In a follow up work to Ref. 48, we discovered that under a different condition described later, the output peak intensity of a matched object may be increased more than two to three times by simply moving that object.[51] This may be potentially useful to those applications in which only the moving object of interest is desired to be identified and tracked. However, we also observe that the output peak intensities of other matched objects that are not moving also go up by approximately the same amount. This last feature could be either an advantage or disadvantage, depending on the applications.

We call this new optical correlator a novelty-filtered optical correlator (NFOC), because functionally it is similar to the combination of an optical novelty filter[37] and an optical correlator. The basic configuration is a VanderLugt[49] correlator consisting of a photorefractive crystal which does both the correlation and novelty filtering. The operation principle is thus very different from actually combining an optical novelty filter with an optical correlator.

A schematic and photograph of a novelty-filtered optical correlator are shown in Fig. 7 and Fig. 8, respectively. The basic configuration is the same as a VanderLugt correlator with a *reflection* grating.

Figure 9 shows a pair of the photographs with the target at rest and in motion, illustrating that the correlation peak intensity of the target in motion is about twice of that at rest. This is an observation of correlation enhancement due to target motion.

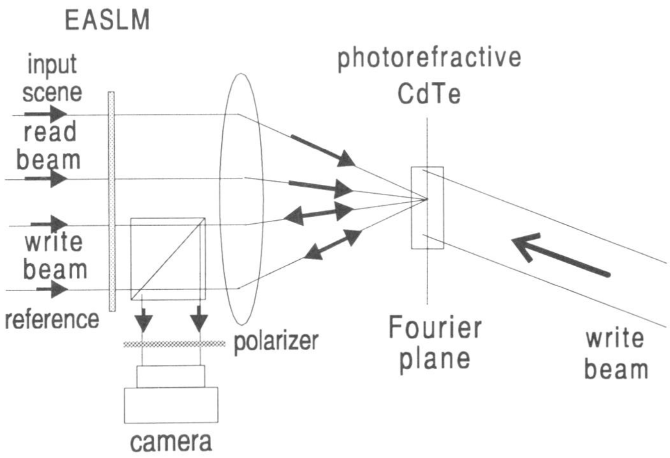

Figure 7. Schematic of a novelty-filtered optical correlator.

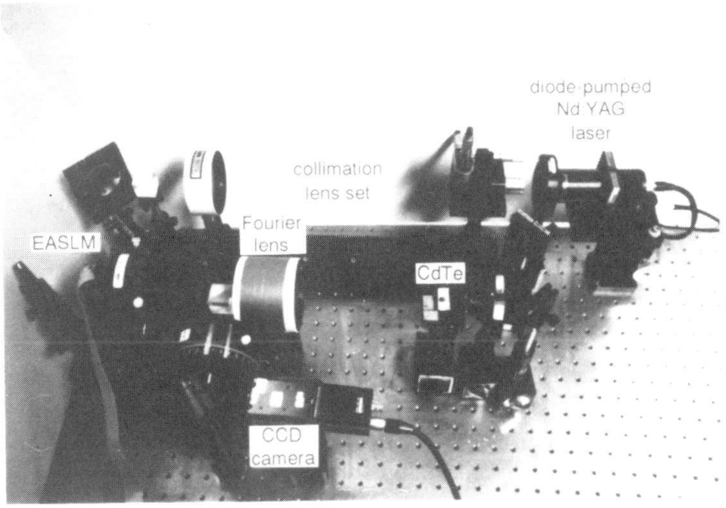

Figure 8. Photograph of a novelty-filtered optical correlator.

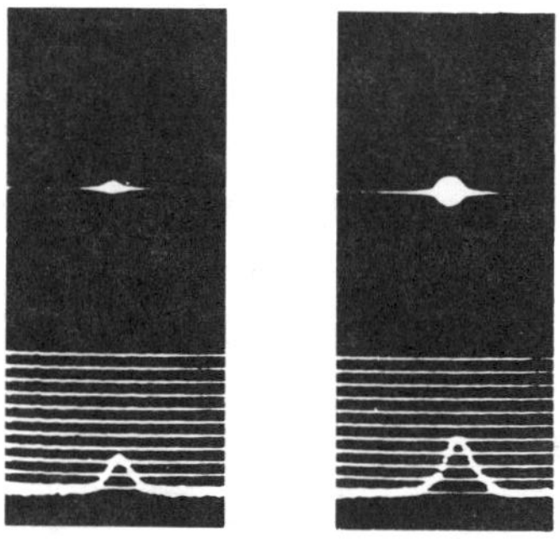

Figure 9. Correlation signal with object moving is about two times stronger than that when object is at rest.

In Ref. 51 a simple qualitative explanation based on the correlation beam geometry in the photorefractive crystal is given to show how the motion enhancement in correlation signal can occur.

7. IMAGE PROCESSING USING PHOTOREFRACTIVE InP

The observation of photorefractive effects in InP and GaAs was simultaneously reported in 1984.[1] Photorefractive Fe-doped InP provides potential advantages in comparison with GaAs, namely, figure of merit for photorefractive effect of InP is about 4.1, whereas that for GaAs is about 3.3. In this section, results from a study on the feasibility of using photorefractive InP for optical image processing are reported.

A polished liquid-encapsulated Czochralski-grown Fe-doped InP crystal was used in the study. This sample did not have anti-reflection coatings because of peeling off anti-reflective aluminum oxide layers from InP. The sample dimensions are 5mm x 6mm x 7mm. The experiment includes imaging by phase conjugation and image edge enhancement. The experimental setup was similar to those used for the GaAs experiment[8]. Images obtained by the phase conjugation process and edge enhancement. has good quality,[50] even though the measured diffraction efficiency in the InP sample was about a factor 10 smaller than those of the GaAs samples obtained previously and the beam intensities needed were considerably higher. A possible reason for the low diffraction efficiency and high intensity requirement in the InP sample could be due to that the energy bandgap of InP (1.27eV) is lower than that of GaAs (1.43eV). Consequently, the InP has a high dark conductivity which reduces the photorefractive effect.

In order to verify this explanation, an experiment to measure the two-beam coupling coefficient as a function of beam intensity and temperature was carried out. Results taken at 0, 6, and 19°C under a total beam intensities from 8 to 700 mW/cm^2, show that the beam coupling coefficient of the InP sample increases with decrease of the temperature and increase of the intensity.[50] The same phenomenon in GaAs was observed previously, but occurs at higher temperatures (60-100°).[7] The photorefractive effect in the InP sample at 19°C was not saturated when the total intensity was about 700 mW/cm^2, whereas the effect in GaAs at the same temperature was saturated when the total intensity was only about 100 mW/cm^2.[7] These observations are consistent with the thought that image processing in InP requires higher beam intensities at room temperature in order to use full potential of the material because of the high dark conductivity due to the smaller bandgap.

8. IMAGE PROCESSING USING SEMICONDUCTOR LASERS

Image processing in photorefractive semiconductors with semiconductor lasers is a critical experiment to demonstrate the feasibility of integrating the photorefractive processor and the semiconductor laser to develop a functional compact processing module. JPL has successfully performed two proof-of-the-concept experiments on image processing in a photorefractive GaAs crystal with two 1.3-μm, 13-mW, DFB single mode GaInAsP/InP lasers supplied by Ortec Corporation. One experiment was imaging by phase conjugation. The other is image correlation. The results demonstrated the feasibility of integrating photorefractive semiconductors and semiconductor lasers together to develop compact image processors.

8.1. Imaging by phase conjugation

One laser provided two coherent beams to form a hologram in the crystal and the beam from the other laser read the information in the hologram.[50] This architecture is necessary, because (1) the combination provides a higher total beam power for the four-wave mixing, and (2) the incoherence between the read beam and the write beams eliminates the mode competition that occurs when all three beams originate from a single-mode laser because of its extremely long coherent length.

A reasonably good phase conjugate image was obtained from the experiment. To our knowledge, this is the first observation of phase conjugation imaging in a four-wave mixing using a photorefractive semiconductor with semiconductor lasers. The result demonstrates the feasibility of integrating the photorefractive processor and the semiconductor laser to develop a functional compact processing module.

Time scans of the phase conjugate signal were taken with the forward pump beam chopped or the read beam chopped. Results show that (1) the grating formation was about 10 ms, which reveals that the optical process can operate at a speed better than the video frame rate, and (2) the read beam can create enough photoexcitation for reducing the grating strength in the crystal and consequently the diffraction by about 60%. The time constant of this process was about 5 ms.[50]

8.2. Image correlation

We have demonstrated, for the first time, image correlation using the four-wave mixing in photorefractive GaAs with two 1.3-μm single-mode, DFB InGaAsP/InP lasers. The result provides further support to the statement in the previousl subsection that the feasibility of developing compact integrated optical processing modules with photorefractive and lasing phenomena in compound semiconductors.

Figure 10 shows a schematic of the experimental setup for the image correlation experiment. As indicated in the figure, wavelengths of the two lasers are slightly different, about 0.0104 μm. This difference makes the four-wave mixing configuration non-degenerate.

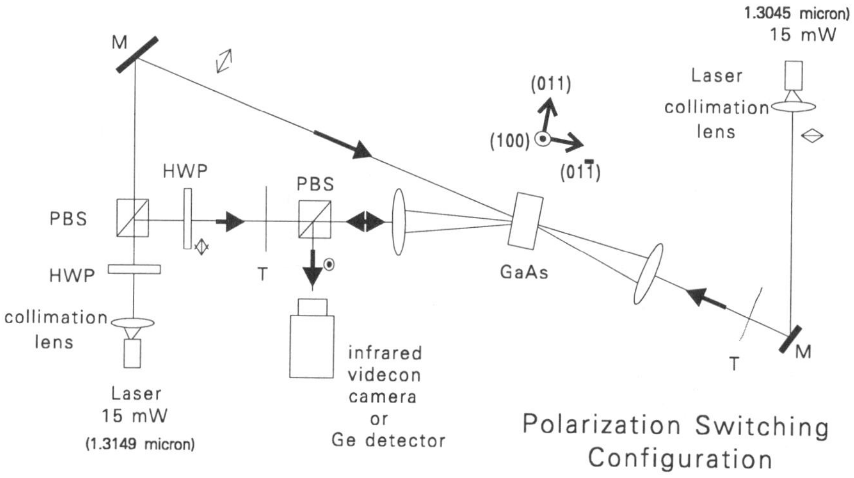

Figure 10. Schematic of the experimental set-up for the image correlation experiment.

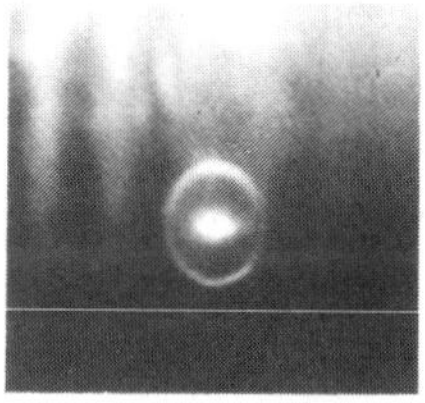

Autocorrelation of a Circle

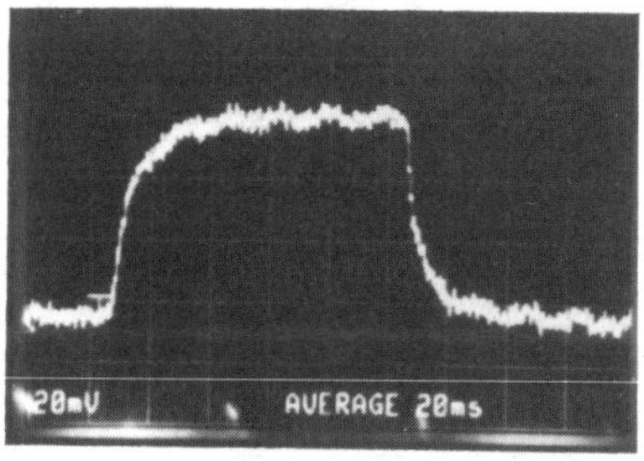

Time Dependance
(with probe beam chopped)

Figure 11. (a) An autocorrelation image of a circle obtained from an experiment using semiconductor lasers. (b) Autocorrelation signal with the probe beam chopped, showing a 5-ms characteristic time for hologram formation in the correlation configuration.

Figure 11(a) gives an autocorrelation image of a circle obtained from the experiment. The correlation signal is reasonably good, but its intensity is weak in comparison with those obtained from the experiments using 1.06-μm beam of similar intensities from a

Nd:YAG laser. This can be attributed to fact that the absorption coefficient of the EL-2 level is about a factor of eight smaller than that at 1.06 μm.[52] If single-mode semiconductor lasers can provide beams of wavelength around 1 μm with power in several tens of mW are available, a compact image correlator with semiconductors can be developed.

Figure 11(b) gives an oscilloscope picture obtained with the probe beam chopped, showing the characteristic times of the hologram formation in the correlation configuration. The formation time and the decay time of the hologram are roughly the same, about 5 ms, illustrating that this correlation system can process input signal at 200 frames/s.

9. TIME CORRELATION

Time-integrating correlator has been implemented using acousto-optic (AO) devices and a linear charge-couple-device (CCD) detector arrays.[53,54] The drawback of using a CCD array is that it detects only the correlation term but extraneous bias terms as well. Recently it has been demonstrated that this problem can be removed by replacing the CCD array with a photorefractive crystal.[55,56]

A potential application of the photorefractive time-integrating correlator is in a real-time radar-jamming-interference rejection system using the adaptive filter method. The response time of the photorefractive crystal determines both the integration time and maximum repetition rate of the time integrator, and thus a fast photorefractive crystal such as GaAs can be beneficial in this application.

A GaAs-based photorefractive time-integrating correlator with the same configuration as that described in Ref. 56 was used to perform the crosscorrelation of signals S_1 and S_2 represented by the following equation,[57]

$$S_{12}(\tau) = C\int_0^\tau S_1(t - x/v)S_2(t + x/v)dt \qquad (1)$$

It carries out this mathematical operation this way. Radio frequency (RF) signals $S_1(t)$ and $S_2(t)$ are applied to excite two acoustic waves propagating along AO devices AO1 and AO2 in opposite directions. Two coherent beams, diffracted by these acoustic waves, are then transmitted through appropriate optics to image and superimpose the patterns of these acoustic waves onto a photorefractive crystal. The resultant refractive index is then read out by the reading beam, yielding a readout beam whose amplitude is spatially modulated according to the correlation function, $S_{12}(\tau)$. The output at a square-law detector array is then proportional to $|S_{12}(\tau)|^2$.[57] For a detailed review of this topic, see the article by Hong on the applications of photorefractors for signal processing in this volume.

The response time of the GaAs crystal in the time-integrating correlator was measured as a function of the total beam intensity. At a total writing beam intensity of 4 W/cm^2, we obtained a response time of better than 80μs. The dynamic ranges of both the AO device and GaAs crystal in the correlator were also measured and both are about equal to 40 dB. The response time and dynamic range can probably be improved by using higher laser intensity and better detector.

Because of the vertical alignment of the Bragg cells in the configuration used in our experiment, we observed an interesting new phenomenon - an interference fringe pattern (see Fig. 12) - in the output of the system under certain conditions.[58] A theoretical explanation in good agreement with experimental data is presented below.

Let us consider the case in which the RF input signal to the Bragg cells are square-wave modulated FM signal, namely, the signal contains two different RF frequencies appearing alternately with time. When the high frequency, f_{rf2}, is applied to the Bragg cells, the diffraction angle is larger than that of the diffracted beam when the lower frequency, f_{rf1}, is applied. The two sets of write beams are in two different planes intersecting at an angle (see Fig. 12). As a result, the index gratings in the photorefractive crystal also intersect with the same angle. The two index gratings may coexist in the crystal if the RF frequency switches between f_{rf1} and f_{rf2} at a rate faster than the reciprocal of the response time, τ_s , of the photorefractive crystal. Under this condition, the two index gratings may interfere with each other and result in a fringe pattern as shown in the thicker solid lines in Fig. 12. The maxima of the fringe occur at where the two index gratings are in phase, while the minima occur at where the two index gratings are out of phase.

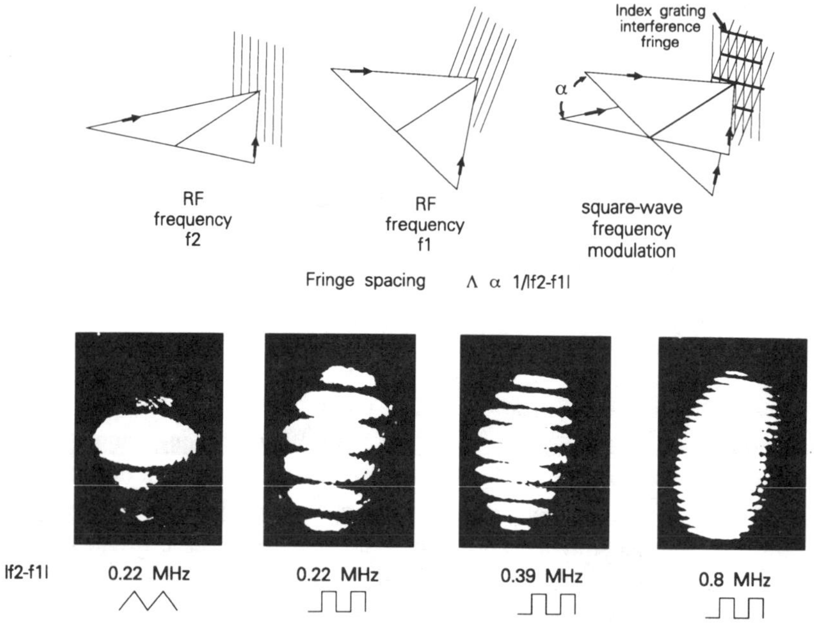

Figure 12. Top: Illustration of the formation of the fringes created by the interference between the index gratings generated by two different RF frequencies. Note that the write beam planes corresponding to frequency 1 and frequency 2 intersect at an angle alpha. As a result, the index gratings corresponding to these two sets of write beams also intersect with the same angle. The maxima of the fringes occur at where the two index gratings are in phase, while the minima occur at where the two index gratings are out of phase. Bottom: Images of the index grating interference fringe patterns. The measured spacing between fringes are consistent with the prediction of Eq. (2).

The fringe spacing, Λ_f, can be derived from the beam geometry which is a function of the RF frequency, the sound velocity v in the AO crystal, and the focal lengths of the lenses. In practice, it is reasonable to assume the diffraction angle of the grating in the Bragg cells is very small. Under this assumption,

$$\Lambda_f = (v/2\Delta f_{rf})\,(f_2/\,f_1) \qquad (2)$$

where $\Delta f_{rf} = f_{rf2} - f_{rf1}$, f_1 is the focal lengths of lenses L_1 and L_3, and f_2 is the focal lengths of lenses L_2 and L_4.

Equation (2) is in good agreement with experimental data. For example, the fringe spacing is proportional to f_{rf} as shown in Fig. 12. For $f_1 = f_2$, v = 616 m/s, and $\Delta f_{rf} = 0.8$ MHz, both the experimental measurement and the calculation given by Eq. (2) yield $\Lambda_f =$ 0.4 mm. The experimental data also show that Λ_f is inversely proportional to f_{rf} and independent of the response time of the photorefractive crystal as predicted by Eq. (2).

The observation of the index grating interference fringe in a photorefractive crystal in a time-integration and correlation system is very important to its operation. For chirped signals, the intercepting angle of two write beams is continuously changing. As a result, the strength of the hologram may not increase linearly with time. Under certain special conditions, the strength could be reduced.

10. IMAGE CORRELATION USING A NEW CLASS OF THIN PHOTOREFRACTIVE DEVICES

Recently an optical image correlator using a 2-μm-thick GaAs/AlGaAs semi-insulating multiple quantum wells (SI-MQW) and diode lasers was successfully demonstrated.[59] The thinness of this type of device could potentially alleviate the space-bandwidth product (the number of resolvable dots discussed in sec. 3.5) limitation of optical correlator systems based on thick (1-5 mm) photorefractive crystals. Although the resolution of the SI-MQW device used in the experiment is ~30 μm, the theoretical limit of this class of devices is ~2.5 μm. The MQW was doped with $10^{16}/cm^3$ of Cr to render it semi-insulating.

The SI-MQW device was reported to show large absorption changes ($\Delta\alpha \sim 6000\ cm^{-1}$ for 20-V applied voltage) and refractive-index changes ($\Delta n \sim 0.06$ for 20-V applied voltage)near the exciton peak at 850 nm. With only 3 mW of power incident upon the SI-MQW, a correlator based on this device showed a rise time of 1 μs in its autocorrelation peak, an erasure time as short as 2 μs and a controllable storage time of 2-25 μs.

Some concerns with this type of SI-MQW are: (1) the undesirability of have Cr in a MBE system in the fabrication process, (2) large grating spacing (20-30 μm) required due to substantial charge migration at the interface. Other potential materials for fabricating this these devices are Fe-doped InGaAs/InP or InGaAsP/InP and low-temperature grown SI AlGaAs/GaAs.

11. CONCLUSION

In this review we have discussed basic properties of photorefractive effects in compound semiconductors relevant to information processing. Experiments demonstrating the feasibility of using these photorefractive crystals as a dynamic holographic medium for information processing have been presented. They include multiple four-wave mixing, parallel image operations, novelty filtering, image correlation, novelty-filtered and time correlation. The results have established distinctive advantages of using photorefractive compound semiconductors to develop fast information processors. The existence of significant difference in intensity requirement between uses of GaAs and InP for image processing is illustrated experimentally. The feasibility of integrating photorefractive semiconductors and semiconductor lasers together to develop compact image processors is demonstrated by using semiconductor lasers as sources in experiments involving imaging by phase conjugation and image correlation, respectively.

More research and development are needed to realize these information processing applications using compound semiconductors. On the application side, development of compact image processing modules by integrating with small-sized lasers, spatial light modulators, and detector array together is desired. In addition, use of novel semiconductor structures such as multiple quantum wells for image processing needs to be further explored. On the materials side, further research to enhance the understanding of crystal defects relevant to photorefractive effects and to improve growth techniques for producing high-quality photorefractive crystals is needed.

12. ACKNOWLEDGMENTS

The work was performed by the Center for Space Microelectronics Technology, Jet Propulsion Laboratory, California Institute of Technology and was sponsored by the Advanced Research Projects Agency through an agreement with the National Aeronautics and Space Administration.

13. REFERENCES

1. A. M. Glass, A. M. Johnson, D. H. Olsen, W. Simpson, and A. A. Ballman, "Fourwave mixing in semi-insulating InP and GaAs using the photorefractive effect," *Appl. Phys. Lett.*, **44**, 948-950 (1984).

2. M. B. Klein, "Beam coupling in undoped GaAs at 1.06 micron using the photorefractive effect," *Opt. Lett.*, **9**, 350-352 (1984).

3. D.D. Nolte, D.H. Olsen, E.M. Monberg, P.M. Bridenbaugh, and A.M. Glass, "Optical and photorefractive properties of InP:Ti : A new photorefractive semiconductor," *Opt. Lett.*, **14**, 1278-1280 (1989).

4. K. Kuroda, Y. Okazaki, T. Shimura, H. Okamara, M. Chihara, M. Itoch and I. Ogura, " Photorefractive effect in GaP," *Opt. Lett.* **15**, 1197-1199 (1990).

5. R.B. Bylsma, P.M. Bridenbaugh, D.H. Olson, and A.M. Glass, "Photorefractive properties of doped cadmium telluride," *Appl. Phys. Lett.* **51**, 889-891 (1987).

6. See, for example, P. Gunter and J.-P. Huignard, "Photorefractive effects and materials," in *Photorefractive Materials and Their Applications I*, ed. by P. Gunter and J.-P. Huignard, Springer-Verlag, Berlin, 1989, p.7; J. Feinberg and K.R. MacDonald, "Phase-conjugate mirrors and resonators with photorefractive materials," in *Photorefractive Materials and Their Applications II*, ed. by P. Gunter and J.-P. Huignard, Springer-Verlag, 1989, p.151; T.J. Hall, "Optical information processing," in *Principles of Modern Optical Systems*, ed. by I. Andonovic and D. Ottamchandani, Artech House, Norwood, MA,1989, p.461.

7. L.J. Cheng and A. Partovi, "Temperature and intensity dependence of photorefractive effect in GaAs," *Appl. Phys. Lett.* **49**, 1456-1458 (1986).

8. G. Gheen and L.J. Cheng, "Image processing by four-wave mixing in photorefractive GaAs," *Appl. Phys. Lett.* **51**, 1481-1483 (1987).

9. G. C. Valley, A. L. Smirl, M. B. Klein, K. Bohnert, and T. F. Boggess, "Picosecond photorefractive beam coupling in GaAs," *Opt. Lett.*, **11**, 647-649 (1986).

10. L. J. Cheng and A. Partovi, "Index grating lifetime in photorefractive GaAs," *Appl. Opt.*, **27**, 1760-1763 (1988).

11. A. M. Glass and J. Strait, "The photorefractive effect in semiconductors," in *Photorefractive Materials and Their Application I*,ed. by P. Grunter and J.P. Huignard,Springer-Verlag, Berlin, 1988, p.237.

12. R. V. Johnson and A. R. Tanguay, Jr., "Fundamental physical limitations of the photorefractive grating recording sensitivity," in *Optical Processing and Computing*, ed. by H. H. Arsenault, T. Szoplik and B. Macukow, Academic Press, New York, 1989, p.59.

13. L. J. Cheng, J. Lagowski, M. F. Rau, and F. C. Wang, "Defects and photorefractive effects in GaAs," *Radiation Effect and Defects in Solids*, **111** & **112**, 37-43 (1989).

14. D. T. H. Liu, L. J. Cheng, M. F. Rau, and F. C. Wang, "Photorefractive gain in GaAs under a dc electric field," *App. Phys. Lett.*, **53**, 1369-1371 (1988); D. T. H. Liu, L. J. Cheng, A. E. Chiou, and P. Yeh, "Two-beam coupling gain in undoped GaAs with applied dc electric field and moving grating," *Opt. Comm.*, **72**, 384-386 (1989).

15. B. Imbert, H. Rajbenbach, S. Mallick, J. P. Herriau and J. P. Huignard, "High photorefractive gain in two-beam coupling with moving fringes in GaAs:Cr crystals," *Opt. Lett.*, **13**, 327-329 (1988).

16. A. Partovi, A. Kost, E.M. Garmire, G. C. Valley and M. B. Klein, "Band-edge photorefractive effect in semiconductors," *Appl. Phys. Lett.*, **56**, 1089-1091 (1990).

17. A. Partovi, J. Millerd, and E. M. Garmire, "Large photorefractive beam-coupling gain using the Franz-Keldysh electrorefractive effect in temperature-stablized InP:Fe," Digest of the First International Meeting on Nonlinear Optics, Kauai, Hawaii, July 16-20, 1990, pp. 212-213.

18. D. L. Staebler and J. J. Amodei, "Coupled-wave analysis of holographic storage in $LiNbO_3$, *J. Appl. Phys.* **43**, 1042-1049 (1971).

19. N. V. Kukhtarev, B. Markov, S. G. Odulov, M. S. Soskin and V. L. Vinetsskii, "Holographic storage in electrooptic crystals. I. Steady state," *Ferroelectrics*, **22**, 949-960 (1979) and " Holographic storage in electrooptic crystals. II. Beam coupling-light amplification," **22**, 961-964 (1979).

20. L. J. Cheng, G. Gheen, T. H. Chao, H. K. Liu, A. Partovi, J. Katz, and E. Garmire, "Spatial light modulation by beam coupling in GaAs crystals," *Opt. Lett.*, **12**, 705-707 (1987); L. J. Cheng, G. Gheen, F. C. Wang, and M. F. Rau, "Image transfer in photorefractive GaAs," *J. Appl. Phys.*, **62**, 3991-3993 (1987).
21. G. Gheen and L. J. Cheng, "Optical correlators with fast updating speed using photorefractive semiconductor materials," *Appl. Opt.*, **27**, 2756-2760 (1988).
22. D. T. H. Liu, L. J. Cheng, T. H. Chao, J. W. Yu, and D. A. Gregory, "Real-time optical correlator using photorefractive GaAs," *Proc. SPIE* **1347**, 156-168 (1990).
23. L. J. Cheng and D. T. H. Liu, "Optical computing and image processing using photorefractive gallium arsenide," *Jap. J. Appl. Phys.*, **29**, L1297-L1300(1990).
24. L. J. Cheng and D. T. H. Liu, "High-speed parallel optical processors of photorefractive semiconductors," Vol. 9 of Technical Digest of the Topical Conference on Optical Computing, Salt Lake City, Utah, February 28-March 2, 1989, p.87.
25. L. J. Cheng and D. T. H. Liu, "Properties and applications of photorefractive GaAs," *Proc. SPIE* **1220**, 1-16 (1990).
26. P. Yeh, "Photorefractive two-beam coupling in cubic crystals," *J. Opt. Soc. Am. B*, **4**, 1382-1386 (1987).
27. L.J. Cheng and P. Yeh, "Cross-polarization beam coupling in photorefractive GaAs crystals," *Opt. Lett.* , **13**, 50-52 (1988).
28. A. Partovi, E. Garmire, and L.J. Cheng, "Enhanced beam coupling modulation using the polarization properties of photorefractive GaAs," *Appl. Phys. Lett.* **51**, 299-301 (1987).
29. M. D. Levenson, "High resolution imaging by wave-front conjugation," *Opt. Lett.* **5**, 182-184 (1980).
30. A. E. Chiou and P. Yeh, "Parallel image subtraction using a phase-conjugate Michelson interferometer," *Opt. Lett.*, **11**, 306-308 (1986).
31. A. E. Chiou, P. Yeh and M. Khoshnevisan, "Coherent image subtraction using phase conjugate interferometry," *Proc. SPIE* **613**, 201-206 (1986).
32. S. K. Kwong, G. Rakuljic, and A. Yariv, "Real time image subtraction and 'exclusive or' operation using a self-pumped phase conjugate mirror," *Appl. Phys. Lett.*, **48**, 201-203 (1986).
33. S. K. Kwong, G. Rakuljic, V. Leyva, and A. Yariv, "Real-time image processing using a self-pumped phase conjugate mirror," *Proc. SPIE*, **613**, 36-42 (1986).
34. P. Yeh, T. Y. Chang, P. H. Beckwith, "Real-time optical image subtraction using dynamic holographic interference in photorefractive media," *Opt. Lett.* **13**, 586-588 (1988).
35. N. V. Vainos, J. K. Khoury, and R. W. Eason, "Real-time parallel optical logic in photorefractive bismuth silicon oxide," *Opt. Lett.*, **13**, 503-505 (1988).
36. D. Z. Anderson, D. M. Lininger, and J. Feinberg, "Optical tracking novelty filter," *Opt. Lett.*, **12**, 123-125 (1987).
37. D. Z. Anderson and J. Feinberg, "Optical novelty filters," *IEEE J. Quant. Electr.* **QE-25**, 635-637 (1989).
38. D. T. H. Liu and L. J. Cheng, "Resolution of a target-tracking optical novelty filter," *Opt. Eng.* **30**, 571-576 (1991).

39. See, for example, J. Goodman, "Introduction to Fourier Optics", MaGraw-Hill, New York (1968).
40. D. M. Pepper, J. Auyeung, D. Fekete, and A. Yariv, "Spatial convolution and correlation of optical fields via degenerate four wave mixing," *Opt. Lett.* **3**, 7-9 (1978).
41. J. O. White and A. Yariv, "Real-time image processing via four-wave mixing in photorefractive materials," *Appl. Phys. Lett.* **37**, 5-7 (1980).
42. L. Pichon and J. P. Huignard, "Dynamaic joint-Fourier transform correlator by Bragg diffraction in photorefractive $Bi_{12}SiO_{20}$ crystals," *Opt. Comm.* **36**, 277-280 (1981).
43. P. D. Foote, T. J. Hall, and L. M. Connors, "High speed two input real-time optical correlation using photorefractive BSO," *Opt. Laser Technol.* **18**, 39-42 (1986).
44. M. G. Nicholson, I. R. Cooper, M. W. McCall, and C. R. Petts, "Simple computational model of image correlation by four-wave mixing in photorefractive media," *Appl. Opt.* **26**, 278-286 (1987).
45. M. G. Nicholson, I. R. Cooper, G. G. Gibbons, and C. R. Petts, "Optimization of an updatable optical image correlator," *Opt. Eng.* **26**, 445-452 (1987).
46. F. T. S. Yu, S. Wu, A. W. Mayers, and S. Rajan, "Wavelength-multiplexed reflection matched spatial filters," *Opt. Comm.* **81**, 343-347 (1991).
47 F. T. S. Yu, S. Wu, S. Rajan, and D. A. Greg "Compact joint transform correlator with a thick photorefractive crystal," *Appl. Opt.* **31**, 2416-2418 (1992).
48. D. T. H. Liu and L. J. Cheng, "Real-time VanderLugt optical correlator using photorefractive GaAs," *Appl. Opt.*, **31**, 5675-5680 (1992).
49. A. VanderLugt, "Signal detection by complex spatial *filtering," IEEE Trans. Information Theory*, **IT-10**, 139-145 (1964).
50. L. J. Cheng, D. T. H. Liu and K. L. Luke, "Photorefractive image processing using compound semiconductors," Int. J. Opt. Comp., in press.
51. D. T. H. Liu, T. H. Chao and L. J. Cheng, "Novelty-filtered optical correlator using photorefractive crystal," *Proc. SPIE*, **1701**, 105-110 (1992).
52. L. Samuelson, "Defect levels in semiconductor alloys," *Proceedings of the 13th International Conference on Defects in Semiconductors*, ed. by L.C. Kimerling and J.M. Parsey, The Metallurgical Society of AIME, Warrendale, PA. 1984, pp. 101-114.
53. W. T. Maloney, "Acoustooptical approaches to radar signal processing," *IEEE Spectrum* **6**, 40-48 (1969) and references therein.
54. W. T. Rhodes, "Acousto-optic signal processing: convolution and correlation," *Proc. IEEE* **69**, 65-79 (1981).
55. D. Psaltis, J. Yu, and J. Hong, "Bias-free time-integrating optical correlator using a photorefractive crystal," *Appl. Opt.* **24**, 3860-3865 (1985).
56. J. H. Hong and T. Y. Chang, "Photorefractive time-integrating correlator," *Opt. Lett.* **16**, 333-335 (1991).
57. D. T. H. Liu, K. L. Luke and L. J. Cheng, "GaAs-based photorefractive time-integrating correlator," *SPIE* **1702**, 205-209 (1992).
58. D. T. H. Liu, K. L. Luke and L. J. Cheng, unpublished results.
59. A. Partovi, A. M. Glass, T. H. Chiu and D. T. H. Liu, "High-speed joint-transform optical image correlator using GaAs/AlGaAs semi-insulating multiple quantum wells and diode lasers," *Opt. Lett.* **18**, 906-908 (1993).

Optical interconnection using photorefraction

Arthur E. Chiou
Rockwell International Science Center
1049 Camino Dos Rios, Thousand Oaks, CA91360

ABSTRACT

This paper surveys various approaches that have been proposed or demonstrated for optical interconnections using photorefractive holograms. Some material issues, technical challenges, and competing technologies are discussed.

1. INTRODUCTION

During the past two decades photorefractive effect[1-3] in a wide variety of materials has been studied extensively for potential applications in optical storage, interconnection, and processing. This paper presents a brief overview of how each of the unique features of photorefractive phenomena has been used to implement optical interconnection networks. To avoid redundancy with other papers in this volume, I have kept the treatment of photorefraction to a minimum. Those who need a comprehensive review of the subject can consult any of the reference papers cited in this or other articles in this volume. This paper is organized as follows: A brief introduction to optical interconnection is given in Section 2, and the photorefractive phenomena that have been demonstrated for potential applications in optical interconnects are outlined in Section 3. Section 4 consists of the main body, where various architectures and approaches are discussed. Section 5 concludes the chapter with a brief summary and a discussion of some of the issues and the competing materials and technologies.

2. OPTICAL INTERCONNECTION

Storage, interconnection, and processing are three basic functional blocks of an information system. Over the last decade, processing speed and storage density in electronic computers based on silicon integrated circuits have increased dramatically.[4] This increment in device speed and density results in new requirements and constraints on data links which cannot be satisfied by the conventional interconnection technology based on metallic conductors. Since the performance of modern super computers is largely limited by data communication rather than processing speed, new technologies for interconnection have been sought. Optical interconnection[5,6] offers several unique advantages compared with its electronic counterpart due to its inherent parallelism, immunity to electromagnetic interference (EMI), and high temporal bandwidth. Optical interconnects capable of switching the data paths among many computing systems or among various components within a system can play an important role in future computing systems.

A generic block diagram for optical interconnection is shown in Fig. 1. In many applications, the input signals or data are likely to be in electrical format (in the form of voltage or current) which can be used to modulate a light source (or sources) either by direct modulation or by driving an external modulator. The modulated optical output is transmitted through some medium (free-space, waveguide or fiber) to an interconnection device (or switching network) which connects each of the input channels to the designated output channels. In the final stage, the signals in optical form are received by

Fig. 1. A generic block diagram of optical interconnection

the detectors which convert the optical signals back into electrical format for further processing or for further transformation into other suitable format. In some cases, however, the original information is naturally in two-dimensional optical format. Parallel readout from a holographic memory[7,8] or from an optically-addressed spatial light modulator (OASLM)[9-11] represents a typical example of such cases. In this paper, I will focus mainly on the optical interconnection devices using photorefractive holograms.

Regardless of the actual implementation scheme, an interconnection network can be represented schematically by a two-dimensional checkerboard pattern illustrated in Fig. 2, where the connections of the input channels (represented by capital letters, A, B, C, and D) and the output channels (represented by numbers 1, 2, 3, and 4) are shown as solid circles. Fig. 2a, b, and c, for example, represent permutation, full broadcasting, and partial broadcasting, respectively. For the case of permutation (or one-to-one interconnection) of N input to N output channels, an optical interconnect can be implemented in a two-dimensional planar geometry with appropriate waveguides and switching devices (at the cross-points).[12] The most common type of such a planar switch array is based on electro-optic directional couplers in lithium niobate.[13,14] Alternatively, an array of localized gratings can be generated optically in a real-time holographic material to diffract the input beam to the designated output channels as shown in Fig. 3.[15] The holographic gratings can be created by the interference of either two guided waves in the plane of the waveguide (Fig. 4a) or two expanded beams in free space (Fig. 4b). In the latter case, a spatial light modulator (SLM) can be used to selectively expose the regions where the gratings are needed. The concept has been demonstrated successfully using photorefractive gratings in iron-doped lithium niobate. The key advantage of this approach (compared with the electro-optics based directional coupler) lies in its relatively smaller size. One major disadvantage is that the time required to change the interconnection pattern is extremely long (on the order of a second) due to the very slow response of lithium niobate. The reconfiguration rate can be improved by using other materials with faster response. For example, an integrated-optical switch array based on electrically controlled dynamic free carrier gratings in GaAs was reported recently.[16] An important drawback shared by all the techniques based on this simple 2-D geometry is that the path lengths are not equal. Consequently, the unequal signal delay may limit the performance of a large network operating in synchronous mode.

In general, an optical interconnection network is often configured to take advantage of the third dimension, as shown schematically in Fig. 5 where a 2-D array of optical transmitters (say, on the xy plane) is connected to a 2-D array of receivers through an interconnection network by transmission and diffraction in the z-direction (normal to the source plane). In this example, the switching is accomplished by diffraction off transmission gratings recorded in a photorefractive crystal. The gratings are usually generated by interfering (inside a photorefractive crystal) two or more coherent light

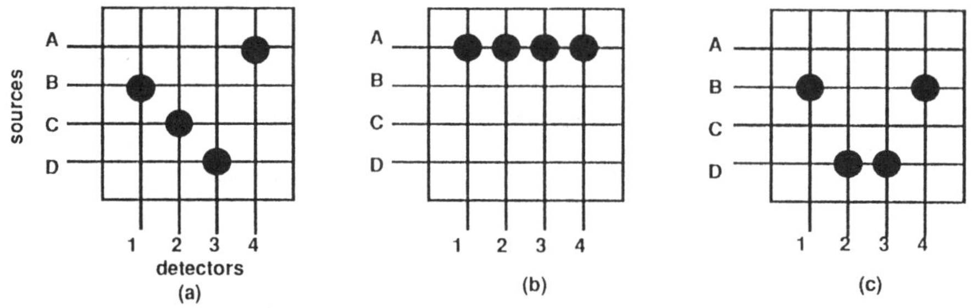

Fig. 2. A schematic representation (for four input and four output channels) of various interconnection functions: (a) permutation, (b) broadcasting, (c) multi-casting.

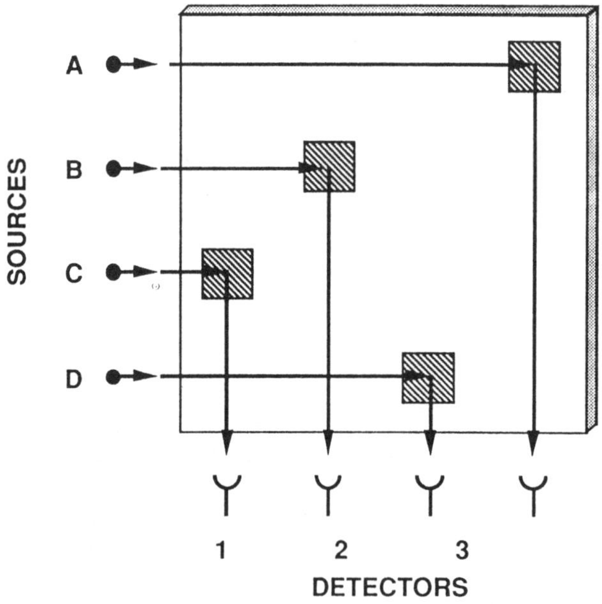

Fig. 3. Two-dimensional planar configuration for permutation using localized diffraction gratings.

sources at appropriate locations. A popular schematic diagram for an optical neural network[17,18] which uses exactly the same concept described above is shown in Fig. 6. The required interconnection holograms are multiplexed in a volume holographic material by the interference of one or more mutual coherent beams from the input and the control planes.

3 PHOTOREFRACTIVE PHENOMENA

Photorefractive phenomena offer many interesting and unique features which can be used for optical interconnection. An extensive discussion of the photorefractive effect and materials is beyond the scope of this paper. The key features which have been proposed or demonstrated for optical interconnection are outlined below:

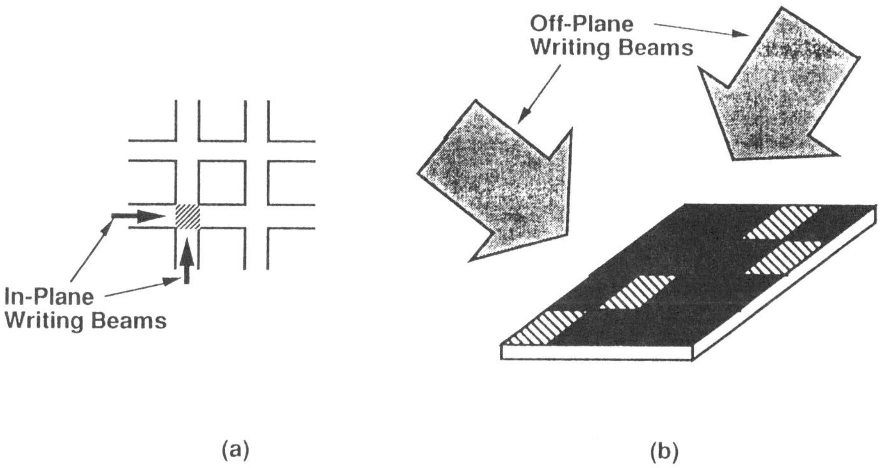

Fig. 4. Generation of localized gratings using (a) in-plane, and (b) off-plane writing beams.

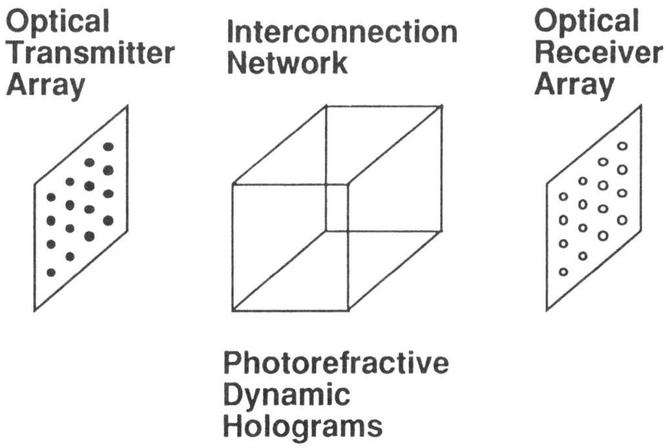

Fig. 5. A generic view of three-dimensional optical interconnection using photorefractive volume hologram.

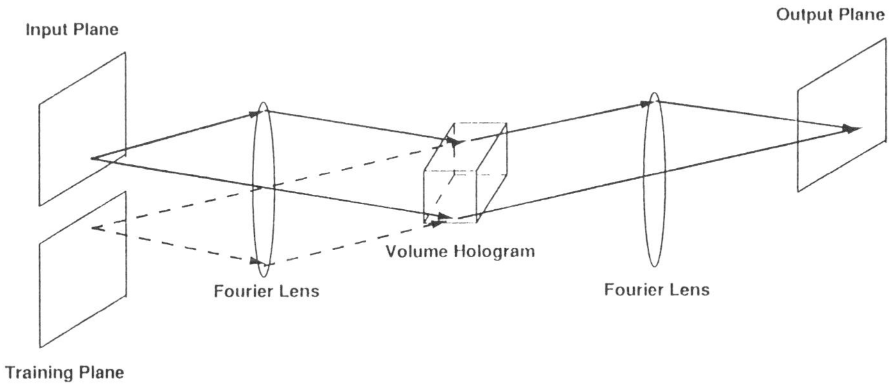

Fig. 6. A schematic illustration of interconnect for optical neural network using volume hologram.

(i) Real-Time Operation: The photorefractive effect[1] involves dynamic charge distribution in an electro-optic material via optical excitation, migration, and trapping of free charge carriers. The combination of these processes generates a non-uniform space charge field which modulates the refractive index (or in other words, creates a phase hologram) via the linear electro-optic effect. The dynamic nature of the photorefractive mechanism gives rise to the formation of new holograms and the simultaneous erasure of the old holograms (with a characteristic time constant) in response to a time varying input pattern. This feature is essential to the implementation of a reconfigurable optical interconnection where the desired interconnection pattern (or holograms) is recorded as needed. The capability to record and erase holograms in real-time is indispensable to the training process of an adaptive optical neural network.[19]

(ii) Multiplexed Volume Holographic Storage: Mok recently demonstrated[20] that as many as five thousand holograms of gray scale images can be stored in (and selectively recalled from) a lithium niobate crystal of about 5 cm^3. The possibility of multiplexing thousands of holograms and selectively retrieving each of them with very little crosstalk allows for a programmable optical interconnection system where the required interconnection patterns are pre-stored in a page-oriented format.

(iii) Coherent Amplification: When two mutually-coherent beams interact in a photorefractive crystal, one beam can gain energy at the expense of the other.[21] In other words, a photorefractive crystal can serve as an adaptive coherent optical amplifier to amplify an image bearing beam. This feature can be used to improve the energy efficiency of a reconfigurable optical interconnect.

(iv) Optical Phase Conjugation: Optical phase conjugation (OPC)[22,23] can take place in photorefractive materials under a wide range of conditions and configurations.[24] The unique capabilities of OPC in aberration correction and self-alignment can facilitate the realization of globally interconnected networks.

Last, but not the least, the most unique aspect of photorefractive materials is that all the features listed above can be realized at relatively low optical power on the order of a few milliwatts. Specific examples of how these unique features have been applied to optical interconnections are discussed in more detail in the next section which constitutes the main body of this paper.

4. A SURVEY OF PHOTOREFRACTIVE OPTICAL INTERCONNECTIONS

4.1 Optical Beam Steering

Beam steering usually refers to the pointing of an optical beam to a designated target. For the purpose of optical interconnection, beam steering can be used to send a signal carrier beams to any one of many receivers. Thus, it is basically a 1-to-N reconfigurable interconnect. In many optical systems, beam steering is accomplished by reflection from a mirror attached to a galvanometer or by diffraction from an acousto-optic (or electro-optic) cell. A method for beam steering using photorefractive dynamic holograms[25] is shown in Fig. 7. In this approach, the two writing beams generate a hologram which diffracts a signal bearing beam to the desired direction. To change the direction of the diffracted output beam, a new set of gratings has to be generated by changing the direction and/or wavelength of the writing beams. For example, two-dimensional beam steering with an angular range of 4° x 7° has been demonstrated[26] in

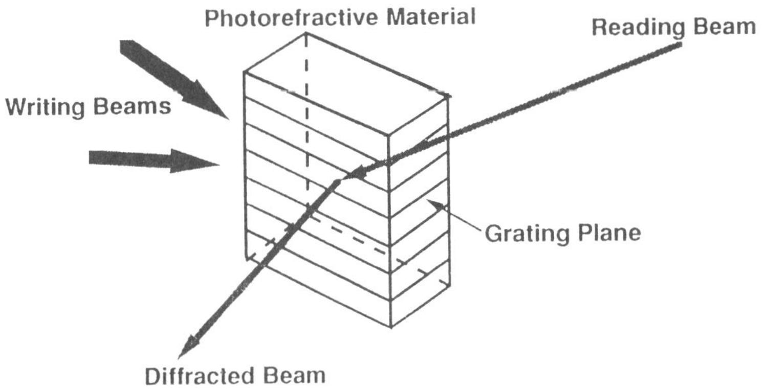

Fig. 7. Optical beam steering using photorefractive dynamic hologram.

bismuth silicon oxide (BSO) by changing both the direction and the wavelength of the writing beams. The experimental configuration and the experimental results are reproduced in Fig. 8. Although sound in principle, this approach suffers from several severe drawbacks including slow reconfiguration time and limited steering-range.

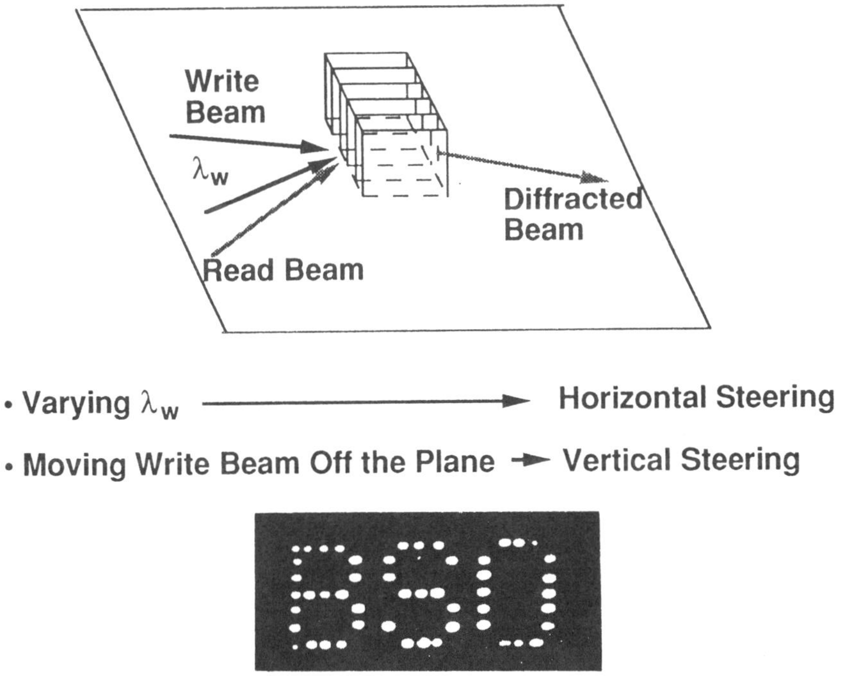

Fig. 8. Two-dimensional optical beam steering in photorefractive dynamic hologram. Direction of the diffracted output beam is controlled by varying both the wavelength and the direction of the writing beams. An experimental result is shown in the lower part of the figure.

A much more attractive method for beam steering using coherent amplification via two-beam coupling in a photorefractive crystal is shown in Fig. 9. In this approach,[27]

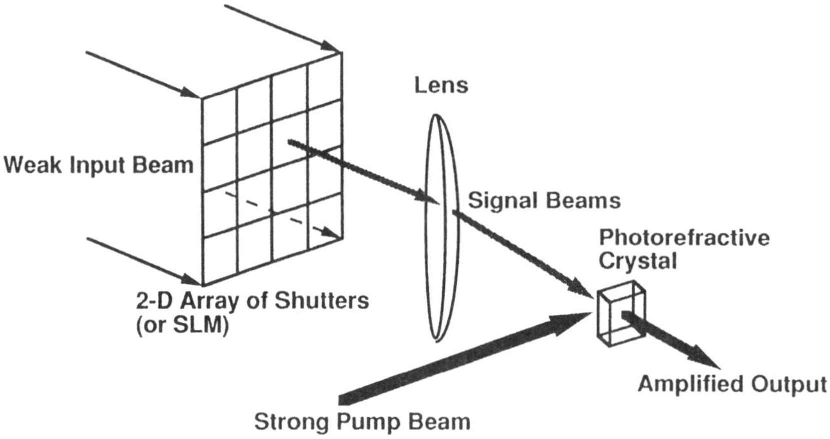

Fig. 9. Optical beam steering based on energy transfer in photorefractive beam coupling.

a small portion of optical energy from the laser source is split off and expanded to illuminate a 2-D array of shutters or a spatial light modulator (SLM). The weak beam transmitted by the shutter (denoted as the signal beam) is subsequently amplified by gaining energy from the main beam (denoted as the strong pump beam) via two-wave mixing in a photorefractive crystal. The net result is that the pump beam is steered to the direction prescribed by the signal beam. When more than one shutter is open at the same instance, the pump beam can simultaneously interact with several signal beams and distribute its energy uniformly (provided that certain conditions are satisfied) to all the signal beams. The dynamic and adaptive nature of the photorefractive hologram is responsible for the reconfigurability. A new hologram is formed and the old one is erased whenever the pattern on the SLM changes. The high energy efficiency is a consequence of the efficient energy transfer in photorefractive beam coupling which allows a large fraction of the beam energy to bypass the inherent fanout loss[28] imposed by the SLM and a redistribution of the reserved energy to the designated channels in a later stage. With some modifications, the basic principle described above can be applied to a NxM crossbar switch which is discussed in the next section.

4.2 NxM Crossbar Switch

A generalized NxM crossbar switch (i.e., a network capable of establishing an arbitrary link of N input channels to M output channels) is best described as a combination of three functional blocks shown in Fig. 10. In the first stage, each input is uniformly split into M outputs (fan-out) resulting in a total of NxM channels each of which is controlled by an ON/OFF switch (or shutter) in the second stage. In the final stage, the output beams from the NxM switches are recombined (fan-in) so that each input can access any one or more (up to M) output channels. For optical implementation of a generalized crossbar switch,[29] the schematic functional diagram shown in Fig. 10 can be conveniently mapped into the optical matrix-vector multiplication scheme[30] in a three-dimensional space shown in Fig. 11. Here, light emitted from each source is projected (fan-out) by appropriate anamorphic optics to a corresponding column of NxM windows whose output is fanned-in on a linear array of detectors. (The anamorphic optics

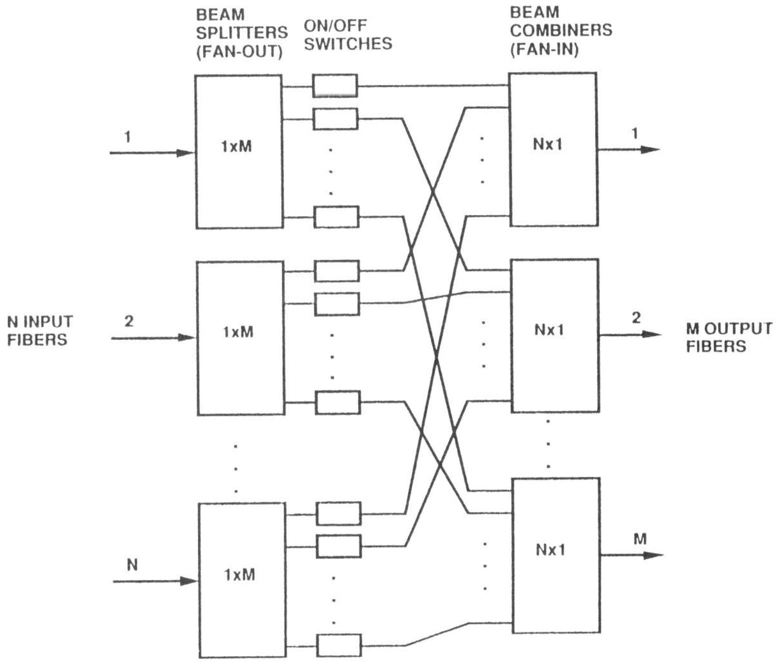

Fig. 10. A schematic block diagram of a NxM generalized crossbar network.

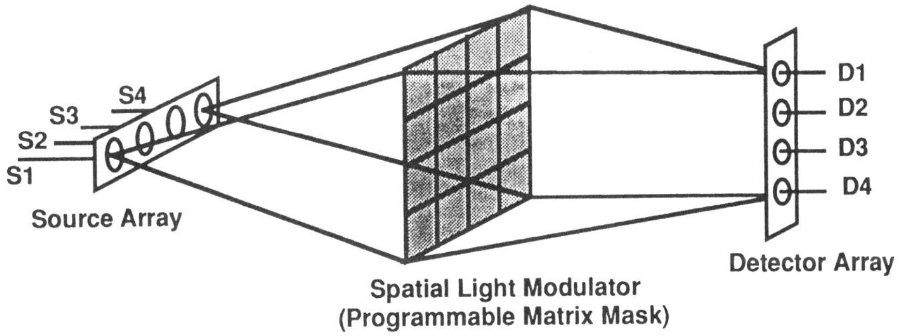

Fig. 11. A generalized crossbar switch based on optical matrix-vector multiplication scheme.

required for fan-out and fan-in are omitted in Fig. 11 for clarity). Connection of a source "i" to a detector "j" is established by opening the window (i,j). A crossbar switch based on this scheme, using optical fibers and fiber star-couplers for beam transmission, fan-out, and fan-in has been developed[31] and is available commercially. Even though this concept has been around for almost two decades, it is still of current interest due to several reasons: (1) It is relatively simple (single stage, straight forward control algorithm) and very flexible (capable of arbitrary interconnection mode ranging from permutation to full broadcasting). (2) Matrix-vector multiplication can be viewed as a basic building block of many mathematical models including the neural network. (3) The technique is readily modified or extended to achieve more complicated operations such as matrix-tensor multiplication[32,33] and optical image crossbar switch,[34] (4) Compact packaging is feasible, for example, by using a matrix-mask sandwiched between linear arrays of elongated sources and detectors[35] as shown in Fig. 12.

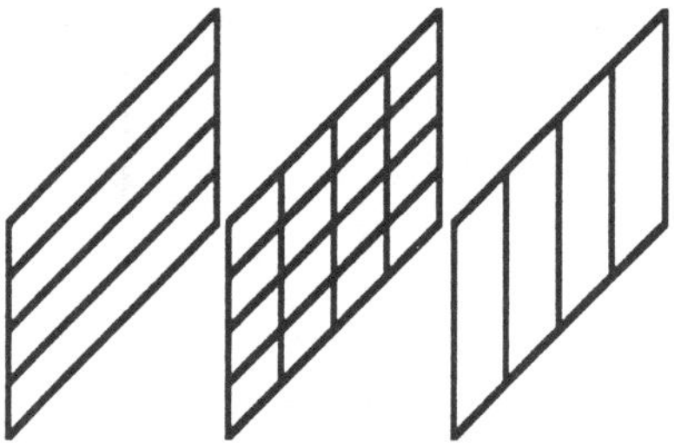

Fig. 12. A compact form of optical matrix -vector multiplier. The three layers (i.e., source plane, matrix-mask, and detector plane) which can be in close contact are shown separately for clarity.

An important drawback of this technique is its inherently low energy efficiency (mainly due to the fan-out and fan-in losses,[28] which may limit the signal-to-crosstalk ratio for switches of considerable size). One approach to overcome this problem is to replace the optical shutters (or SLM) with semiconductor laser amplifiers or erbium-doped fiber amplifiers to significantly improve the contrast, and the signal-to-crosstalk ratio. For example, an 8 x 8 crossbar switch with 64 semiconductor optical amplifiers has been demonstrated by Optivision. Another elegant approach is to use a photorefractive coherent amplifier similar to the one described in the last paragraph of Section 4.1.

The basic principle of an energy efficient photorefractive crossbar network[36] is illustrated in Fig. 13 for a 4 x 4 geometry. A beam splitter (BS) is used to reflect a small fraction of the optical energy from the source array (V) to the signal arm where the output from each individual source is fanned out (via anamorphic optics) and is directed to a corresponding column of windows on an SLM. In contrast with the conventional approach illustrated in Fig. 11, the signal beams transmitted by the SLM are recombined in a photorefractive crystal with the pump beams transmitted by the beam splitter prior to fanning-in to a detector array. Under proper conditions, efficient energy transfer from the pump to the signal beams takes place in the crystal due to photorefractive two-beam coupling. The transferred energy is equally distributed to the designated detectors prescribed by the interconnection pattern on the SLM. Critical performance parameters of this technique, including energy efficiency, reconfiguration time, and signal-to-crosstalk ratio, have been investigated and reported.[37-39] In what follows, we briefly discuss some techniques that have been developed to improve the reconfiguration rate and the signal-to-noise ratio.

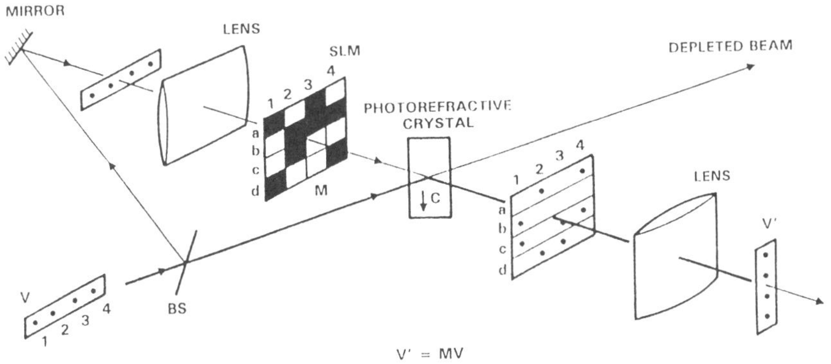

Fig. 13. An energy efficient optical crossbar network based on energy transfer in photorefractive beam coupling

4.2.1 Reconfiguration Rate

Despite its high energy efficiency,[38] the photorefractive crossbar interconnect described above suffers from one major disadvantage; namely, relatively long reconfiguration time (~ 1 sec) which is limited by the buildup (and erasure) time of the photorefractive hologram at low optical intensity (~ 100 mW/cm^2). A technique with the potential to improve the reconfiguration rate by several orders of magnitude has been developed and demonstrated.[40] In principle, the response time can be reduced by increasing the optical power since the photorefractive response time is approximately proportional to $1/I^x$, where I is the total beam intensity and the exponent x can range from 0.4 to 1 depending on the specific material. Photorefractive grating formation in a barium titanate crystal on a time scale of the order of 100 picoseconds using 43 picosecond pulses (~ 10 mJ/cm^2) had been reported.[41,42] The use of high power coherent sources as the signal carrier may be unacceptable because of various constraints imposed by other system requirements such as the data rate, duty cycle, and heat dissipation rate. A separate source, however, can be incorporated into a photorefractive reconfigurable interconnect (Fig. 14) to generate an auxiliary pulse which is flashed momentarily only when reconfiguration is desired. Such an arrangement can, in principle, reduce the reconfiguration time by several orders of magnitude. Preliminary experimental results[40] demonstrating the reduction in reconfiguration time are shown in Fig. 15. The two oscillographs at the top show the long switching time (~ 6 sec) without the aid of the auxiliary pulse. When the auxiliary pulse is turned on (for 500 ms) simultaneous to the switching of the SLM window, the switching time (for both on and off) is reduced to ~ 500 ms as shown in the lower oscillographs. Based on the response time of a fast barium titanate crystal reported recently,[43] we estimate that the reconfiguration time of the order of a millisecond or less should be feasible using an auxiliary nanosecond pulse of sub-mJ/cm^2.

4.2.2 Techniques for Noise Reduction

Signal-to-noise ratio is an important parameter that characterizes the overall performance of an interconnection network. In optical interconnects, crosstalk noise due to scattering from various optical components and optical aberration can be minimized by using high quality components. In photorefractive interconnects, nonuniform

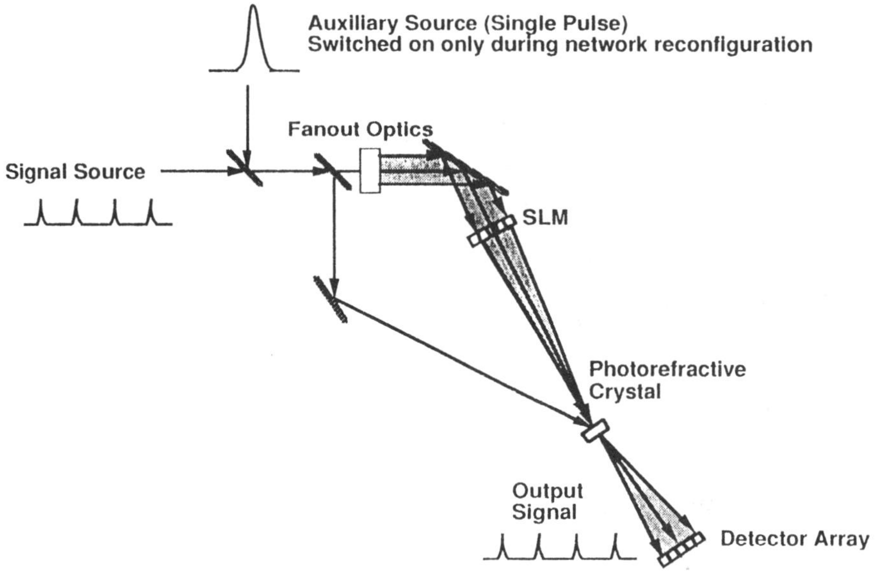

Fig. 14. A schematic illustration of a technique for improving reconfiguration rate of a photorefractive crossbar switch

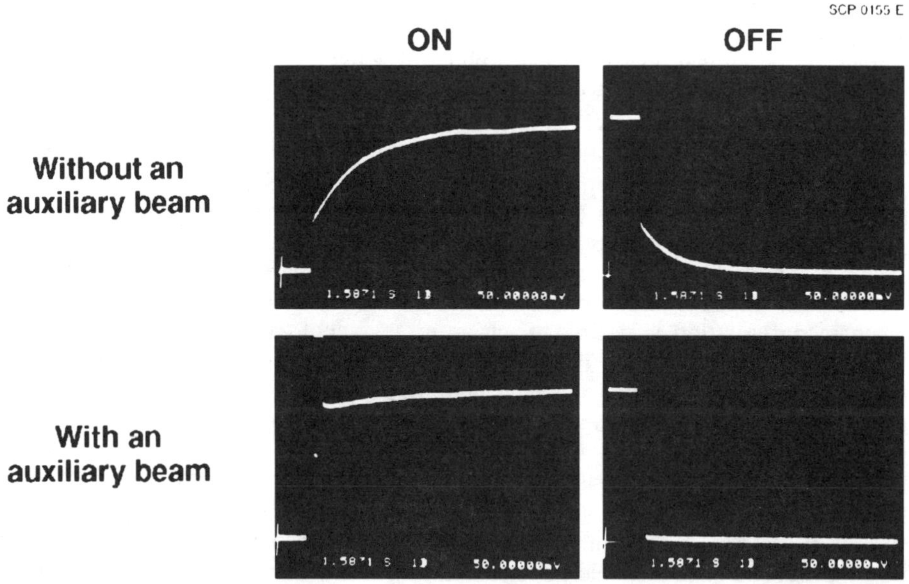

Fig. 15. Experimental results showing the reduction in reconfiguration time using the technique depicted in Fig. 14.

amplification[44] and beam fanning[45] may generate additional noise components that need to be minimized.

4.2.2.1 Reduction of Crosstalk Noise Due To Nonuniform Amplification

In the energy efficient photorefractive interconnect based on two-wave mixing, the signal beams and the corresponding pump beam must overlap completely independent of the interconnection pattern prescribed by the SLM so that maximum energy efficiency is always achieved. This is accomplished by placing the SLM at the front focal plane of a Fourier transform lens with the photorefractive material at the back focal plane. For one-to-one interconnection in the beam steering configuration (Fig. 9) and the permutation in a NxN crossbar configuration (Fig. 13), the shift invariance property of Fourier transformation ensures maximum beam overlap of any selected signal beam and the corresponding pumping beam inside the crystal (Fig. 16). For one-to-many interconnection, more than one window in a column of the SLM will be switched "ON". The Fourier transform of a column of the signal beam consists of a linear superposition of identical patterns each with a different phase factor which is dictated by the relative positions of the windows. As a consequence, the resulting intensity distribution of the signal beam will not be identical with that of the pumping beam at the Fourier plane. Because of the difference in the intensity distribution, the beam overlap is no longer perfect (in contrast to the case of single signal beam illustrated in Fig. 16). This imperfect overlap of the pump and signal pattern at the Fourier plane may reduce the efficiency of energy transfer. In addition nonuniform amplification in photorefractive two-wave mixing involving these interference structures in the Fourier domain can generate undesirable crosstalk noise at the output plane. This crosstalk noise can be reduced by moving the crystal slightly away from the exact Fourier plane. Theoretical (or experimental) results for the optimum position of the crystal along the optical axis are not yet available. An analysis of the crosstalk noise due to nonuniform amplification at the Fourier plane and the preliminary experimental verification can be found in Ref. 46.

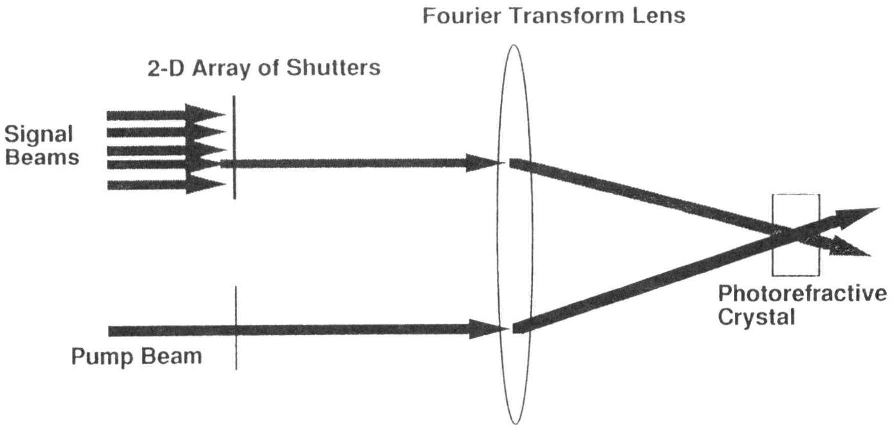

Fig. 16. Beam overlap in photorefractive crystal due to the shift-invariance property of Fourier transformation.

4.2.2.2 Crosstalk Noise Due To Beam Fanning

In the photorefractive crossbar interconnect, the phenomenon of beam fanning[45] (due to coherent amplification of the scattered beams by the main beam) has to be minimized to avoid intolerable crosstalk. Although several techniques to reduce the beam fanning effect have been reported,[47,48] they may not be appropriate for the optical crossbar, and the issue requires further studies. Since beam fanning is essentially along the horizontal direction, however, its contribution to crosstalk can be significantly reduced by choosing an appropriate configuration where the linear array of detectors is oriented vertically. A qualitative explanation is given below.

In a conventional optical crossbar, one can use a horizontal (or vertical) array of sources with a vertical (or horizontal) array of detectors as illustrated in Fig. 11 or vice versa. The two geometries are formally equivalent. This is, however, not true in a photorefractive crossbar interconnect because the contribution of beam fanning to crosstalk noise is much more serious in one geometry than in the other. In the configuration where the array of detectors is horizontal (Fig. 17a), the fanned beams collected by the anamorphic lens are directed into the "wrong" detectors, and thus results in strong crosstalk. In the alternative (orthogonal) geometry (Fig. 17b), the fanning pattern of each beam, which lies essentially in a horizontal plane, is collected by the anamorphic optics and directed into the designated detector. Quantitative assessment of the advantage of the second geometry over the first in terms of crosstalk noise has yet to be determined.

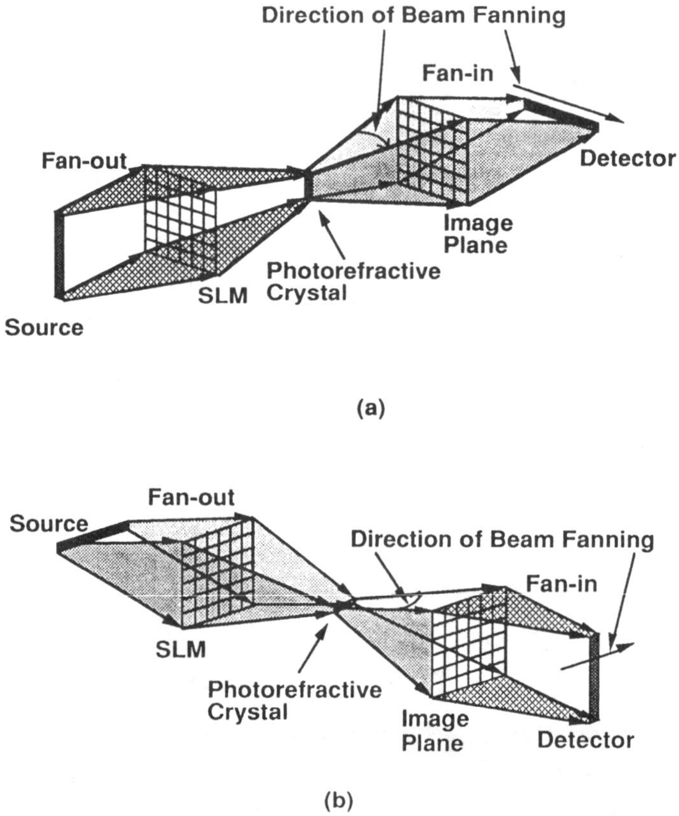

Fig. 17. Effect of beam fanning on signal-to-noise ratio (S/N) in a photorefractive crossbar switch. A much better S/N can be expected from the arrangement shown in (b) with vertically oriented detector array than that of the geometry shown in (a).

4.3 Programmable Interconnection Using Multiplexed Volume Holograms

The conventional recording of a hologram and subsequent recall, as shown in Fig. 18(a) and (b), can be regarded as a fixed one-to-many fanout interconnect because the reading beam is diffracted into many directions prescribed by the object transparency used in the recording process. In other words, a specific interconnection pattern can be recorded as a hologram. Mok recently demonstrated[20] that 5000 holograms can be recorded and fixed in a photorefractive crystal (of about 1 cm x 1 cm x 1 cm) by an angular multiplexing technique and can be selectively retrieved with very little crosstalk. A programmable interconnection scheme where all the desirable interconnection patterns are pre-recorded and fixed in a photorefractive crystal can thus be realized. In addition to the angular multiplexing technique in which each page of input is identified by a unique angular orientation of the reference beam, other multiplexing techniques such as those listed in Fig. 19, and various combinations of these techniques, have also been demonstrated.[49-52]

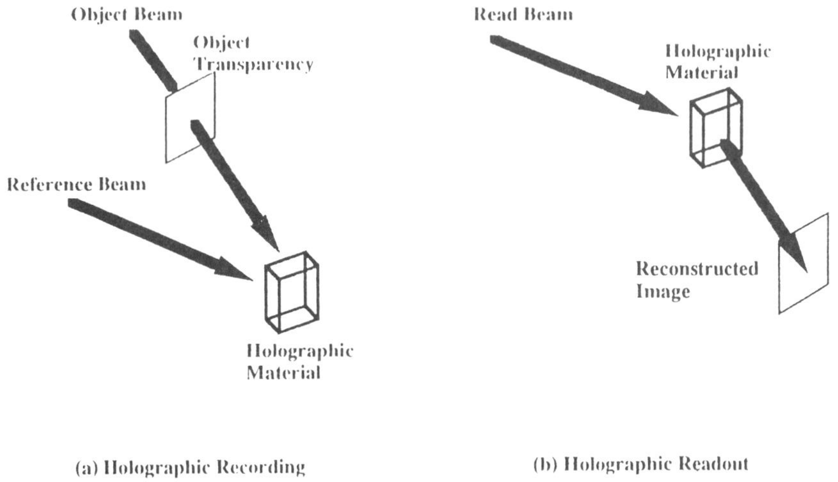

Fig. 18. A conventional holographic process of (a) recording, (b) recall as a 1-to-N fan-out interconnection.

4.4 Optical Interconnection Based on Phase Conjugation

Photorefractive materials are unique in exhibiting efficient phase conjugation at low optical power on the order of a few milliwatts. Many different configurations for self-pumped phase conjugation and mutually-pumped phase conjugation have been demonstrated. Their applications for optical interconnection are discussed below.

4.4.1 Optical Interconnect Based on Self-Pumped Phase Conjugation (SPPC)

A very popular scheme for self-pumped phase conjugation (which is based on total internal reflection at the crystal/air interface to provide the optical feedback),[53] is illustrated schematically in Fig. 20. This simple arrangement has been shown to provide global connectivity as follows.[54] When a coherent beam illuminating an input transparency is weakly focused in a photorefractive crystal at proper orientation, a stable phase conjugate replica of the input object can be observed at the output plane after the

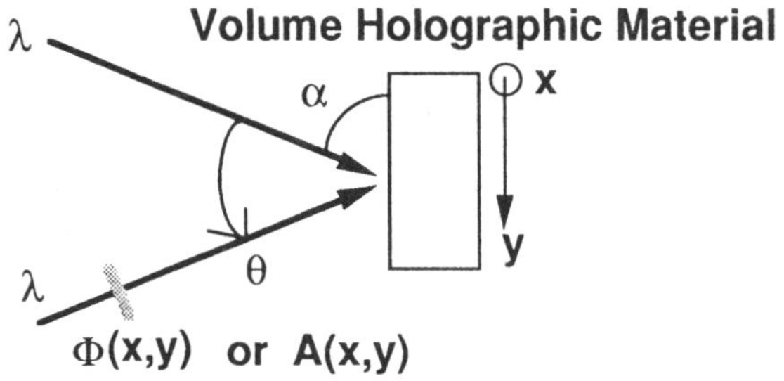

- Spatial Multiplexing: vary (x, y)
- Angular Multiplexing: vary θ and/or α
- Wavelength Multiplexing: vary λ
- Phase Encoding : vary Φ(x,y)
- Amplitude Encode: vary A(x,y)
- Field Multiplexing: vary external electric field

Fig. 19. Various multiplexing techniques in volume hologram.

self-generated gratings reach their steady state. To demonstrate the global connectivity, a two-dimensional array of dots was used as the input pattern. When the input pattern was suddenly switched from the array to any one of the dots from the original array (as shown in Fig. 20), the original output (i.e., the whole array of dots) can be observed within the decay time of the gratings. In other words, the input (i.e., light transmitted through the single dot) is globally connected via diffraction through the self-generated gratings to all the points in the output plane. Global connectivity is a very important aspect of neural networks. Major issues in this approach for neural network applications include the nonuniformity and the difficulty in achieving independent control of the individual interconnection weights.

4.4.2 Optical Interconnect Based on Mutually-Pumped Phase Conjugation (MPPC)

An optical interconnection scheme based on mutually-pumped phase conjugation is shown in Fig. 21 in which a beacon source illuminating a 2-D SLM is used to interfere with an array of sources inside a photorefractive crystal to generate the interconnection holograms.[55,56] The recorded holograms are then used to diffract the signal beams (i.e., the message-bearing beams from the same array) to the receiving channels prescribed by the SLM. In principle, techniques such as time division multiplexing,[57] wavelength division multiplexing, and space division multiplexing can be incorporated to achieve point-to-point interconnection, broadcasting, and crossbar switch.[58]

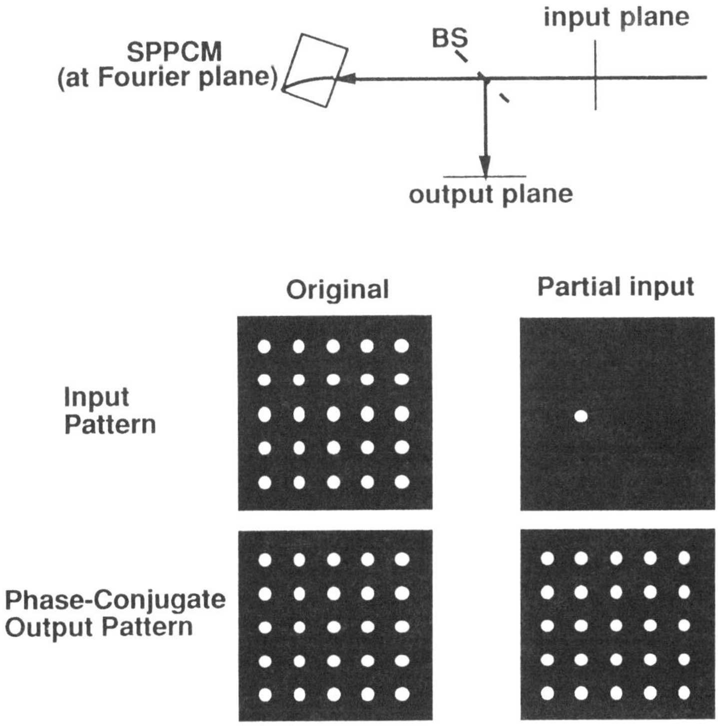

Fig. 20. A demonstration of global interconnection using holograms formed in self-pumped phase conjugation.

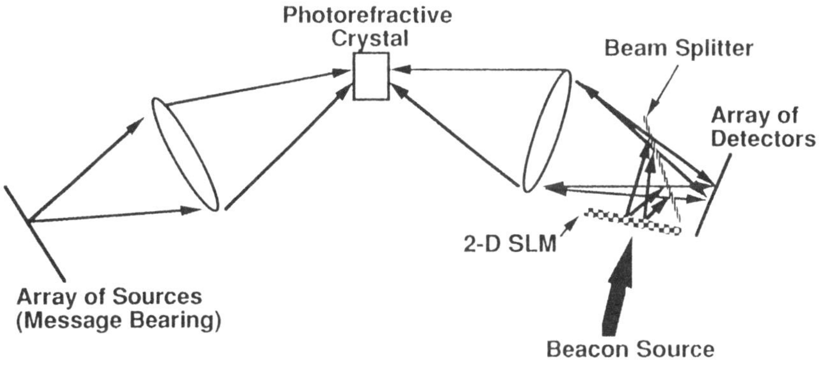

Fig. 21. A schematic illustration of optical interconnect using mutually-pumped phase conjugation.

4.4.3 Applications Based On Aberration Correction and Tracking

The unique features of optical phase conjugation (OPC), such as aberration correction and tracking[22] can be very useful for point-to-point optical links as well as image transmission. In the context of optical interconnects, the former represents the most primitive form of optical interconnect whereas the latter can be regarded as a special form of parallel optical interconnect which is highly structured and regular.

A point-to-point optical link for signal transmission from one station to another remote station can benefit from the aberration correction and tracking properties of OPC as illustrated in Fig. 22.[59,60] Station No. 1 first sends out a broadcasting beacon beam which is accepted by station No. 2. Signals or data to be transmitted are encoded by modulating the amplitude of the retroreflecting beam which is sent back to station No. 1 via OPC. The transmitting medium can be either free space or any guiding structure such as a fiber. Phase aberration introduced by the medium will be corrected as the beacon beam traverses back to station one, provided that the temporal variation of aberration is slower than the response time of the phase conjugator. Note that although the beam is transmitted back and forth (i.e., double pass) through the medium, the signal is transmitted one way (from station 2 to station 1). The tracking nature of OPC also ensures alignment-free operation.

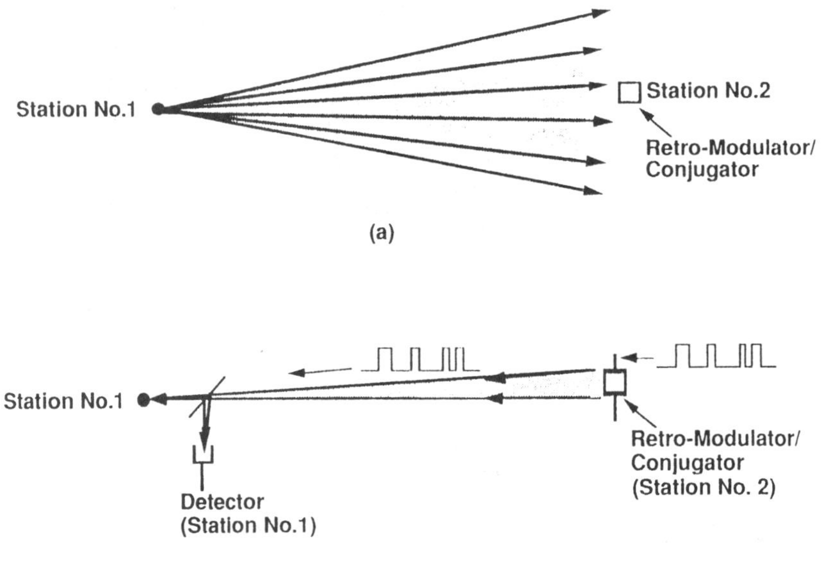

Fig. 22. An optical link with aberration correction and self-tracking capabilities based on a retro-modulator/conjugator.

Parallel image transmission from a transmitting plane to a remote receiving plane through a thick aberrating medium, such as the atmosphere or a multi-mode fiber has been difficult. A solution to this problem was recently demonstrated[61] by angular-multiplexed volume holograms in a photorefractive crystal (Fig. 23). In the recording process, each pixel (i,j) in the input SLM is turned on, one at a time, to allow the transmitted beam to interfere (inside a photorefractive crystal) with a reference plane wave incident at a proper angle Q_{ij} The recorded grating thus encodes the aberration (introduced by the fiber) associated with each input point in terms of the specific orientation of the reference plane wave. When the grating is illuminated by light originating from the same point, it will transform (by diffraction) the incoming beam into a plane wave at a proper orientation such that the resulting plane wave is focused to a corresponding point image at the receiving plane. An experimental result obtained by using this technique is shown in Fig. 23.

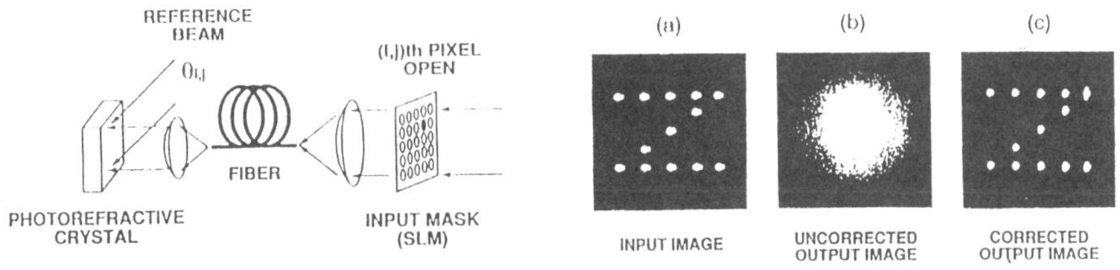

Fig. 23. A technique for one-way image transmission through thick aberrating media. A schematic illustration of the experimental configuration is shown on the left, and the experimental result is shown on the right.

A different approach using a surface emitting laser diode array (SELDA) with wavelength division multiplexing (WDM) technique to achieve parallel image transmission through a single mode fiber can be found in a paper by Paek in this volume.

5. SUMMARY AND CONCLUSION

In this paper, many configurations for optical interconnection using photorefractive holograms are briefly reviewed. The photorefractive phenomena involved in this application include real-time operation, hologram multiplexing, coherent amplification via beam coupling, and optical phase conjugation. The mapping of these phenomena to various interconnection topologies or functions such as beam steering, NxN permutation, 1-to-N^2 fanout, NxN generalized crossbar switch, and optical link or image transmission through a thick aberrating medium or a multimode fiber is discussed.

Although detailed discussions of photorefractive and other real-time holographic materials are beyond the scope of this paper, I conclude this paper with a few lines about some competing materials and technologies. Many photorefractive materials such as strontium barium niobate, lithium niobate, and barium titanate, can now be fabricated in the form of a fiber (or a mini-rod). Assuming that the fabrication technique becomes mature and good quality photorefractive fibers are readily available in many sizes, one has the option of using a photorefractive material either in the form of a single bulk crystal or in the form of a bundle of closely packed fibers. In addition, photorefractive polymers and multiple quantum well (MQW) structures represent two new classes of materials which can be used either in the waveguide mode or in the normal incidence mode. MQW structures are particularly attractive for applications that require extremely small thickness (on the order of or less than a micron). Last, but not the least, bacteriorhodopsin and liquid crystal have also been demonstrated for real-time holography and optical phase conjugation. Although a wide range of materials and technologies, each with its own strengths and weaknesses, is available for concept demonstration of optical interconnects, integration of any one or more of these options (such as bulk vs waveguide or fiber geometry, single crystal vs polymer vs multiple quantum well structures, photorefractive vs bacteriorhodopsin vs liquid crystal) into a practical system has yet to be demonstrated.

6. ACKNOWLEDGMENTS

I thank J. Norman for critical review of this paper. My own research in the photorefractive crossbar switch was supported by the Defense Advanced Research Projects Agency and Air Force Office of Scientific Research under contract No. F49620-90-C-0006.

7. REFERENCES:

1. P. Yeh, *Introduction to Photorefractive Nonlinear Optics* (John Wiley & Sons, Inc., New York, 1993).
2. P. Gunter and J.-P. Huignard, Ed., *Photorefractive Materials and Their Applications I* (Springer-Verlag, Berlin, 1988).
3. P. Gunter and J.-P. Huignard, Ed., *Photorefractive Materials and Their Applications II* (Springer-Verlag, Berlin, 1989).
4. C. Mead, "Potential and limitation of VLSI," in *Topical Meeting on Optical Computing*, 1985 (Optical Society of America, Washington, DC 1985), MA2-1.
5. J.W. Goodman, F.I. Leonberger, S.Y. Kung, and R.A. Athale, "Optical Interconnections for VLSI Systems," Proc. IEEE **72**, 850-866 (1984).
6. J.W. Goodman, "Optics as an Interconnection Technology," in *Optical Processing and Computing*, H.H. Arsenault, T. Szoplik, B. Macukow, Eds. (Academic Press, Inc., Boston, 1989), pp. 1-32.
7. R.G. Zech, "Volume holographic optical memories," Opt. & Photonics News **3**, 16-25 (1992).
8. D. Chen and J.D. Zook, "An overview of optical data storage technology," Proc. IEEE **63**, 1207-1230 (1975).
9. T.D. Hudson and D.A. Gregory, "Optically addressed spatial light modulators," Opt. & Laser Technol. **23**, 297-302 (1991).
10. A.D. Fisher and J.N. Lee, "The current status of two-dimensional spatial light modulator technology," SPIE Proc. **634**, 352-371 (1986).
11. J.A. Neff, R.A. Athale, and S.H. Lee, "Two-dimensional spatial light modulators: a tutorial," Proc. IEEE **78**, 103-132 (1990).
12. H.S. Hinton, "Photonic switching fabrics," IEEE Commun. Mag. 71-89 (1990).
13. W.A. Payne and H.S. Hinton, "Design of lithium niobate based photonic switching systems," IEEE Commun. Magazine **25**, 37-41 (1987).
14. H.S. Hinton, "Photonic switching using directional couplers," IEEE Commun. Magazine **25**, 16-26 (1987).
15. L.B. Aronson and L. Hesselink, "Photorefractive integrated-optical switch arrays in $LiNbO_3$," Opt. Lett. **15**, 30-32 (1990).
16. L.B. Aronson and L. Hesselink, "Integrated-optical switch arrays in GaAs based on electrically controlled dynamic free carrier gratings," Appl. Phys. Lett. **62**, 449-451 (1993).
17. H.J. Caulfield, J. Kinser, and S.K. Rogers, "Optical neural networks," Proc. IEEE **77**, 1573-1583 (1989).
18. D. Psaltis, D. Brady, X.-G. Gu, and S. Lin, "Holography in artificial neural network," Nature **343**, 325-330 (1990).
19. D. Psaltis, D. Bradley, and K. Wagner, "Adaptive optical network using photorefractive crystals," Appl. Opt. **27**, 1752-1759 (1988).
20. F.H. Mok, "Angle-multiplexed storage of 5000 holograms in lithium niobate," Opt. Lett. **18**, 915-917 (1993).

21. N.V. Kukhtarev, V.B. Markov, S.G. Odulov, M.S. Soskin, and V.L. Vinetskii, "Holographic storage in electro-optic crystals. II. beam coupling - light amplification," Ferroelectrics **22**, 961-964 (1979).
22. R.A. Fisher, Ed., *Optical Phase Conjugation* (Academic Press, New York, 1983).
23. A. Yariv, "Phase conjugate optics and real-time holography," IEEE J. Quantum Electron. **QE-14**, 650-660 (1978).
24. J. Feinberg and K.R. MacDonald, "Phase-conjugate mirrors and resonators with photorefractive materials," in *Photorefractive materials and their applications II: Survey of Applications*, P. Gunter, J.-P. Huignard, Eds. (Springer-Verlag, Berlin, 1989), pp. 151-203.
25. P.D. Henshaw, "Configurations for photorefractive laser beam steering," Proc. SPIE **464**, 21-28 (1984).
26. J.-P. Herriau, A. Delboulbe, J.-P. Huignard, G. Roosen, and G. Pauliat, "Optical-beam steering for fiber array using dynamic holography," J. Lightwave Tech. **LT-4**, 905-907 (1986).
27. D. Rak, I. Ledoux, and J.P. Huignard, "Two-wave mixing and energy transfer in $BaTiO_3$ application to laser beam steering," Opt. Commun. **49**, 302-306 (1984).
28. J.W. Goodman, "Fan-in and Fan-out with optical interconnection," Optica Acta **32**, 1489-1496 (1985).
29. A.A. Sawchuk, B.K. Jenkins, C.S. Raghavendra, and A. Varma, "Optical crossbar network," Computer **20**, 50-60 (1987).
30. J.W. Goodman, A.R. Dias, and L.M. Woody, "Fully parallel, high-speed incoherent optical method for performing discrete Fourier transforms," Opt. Lett. **2**, 1-3 (1978).
31. A.R. Dias, R.F. Kalman, J.W. Goodman, and A.A. Sawchuk, "Fiber-optic crossbar switch with broadcast capability," Opt. Eng. **27**, 955-960 (1988).
32. J.-S. Jang, S.-Y. Shin, and S.-Y. Lee, "Parallel N^4 weighted optical interconnections: comments," Appl. Opt. **27**, 4364 (1988).
33. H.J. Caufield, "Parallel N^4 weighted optical interconnections," Appl. Opt. **26**, 4039-4040 (1987).
34. M. Fukui and K. Kitayama, "High-throughput optical image crossbar switch that uses a point light source array," Opt. Lett. **18**, 376-378 (1993).
35. Y. Nitta, *et al.*, "GaAs/AlGaAs optical interconnection chip for neural network," JJAP **28**, L2101-L2103 (1989).
36. P. Yeh, A.E. Chiou, and J. Hong, "Optical Interconnects Using Photorefractive Dynamic Holograms," Appl. Opt. **27**, 2093-2095 (1988).
37. A. Chiou, P. Yeh, S. Campbell, and J. Hong, "Reconfigurable optical interconnection using photorefractive holograms," SPIE Proc. **1151**, 24-32 (1989).
38. A. Chiou and P. Yeh, "Energy efficiency of optical interconnections using photorefractive dynamic holograms," Appl. Opt. **29**, 1111-1117 (1990).
39. A. Chiou and P. Yeh, "2x8 photorefractive reconfigurable interconnect with laser diodes," Appl. Opt. **31**, 5536-5541 (1992).
40. A. Chiou and P. Yeh, "Photorefractive optical interconnect with high-speed reconfiguration," in *Conference on Lasers and Electro-Optics , 1992*, Vol. 12, OSA Technical Digest Series, 1992 (Optical Society of America, Washington, DC), pp.60.
41. A.L. Smirl, K. Bohnert, G.C. Valley, R.A. Mullen, and T.F. Boggess, "Formation, decay, and erasure of photorefractive gratings written in barium titanate by picosecond pulses," J. Opt. Soc. Am. B, **6**, 606-615 (1989).
42. A.L. Smirl, *et al.*, "Picosecond photorefractive effect in $BaTiO_3$," Opt. Lett. **12**, 501-503 (1987).

43. M.H. Garrett, J.Y. Chang, H.P. Jenssen, and C. Warde, "High photorefractive sensitivity in an n-type 45°-cut $BaTiO_3$ crystal," Opt. Lett. **17**, 103-105 (1992).

44. Y. Fainman, E. Klancnik, and S.H. Lee, "Optimal coherent image amplification by two-wave coupling in photorefractive $BaTiO_3$," Opt. Eng. **25**, 228-234 (1986).

45. J. Feinberg, "Asymmetric self-defocusing of an optical beam from the photorefractive effect," J. Opt. Soc. Am. **72**, 46-51 (1982).

46. C. Gu, A. Chiou, and J. Hong, "Crosstalk noise in photorefractive interconnection," Appl. Opt. **32**, 1437-1440 (1993).

47. H. Rajbenbach, A. Delboulbe, and J.-P. Huignard, "Noise suppression in photorefractive image amplifiers," Opt. Lett. **14**, 1275-1277 (1989).

48. W.S. Rabinovich, B.J. Feldman, and G.C. Gilbreath, "Suppression of photorefractive beam fanning using achromatic grating," Opt. Lett. **16**, 1147-1149 (1991).

49. H. Lee and S.K. Jin, "Experimental study of volume holographic interconnects using random patterns," Appl. Phys. Lett. **62**, 2191-2193 (1993).

50. A. Kewitsch, M. Segev, A. Yariv, and R.R. Neurgaonkar, "Electric-field mutiplexing/demultiplexing of volume holograms in photorefractive media," Opt. Lett. **18**, 534-536 (1993).

51. S. Tao, D.R. Selviah, and J.E. Midwinter, "Spatio-angular multiplexed storage of 750 holograms in an Fe-doped lithium niobate crystal," Opt. Lett. **18**, 912-914 (1993).

52. S. Wu, Q. Song, A. Mayers, D.A. Gregory, and F.T. S. Yu, "Reconfigurable interconnections using photorefractive holograms," Appl. Opt. **29**, 1118-1125 (1990).

53. J. Feinberg, "Self-pumped, continuous-wave phase conjugator using internal reflection," Opt. Lett. **7**, 486-488 (1982).

54. Y. Owechko and B.H. Soffer, "Optical interconnection method for neural networks using self-pumped phase-conjugate mirrors," Opt. Lett. **16**, 675-677 (1991).

55. M.P. Schamschula, H.J. Caulfield, and C.M. Verber, "Adaptive optical interconnection," Opt. Lett. **16**, 1421-1423 (1991).

56. B. Fischer and S. Sternklar, "Self Bragg-matched beam steering using the double color pumped photorefractive oscillator," Appl. Phys. Lett. **51**, 74-75 (1987).

57. M. Cronin-Golomb, "Dynamically programmable self-aligning optical interconnect with fan-out and fan-in using self-pumped phase conjugations," Appl. Phys. Lett. **54**, 2189-2191 (1989).

58. S. Weiss, M. Segev, S. Sternklar, and B. Fischer, "Photorefractive dynamic optical interconnects," Appl. Opt. **27**, 3422-3428 (1988).

59. M. Minden, D. Pepper, and M.B. Klein, "Self-pumped phase conjugator/retromodulator using InP:Fe at 1.3mm," in *Conference on Lasers and Electro-Optics , 1992*, Vol. 12, OSA Technical Digest Series, 1992 (Optical Society of America, Washington, DC), pp.58-60.

60. M. Minden, R.A. Mullen, and D. Pepper, "Coherent communication through water using a photorefractive retromodulator/conjugator," in *Conference on Lasers and Electro-Optics , 1990,* Vol. 7, OSA Technical Digest Series, 1990 (Optical Society of America, Washington, DC), pp.506.

61. T.Y. Chang and J. Hong, "One-way image transmission through thick aberrating media using volume holography," in *Conference on Lasers and Electro-Optics , 1993,* Vol. 11, OSA Technical Digest Series, 1993 (Optical Society of America, Washington, DC), pp. 630-632.

Optical neural networks using photorefractive crystals and microlasers

Eung Gi Paek

Bellcore, Red Bank, New Jersey 07701

ABSTRACT

Holographic implementations of neural networks using photorefractive crystals and vertical-cavity surface-emitting microlaser arrays are described. This paper concentrates on the nonlinear thresholding elements (optical neurons), an associative memory using time-division multiplexing, and image transmission through a single mode fiber.

1. INTRODUCTION

Over the past ten years, many optical neural network systems have been demonstrated.[1] Especially, holographic systems employing photorefractive crystals allow high-capacity adaptive neural networks. However, even with its excellent match with neural networks and the immense computational capability due to the wave nature of optics, the optical systems require bulky components and critical alignment, making their practical application difficult.

Recent advances in vertical-cavity surface-emitting microlaser diode arrays (VC-SELDAs)[2-7] can alleviate many of these problems, making the realization of compact and robust holographic neural networks feasible. Some of the basic elements of a neural network system - pattern recognition and holographic memory readout parts - have been demonstrated using the devices.[8,9] These implementations demonstrate the advantages of the SELDA's over conventional lasers in implementing compact and robust systems.

In this paper, the rest of a neural network - nonlinear thresholding elements, an associative memory using TDM (time-division multiplexing) and a communication network using WDM (wavelength-division multiplexing) - using the SELDA will be described.

2. SURFACE-EMITTING LASER DIODE ARRAYS (SELDA's)

The SELDA's (shown in Figure 1) have many features that make them highly desirable for use in optical neural networks and information processing. Some of the features include:

First, the individual lasers can be as small as a few microns (often called "microlasers"). Also, the cavity of a VC-SELDA is formed along the direction normal to a wafer surface, as the name "VC (vertical cavity)" implies. Therefore, a high density stacking of over one million microlasers on a 1 cm^2 chip in a two-dimensional configuration is possible. In addition, the threshold current of the laser is low, near 1 mA for lasers less than 5 μm in diameter, due to the small active volume of a SEL (surface-emitting laser). Recently, a microlaser with threshold current as low as 1 μA has been predicted.[10] This low threshold current will allow operation of many lasers at the same time without significant power consumption.

Second, the wavelength of the light from the microlaser ranges discretely from 0.7 - 1.5 μm, with a spectral linewidth of about 0.001 nm due to its small cavity length that supports only one longitudinal laser mode, as well as the high wavelength selection property of the distributed Bragg reflector (DBR) mirrors. Recent measurements show the spectral linewidth of 50 MHz, corresponding to a coherence length of a few meters (longer than that of a HeNe laser) and the spectral resolution of several millions. Therefore, a hologram of a high quality image can be recorded and reconstructed by using a microlaser. However, the microlasers are mutually incoherent meaning that phases of the light from each of the microlasers are not locked with each other.

An electrically pumped visible SELDA has been recently demonstrated.[11] As shown in Figure 2, the device uses InAlGaP as an active medium and laser light is emitted from the top surface through AlAs/AlGaAs DBR mirrors. The wavelength of the SELDA ranges between 639-661 nm, depending on temperature and it operates at room temperature (300 °K) in a pulsed mode with peak optical power of about 3.5 mW. However, duty cycle of the device can be as high as 40 %, showing promise for future CW operation. Such visible operation of a SELDA is extremely important for optical neural networks because most of holographic recording materials including photorefractive crystals are sensitive only in the visible wavelength range. Moreover, the shorter wavelength in an optical memory allows the larger storage capacity because each bit takes a space of λ^2 and λ^3 (λ is wavelength) for plane and volume recording materials, respectively.

Third, the output light power from each microlaser is bright enough (typically, several mW, CW) to reconstruct a hologram which can be easily detected by a normal 2-D image sensor. Previously, SEL's have suffered from low power levels and low electrical efficiency, compared with edge-emitting lasers. Recently, a group at UCSB demonstrated a SEL that produces CW output of more than 100 mW, a tenfold increase over previous records with wall-plug efficiency of up to 20 % - equivalent to edge-emitting devices.[12]

Finally, the switching speed of the lasers is very high (much less than one billionth of a second) and contrast is extremely high since a laser generates no light when it is off. Besides, heat sinking, matrix addressing (shown in Figure 3)[13], and integration with silicon electronics using flip-chip bonding are the areas to be improved in the future before the devices become practical.

In the following, some of the applications of the SELDA for nonlinear thresholding and decision making, a TDM-based associative memory and WDM-based image transmission through a single mode fiber will be described.

3. NONLINEAR THRESHOLDING ELEMENTS ("NEURONS")

Recently, an integrated vertical-cavity surface-emitting microlasers with GaAs/AlGaAs heterojunction phototransistors (HPT) to yield optically controlled lasers has been demonstrated.[14] As shown in Fig. 4 (a), incident light is absorbed by the HPT which multiplies the photocurrent to exceed the laser threshold current. The output laser light is emitted along the same direction through the substrate. Figure 4 (b) shows the input vs. output optical power relationship for various bias voltages. The device displays a nonlinear sigmoidal thresholding function that is useful in implementing multilayer perceptron networks. The threshold input optical power of the device is 0.1 mW (for the wavelength of 855 nm) and optical gain of approximately 5 may be obtained. Therefore, it can be used as an efficient optical neuron.

Besides its nonlinear thresholding function, the output light from the optical neuron is coherent. It's a laser! Therefore, it can reconstruct holograms. In Fig. 5 (a), the neuron with input power above threshold is "fired" to reconstruct the corresponding memory recorded in a three dimensional photorefractive crystal. Figure 5 (b) shows an experimental result. If the input light signal is below threshold, nothing appears in the output plane (left). On the other hand, as soon as input light is increased above threshold level, the neuron is fired and the corresponding holographic memory is retrieved, due to the phase matching selection property of a volume hologram (right).

4. A HOLOGRAPHIC ASSOCIATIVE MEMORY USING TDM

Previously, we have shown that the holographic memory using a SELDA is capable of retrieving a 2-D information with about 10^5 bits within one nanosecond with the total storage capacity of around 10^{10} bits.[8] However, the question is how to detect such an ultrafast information, 10^{14} bits/sec or 100 terabits /sec. This is difficult first because the speed of current 2-D image sensor is much slower than 1 Giga frames/sec and second, the light emitted from each microlaser for only 1 nsec has to be distributed over a 2-D image, raising a light budget problem.

However, whether one can detect or not, the retrieved information exists in the output plane. So the next question is how to utilize the fast information. Figure 6 shows one possible example to utilize the ultrafast information using currently available devices. In the Figure, an input transparency is placed on the plane where memories are retrieved and the light passing through the input is collected by a single detector using a lens to obtain the inner product of the input and the retrieved memory. The speed of the single detector normally matches that of a microlaser. Furthermore, since a significant portion of the light from a microlaser is focused to the single detector, one can be free from the serious light budget problem.

Such a TDM system can be extended to an associative memory by adding a feedback loop as in the Figure. Assuming that a whole encyclopedia is recorded in the volume holographic memory and a partial information on one of the pages is presented in the input plane, as each microlaser is turned on one by one, the corresponding page is retrieved and is instantly compared with the input. When the memory that matches the input appears, the inner product value reaches above a certain threshold level, and a control signal latches the microlaser. The light from the latched microlaser reconstructs the complete information on the output plane. In this way, advantages of optics can be fully utilized with currently available optoelectronic devices.

5. 2-D IMAGE TRANSMISSION THROUGH A SINGLE-MODE FIBER (2-D WDM)

Although the 2-D coherent optical processor is capable of performing correlation and convolution at the speed of light, it is difficult to preserve these merits if one wants to interface (or cascade) it with other systems (or layers) because this normally requires a parallel-to-serial electronic conversion that negates the advantages of optics.

As a means of alleviating this problem, an all optical 2-D image transmission through a single-mode fiber is presented.[15] The communication link is conducted in parallel between 2-D optical processors. To provide the necessary spatial information within a single mode fiber, WDM is employed. This is another big advantage of optics -allowing the use of multiple colors in a system. Such a WDM is becoming increasingly important nowadays due to the EDFA's (erbium doped fiber amplifiers) which can amplify light signals without losing color information.

In Figure 7 (a), an input image f(x,y) is encoded to spectral information, f(λ), and is transmitted through a single-mode fiber. In the decoding process, the spectral information is retransformed back to spatial information, f(x,y). This 2-D WDM involves two difficult problems : first, how to efficiently couple all the light from the input image into a very small single-mode fiber and second, how to decode the spectral information to reconstruct the original 2-D image.

These problems can be tackled by combining a 2-D multicolor surface-emitting laser diode array (MC-SELDA)[16] and a volume hologram as in Fig. 7 (middle and bottom). The 2-D MC-SELDA is a 2-D array of microlasers, each with its own unique wavelength. In this system, efficient light coupling is achieved by an in-situ self-aligned light path provided by holography. In the decoding process, the spectrally transformed signal f(λ) from the single-mode fiber is diffracted by the same volume hologram and is redistributed to its original position to reconstruct the original image f(x,y). A volume grating responds to its own specific wavelength, due to the phase matching selection property. Therefore, each wavelength of light can pass through a volume hologram as if there were only one grating created by itself, without being affected by other gratings. Moreover, diffraction efficiency of a volume grating is ideally close to 100 % and is independent of the number of gratings within the available index of the recording material. These advantages permit high coupling efficiency in encoding and a clear image reconstruction in decoding. Also, the system can cover a wide range of wavelengths, each of which does not need to be uniformly separated.

Figure 8 shows how each of the gratings recorded as above can be independent, ensuring no crosstalk noise. For the same reference beam direction, R_1 and R_2, the grating G_1 on k_1 surface which was recorded using a wavelength λ_1, cannot sit on the surface of the k_2 sphere. Therefore, the light signal with a wavelength λ_2 is not affected by the grating G_1, avoiding the crosstalk problem..

A monolithic 2-D MC-SELDA was recently demonstrated.[16] The array has 7X20 (total 140) microlasers as shown in Figure 9. Each laser has its unique wavelength that is uniformly separated from its neighbors by 0.3 nm, spanning the total wavelength of 43 nm. Such a wavelength variation was obtained by varying the laser cavity lengths when growing the wafers as in Fig. 9 (b). By simulating such a 2-D MC-SELDA using an Argon laser and an array of silicon mounted v-grooves, transmission and reception of a simple 2-D image through a single-mode has been successfully demonstrated over one mile.[15]

6. CONCLUSION

We have shown that vertical-cavity SELDA's allow a variety of new multiplexing (TDM and WDM) schemes in a system that can be useful in implementing holographic neural networks. Also, an array of optical neurons that can perform nonlinear decisions, are used to demonstrate a selective data retrieval capability from a three dimensional volume hologram.

However, there still remain a few problems that need to be addressed before the SELDA's become practical. Some examples are lower threshold current, heat dissipation and packaging, and visible wavelength operation to permit recording holograms.

One of the biggest bottlenecks in these holographic neural networks is the photorefractive crystal. This technology has developed incrementally for the past 20 years. The requirements include : high sensitivity to allow real-time recording of multiple volume holograms using a microlaser, rapid permanent storage[17] and erasure for reuse, large refractive index change to allow large storage capacity and efficiency, ideal linear recording without decay and high optical quality without scattering noise. When all these requirements on both photorefractive crystals and SELDA's are satisfied in the near future (within three to five years), optical signal processing and neural networks will be very promising in the future even with strong competition from exponentially growing electronics.

7. ACKNOWLEDGMENTS

The author would like to express his appreciation to many colleagues : P.F. Liao and P.W. Smith for encouraging him to initiate investigating the possible use of the SELDA's; J. Jewell (Photonic Research Inc.), Y.H. Lee (now at KAIST in Korea) and A. Scherer (Caltech) for providing him one of their first SELDA's that was used for holographic memory experiments; W. Chan for an array of optical neurons; C.E. Zah, K.W. Cheung, K. Rastani, S. Etemad, L. Curtis and C. Chang-Hasnain (presently at Stanford) for their stimulating discussions and collaborations; A. Von Lehemn and J. Wullert for an independently addressable SELDA and its driver circuit that were essential in implementing the incoherent correlator; H.J. Yoo (Hyundai in Korea), M. Jain (Stanford), J. Harbison, L. Florez, N. Stoffel, J. I. Song and M. Orenstein (Technion, Israel) for providing him various SELDA devices that were used in these experiments.

8. REFERENCES

1. For instance, see the recent special issue of Applied Optics on Optical Implementations of Neural Networks edited by K. Wagner and D. Psaltis, vol. 32, no. 8, 1993.

2. J. L. Jewell, A. Scherer, S. L. McCall, Y. H. Lee, J. P. Harbison, and L. T. Florez, "Low threshold electrically-pumped vertical cavity surface emitting microlasers," Electron. Lett., 25, 1123-1124, 1989.

3. Y. H. Lee, J. L. Jewell, A. Scherer, S. L. McCall, J. P. Harbison, and L. T. Florez, "High efficiency (1.2 mW/mA) top surface-emitting GaAs quantum well lasers," Electron. Lett., 26, 1990, 1308-1309.

4. J. L. Jewell, J.P. Harbison and A. Scherer, "Microlasers," Sci. Amer., Nov., 86-94, 1991.

5. R. E. Slusher, "Semiconductor Microlasers and their applications," Optics & Photonics News, 4, 8-17, 1993.

6. H. Soda, K. Iga, and Y. Suematsu, "GaInAsP/InP surface emitting injection lasers," Japan. J. Appl. Phys., 18, 2329-2330, 1979.

7. K. Iga, F. Koyama, and S. Kinoshita, "Surface emitting semiconductor lasers," IEEE J. Quant. Electron., 24, 1845-1855, 1988.

8. E. G. Paek, J.R.Wullert, M.Jain, A.VonLehmen, A.Scherer, J.Harbison, L.T.Florez, H.J. Yoo, R. Martin, J.L. Jewell and Y.H. Lee., "Compact and ultrafast holographic memory using a surface-emitting microlaser diode array," Opt. Lett. 15, 341-343, 1990.

9. E. G. Paek, A. Voh Lehmen, J. Wullert and R. Martin, "Compact and robust incoherent holographic correlator using a surface-emitting laser-diode array," Opt. Lett. 16, 937, 1991.

10. K. Iga, "Surface emitting lasers - How small is small and how large is large enough?-," 1993 Technical Digest Series Vol. 10, 318/ITuG1-1, Proc. Integrated Photonics Research held at Palm Springs, CA, March 22-24, 1993.

11. J. A. Lott and R. P. Schneider, :Electrically injected visible *639-661 nm) vertical cavity surface emitting lasers," Electron. Lett. 29, 830-832, 1993.

12. C. T. Troy, "Vertical cavity laser hits new power levels high-temperature operation broadens applications," Photonics Spectra, March, 24,, 1993.

13. M. Orenstein, A. C. Von Lehmen, C. Chang-Hasnain, N. G. Stoffel, J. P. Harbison, and L.T. Florez, "Matrix addressable vertical cavity surface emitting laser array," Electron. Lett. 27, (5), 437-438, 1991.

14. W. K. Chan, et al., "Optically controlled surface-emitting lasers," Appl. Phys. Lett. 58, 2342-2344, 1991.

15. E. G. Paek, C. E. Zah, K. W. Cheung, and L. Curtis, "All-optical image transmission through a single-mode fiber," Opt. Lett. 613-615, 1992.

16. C. J. Chang-Hasnain, J. P. Harbison, C. E. Zah, M. W. Maeda, L. T. Florez, N. G. Stoffel, and T.P. Lee, "Multiple wavelength tunable surface-emitting laser arrays," IEEE J. Quant. Electron. 27 (6), 1368-1376, 1991.

17. Y. Qiao, S. Orlov, and D. Psaltis ,"Electrical fixing of photorefractive holograms in SBN," Opt. Lett. 18, 1004-1006, 1993.

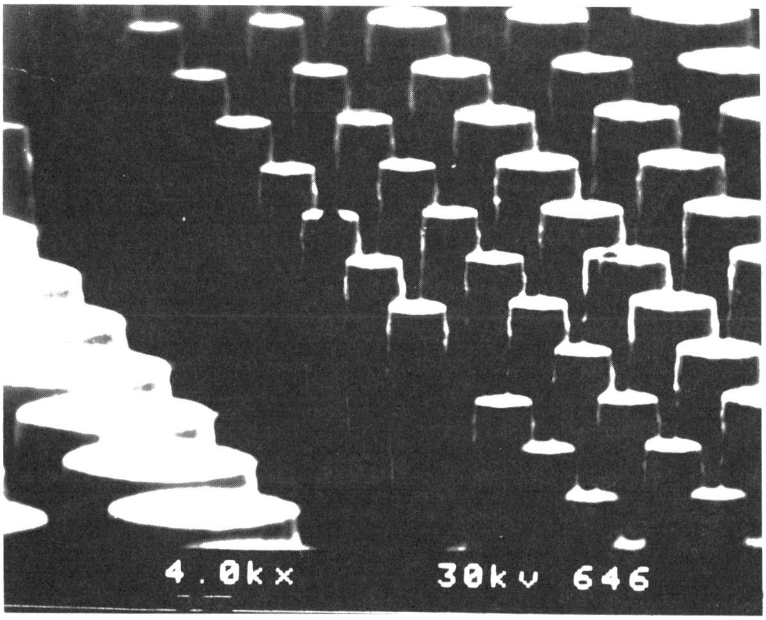

Fig. 1. Scanning electron micrograph of a vertical cavity surface-emitting laser diode array (SELDA)

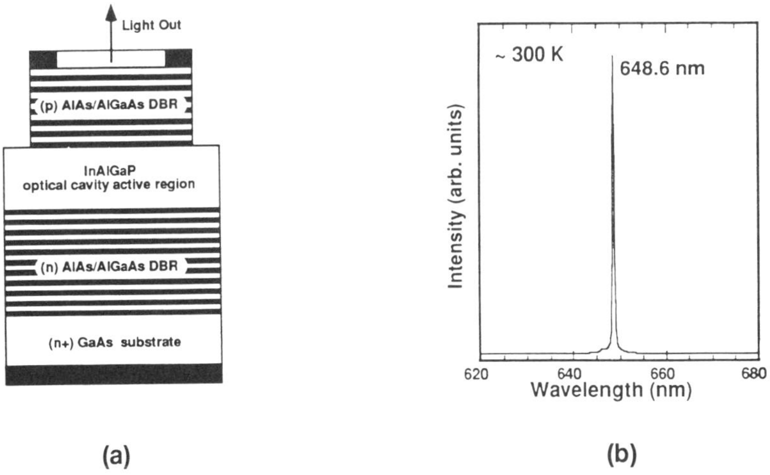

Fig. 2. Visible (639 nm < λ < 661 nm) vertical cavity surface-emitting laser diodes. (a) structure of the device, (b) lasing spectrum. (After J.A. Lott. et. al[11])

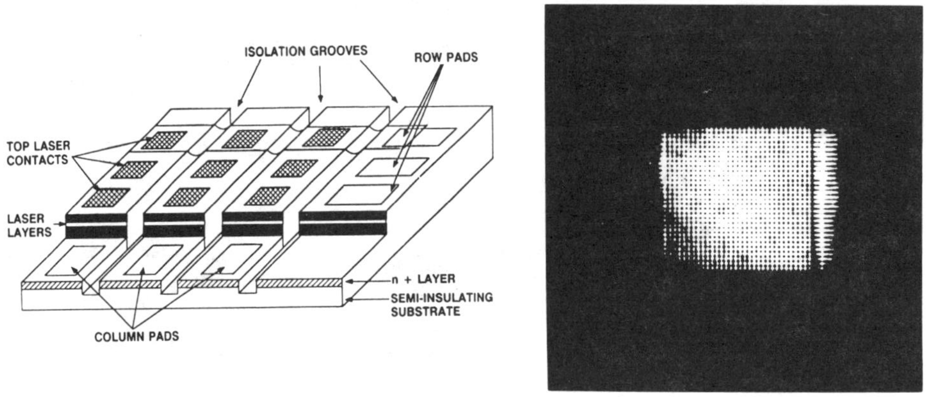

(a) (b)

Fig. 3. An integrated matrix addressable SELDA. (a) structure, (b) devices. (After M. Orenstein et al[13])

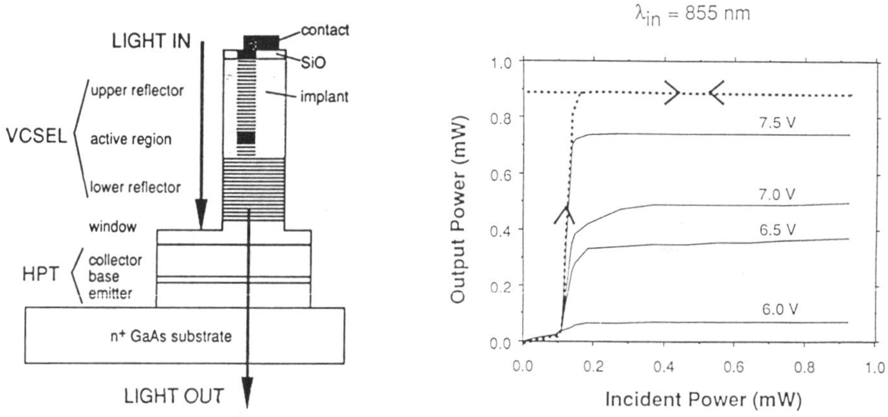

(a) (b)

Fig. 4. An integrated array of optical neurons using microlasers. (a) structure, (b) input vs. output optical power relationship for various bias voltages. (After W. Chan[14])

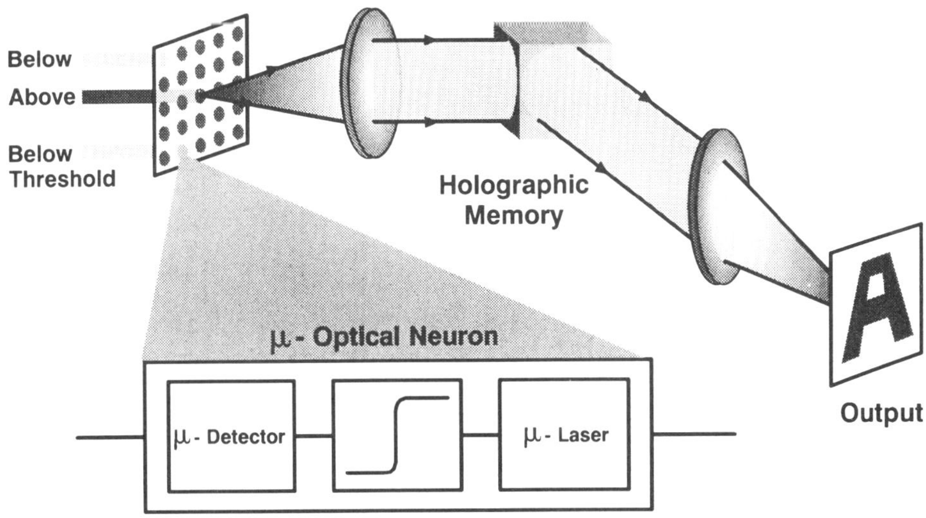

(a)

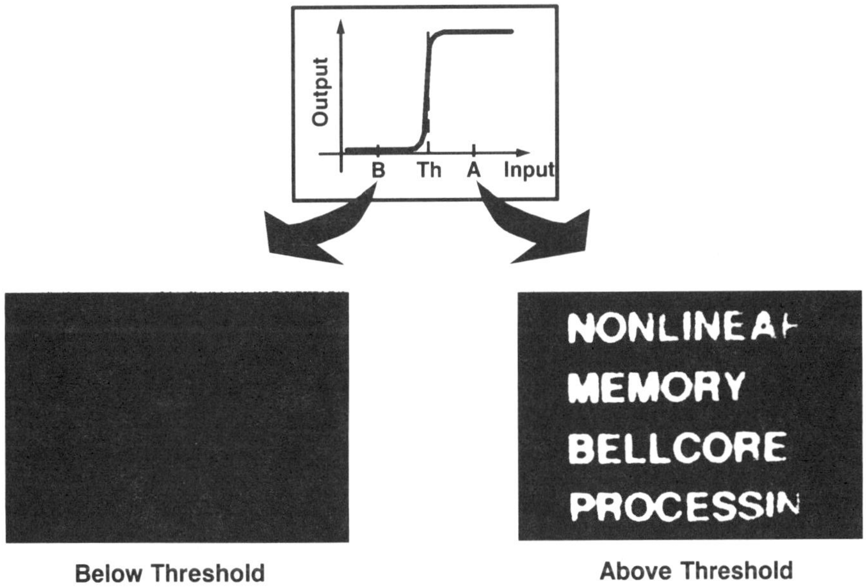

(b)

Fig. 5. (a) An array of optical neurons retrieving corresponding memory from a three dimensional volume hologram; (b) an experimental demonstration.

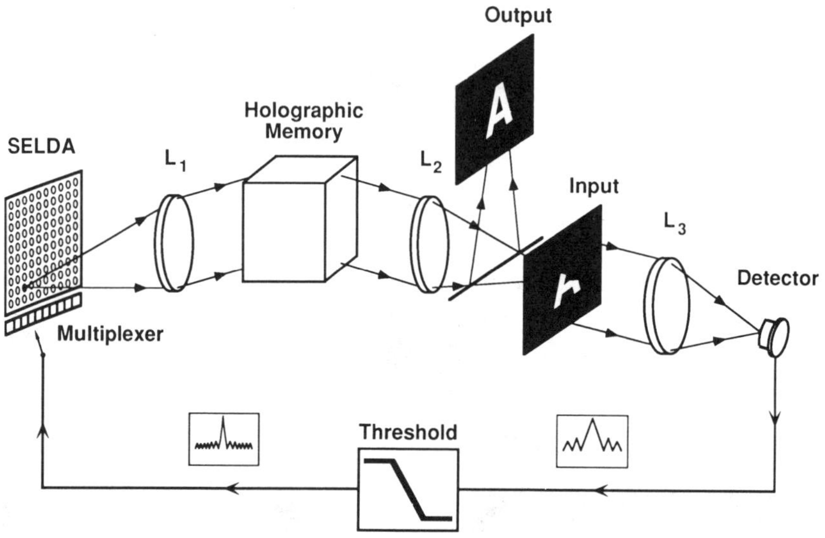

Fig. 6. Holographic associative memory using TDM (time-division multiplexing).

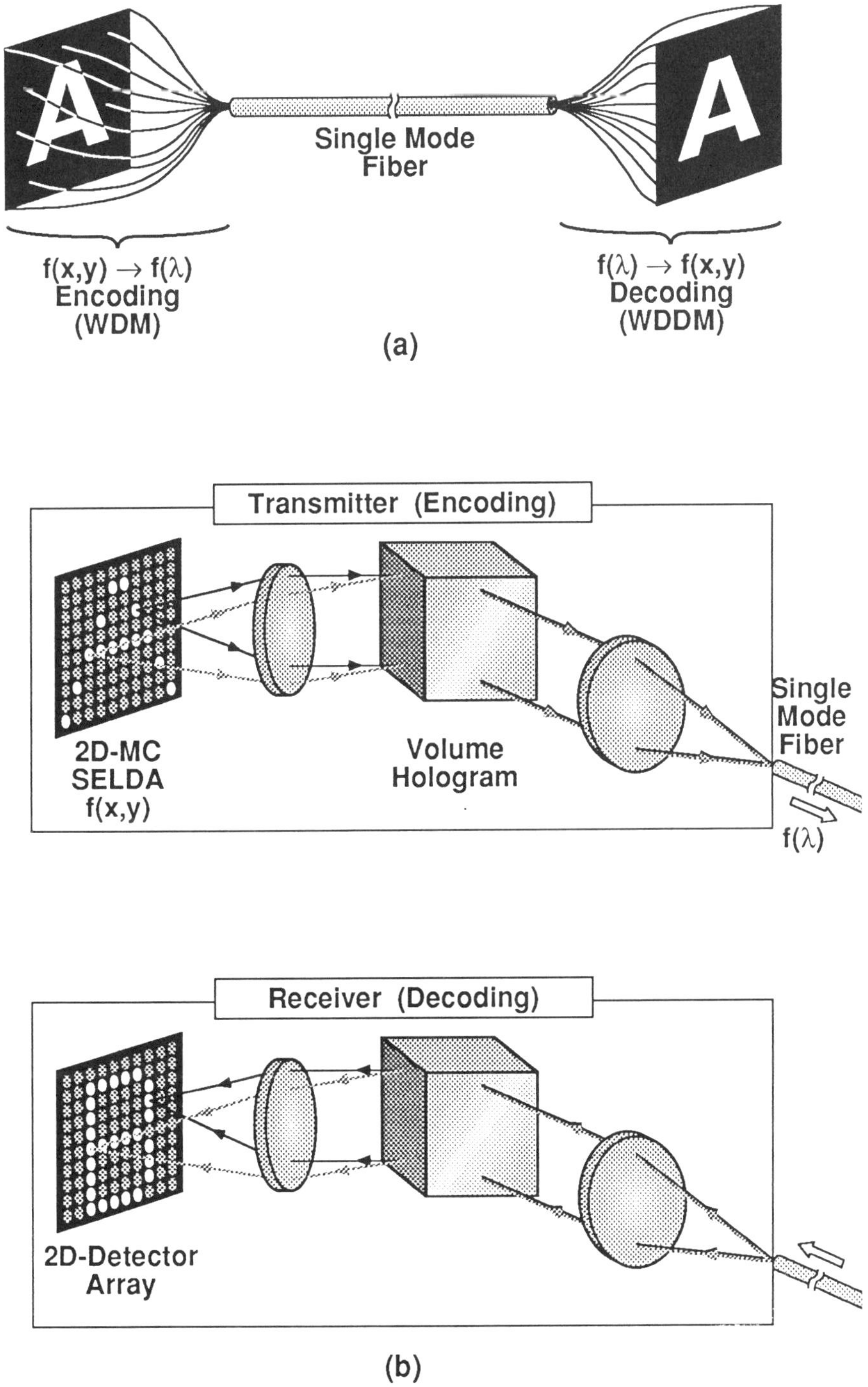

Fig. 7. (a) Image transmission through a single-mode fiber. (b) Both encoding (top) and decoding (bottom) are achieved by combining a 2-D MC-SELDA (two-dimensional multi-color surface-emitting laser diode array) and a volume hologram.

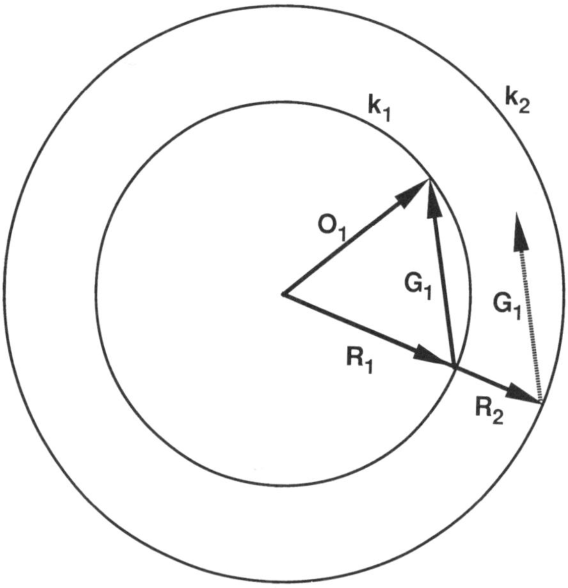

Fig. 8. Crosstalk noise in the 2-D WDM system. The light with wavelength λ_2 is not affected by the grating G_1.

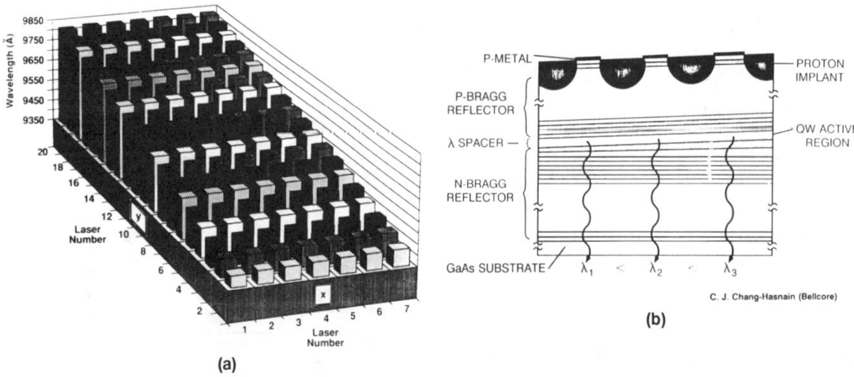

Fig. 9. An integrated 2-D MC-SELDA (two-dimensional multi-color surface-emitting laser diode array) (After C. Chang-Hasnain[16])

Applications of Photorefractors for Signal Processing

John H. Hong
Rockwell International Science Center
1049 Camino dos Rios, A25A
Thousand Oaks, CA 91360

Abstract

Applications of photorefractive devices for radio frequency signal processing problems are reviewed. The problems of signal correlation, filtering and adaptive filtering are addressed using the time integrating property of grating formation in photorefractive crystals. The dynamic range of photorefractive crystals is sufficiently large for critical applications such as adaptive signal processing. New system architectures that use photorefractive two-beam coupling to substantially simplify overall complexity are shown and demonstrated for the problems of tunable notch filtering and adaptive notch filtering.

1. Introduction

Dynamic holography and its associated nonlinear optical phenomena using photorefractive crystals have been investigated for optical information processing applications ranging from volume holographic memory [1,2] to optical neural network implementations [3-5]. The ability to record an efficient hologram in real time without the need for chemical or thermal processing and then to store it for extended periods of time is a unique property of photorefraction which holds promise for information processing tasks.

A characteristic of photorefractive holography that is problematic in some applications is the time-integrative response of photorefractive materials to coherent laser beams that write the holographic grating. This aspect of photorefraction, which derives from the fact that photons must be absorbed to generate the charged carriers that are responsible for setting up the gratings in the crystal, has been used to advantage in applications which require time-integration as a processing tool. In particular, photorefractive crystals have been used in conjunction with acoustooptic devices to implement time-integrating signal correlators for RF (radio frequency) signals [6]. The correlator is a basic signal processing module that can be inserted in a more complex system such as adaptive filters [7,8], which use such devices to estimate the statistics of the signal environment. More recently, a nonlinear optical phenomenon known as two-beam coupling that is associated with photorefraction has been used to greatly simplify the implementation of such adaptive filters [9,10]. In this paper, we present a review of such photorefractive signal processing systems and an important quantitative discussion of the dynamic range of photorefraction which directly affects the overall performance achieveable in photorefractive systems.

2. The Photorefractive Effect

Although a complete description of the photorefractive effect is beyond the scope and intent of this paper, a description of the basic mechanisms that comprise the effect in simple terms will be helpful. The photorefractive effect in electrooptic crystals generally involves photoexcitation of charged carriers from impurity levels and their transport resulting in nonuniform space charge densities whose associated electric fields modulate the index of refraction via the linear electrooptic effect . The complete process is illustrated by the steps shown in Figure 1. First, two coherent optical waves intersect within an electro-optic crystal to form an intensity interference pattern. Electronic carriers are

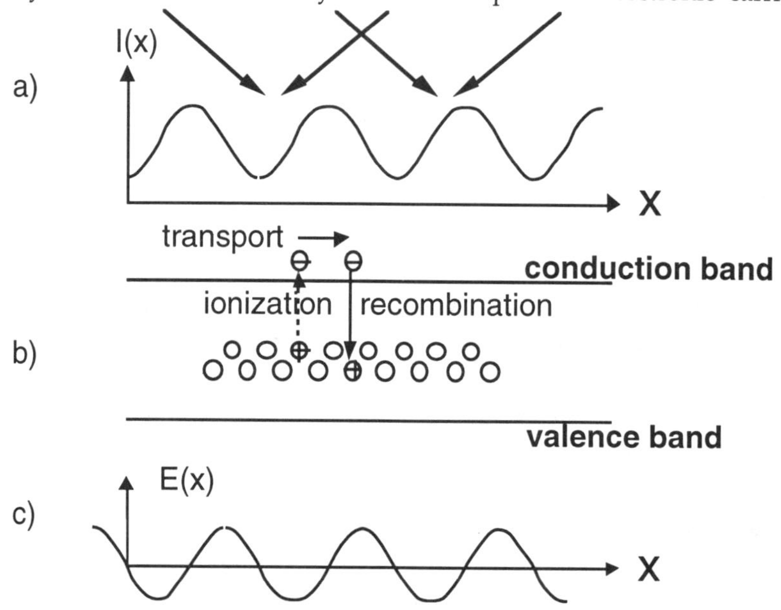

Figure 1.
Hologram Formation in Photorefractive Materials
a) a nonuniform intensity pattern illuminates the crystal, b) charged carriers are photo-excited to the conduction band and are transported via drift and diffusion to be re-trapped at different sites, leaving behind a nonuniform space charge density, c) the electric field associated with the space charge modulates the local refractive index via the electrooptic effect.

produced in the bright fringes, and by drift and diffusion transport mechanisms, they are swept into the dark regions in the crystal to be trapped. This process creates a space charge distribution and its associated electric field that locally modulates the index of refraction via the electro-optic effect to create a holographic phase grating whose amplitude is given by [9]

$$\Delta n(t) = \Delta n_{sat} \int_{-\infty}^{t} \frac{A_1(t')A_2^*(t')e^{(t'-t)/\tau}dt'}{I_0} , \tag{1}$$

where $A_1(t')$, $A_2(t')$ are the amplitudes of the optical waves, $I_0 = |A_1|^2 + |A_2|^2$, and the saturation index amplitude Δn_{sat} is a material dependent parameter. The index perturbation amplitude is a time integrated form of the modulation depth of the interference pattern in the crystal with the time constant given by the approximate relationship

$$\tau \propto \frac{1}{I_0}, \tag{2}$$

where the constant of proportionality is a material dependent parameter. That the time constant is inversely proportional to the total optical intensity is a manifestation of the energy dependent nature of photorefraction: photons must be absorbed to initiate the process.

Because of the large saturation amplitudes Δn_{sat} that can be achieved with ferroelectric crystals such as SBN (Strontium Barium Niobate), $BaTiO_3$ (Barium Titanate) and $LiNbO_3$ (Lithium Niobate), such crystals are preferred for applications requiring large nonlinear optical effects. While the time constants of ferroelectric crystals that are available today lie near the 1 second range at intensity levels of ~ 1 W/cm^2, a simple quantum mechanical calculation yields fundamental limits at least three orders of magnitude faster [10]. Recent breakthroughs in the growth and preparation of reduced ferroelectric crystals [11,12] have yielded results that are coming to within one order of magnitude of this fundamental limit. The temporal characteristics of the optical systems that have been discussed can therefore be expected to improve in proportion.

3. Photorefractive Correlator

The processing of radio frequency (RF) signals has been realized in many forms with the use of acoustooptic deflectors (AOD) to impress the information both spatially and temporally onto optical carriers. For the particular application of signal correlation, time integration techniques using detectors with integrative temporal response have been successful where long integration times are necessary [13]. The conventional approach uses charge-coupled device (CCD) detector arrays where photogenerated charge is accumulated in potential wells formed for each detection element to perform the necessary integration. One of its drawbacks is the undesirable background bias that builds up in each detection element, which limits both the integration time and the dynamic range of the system. These problems can be circumvented by using a photorefractive crystal as a holographic, time-integrating detector [6]. A schematic diagram of that technique is shown in Figure 2; the system consists of two AODs (one for the input signal and one for the reference signal), imaging lenses, a photorefractive crystal and a linear detector array. The diffracted light from the two AODs is imaged onto the photorefractive crystal, each incident at a different angle so that a holographic grating is formed within the crystal. Time-integration is achieved since the hologram forms with a characteristic time constant that is proportional to the incident intensity. Mathematically, the amplitude of the index grating that is formed is given by

$$\Delta n(x,t) = e^{i\frac{2\pi \sin\theta}{\lambda}x} \int_0^t \frac{s_1(t'+x/v)s_2^*(t'-x/v)}{|s_1(t'+x/v)|^2 + |s_2(t'-x/v)|^2}\,dt' \tag{3}$$

where $s_1(t)$ and $s_2(t)$ are the electrical signals driving the two AODs, v is the velocity of acoustic waves travelling in the x-direction inside the AODs, τ is the photorefractive material response time, λ is the wavelength of light and 2θ is the full angle between the two writing beams incident on the crystal. A weak readout beam is Bragg-tuned to read out the hologram, and the diffracted output that carries the correlation information is imaged onto a linear detector array. Since the readout beam is coherent, further optical processing of the ouput can be performed directly if so desired.

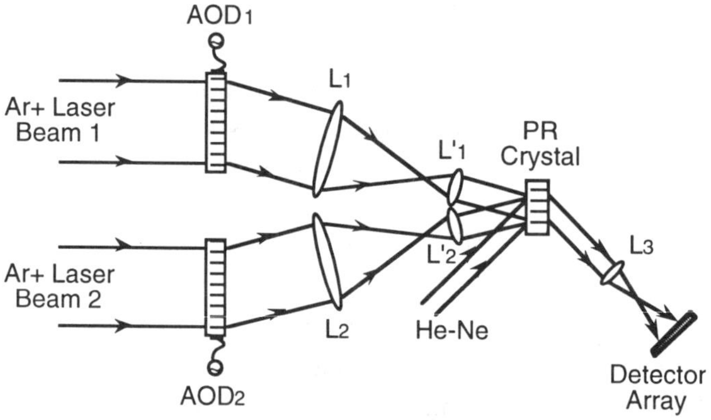

Figure 2
Photorefractive Time Integrating Correlator

An example output of one such time integrating correlator [8] is shown in Figure 3 where the input signal applied to both Bragg cells is a binary pseudo-noise (PN) sequence so that the correlator output is the autocorrelation function of the PN sequence. Figure 3a shows the input signals at 1 MHz and 5 MHz sampling rates, respectively, and the corresponding correlation results are shown in Figure 3b. The complete lack of a background bias in the output correlations should be noted.

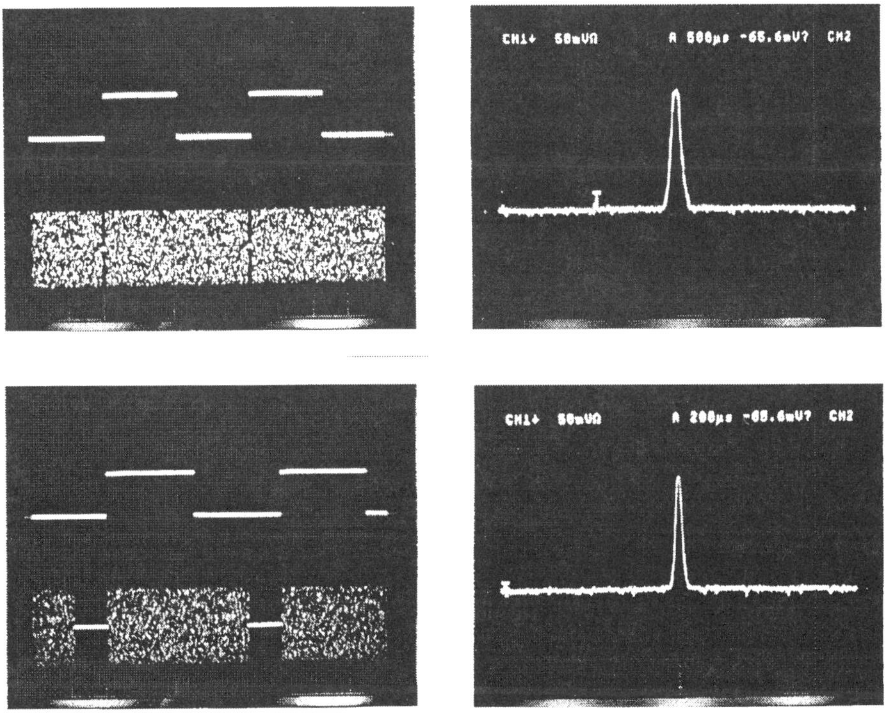

(a) (b)

Figure 3

a) pseudo noise (PN) sequence input signal waveforms (upper at 1 MHz and lower at 5 MHz sampling rates), b) corresponding correlation outputs from photorefractive correlator

4. Adaptive Filter Application

The time integrating correlator just described forms the heart of a more complex system that performs an adaptive filtering task. The adaptive filtering problem is defined in Figure 4 where an input signal is received after being contaminated by an additive jamming noise source. The limited a priori information that is available is that the signal is characterized by a broadband spectrum and the jamming noise consists of a set of narrowband interference waveforms with unknown frequencies. Typically, the input signal-to-noise ratio is very small, the signal being buried in the jamming background. The task for the adaptive filter is to track the interference sources and to suppress the interference noise favoring the broadband signal. A generalized version of this problem exists in the space-time domain for phased array applications [14], but this paper concentrates on the temporal filtering aspect.

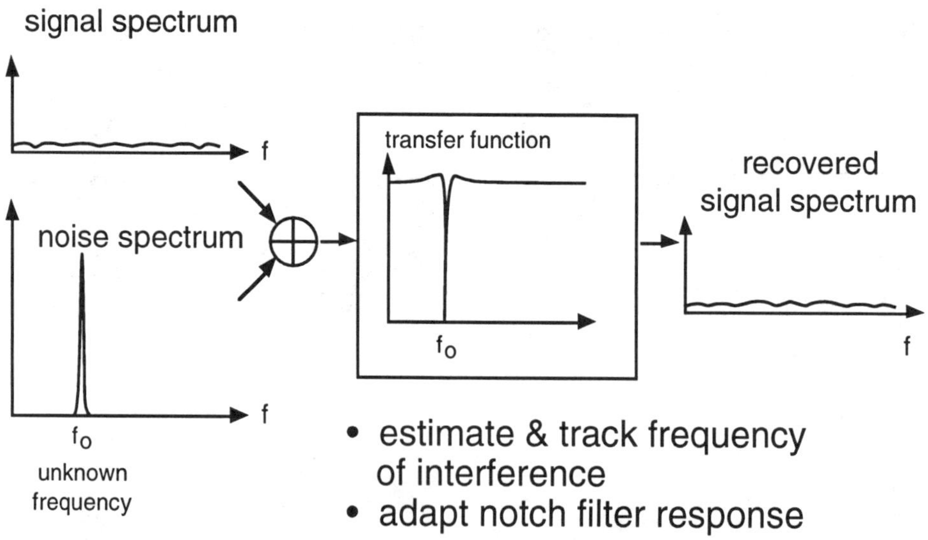

Figure 4
Adaptive Filtering Problem

If the input signal and noise statistics were known, optimum filters could be designed to satisfy a number of optimization constraints yielding systems such as the Wiener filter [15]. Since the adaptive problem does not make such information available a priori, the statistics must be estimated and filters be implemented in real time to approach the performance of such optimum filters. An adaptive system capable of performing this task is described in Figure 5 [16]. The two necessary devices in the implementation of this nonlinear filter are the convolver and correlator, which are placed in the feedback path as shown. The output signal is the difference between the input and the feedback signal, which is an adaptive reconstruction of the interference component of the input. The feedback signal is formed by first computing the correlation function of the output signal and the input signal. The correlation output is then convolved with the input signal to produce the feedback signal of the appropriate phase required for stable nulling of the intereference components in the input signal. The correlation can be performed using the photorefractive approach described in the previous section, while the convolver can be implemented with well-known space-integrating methods using an acoustooptic device and a detector [7].

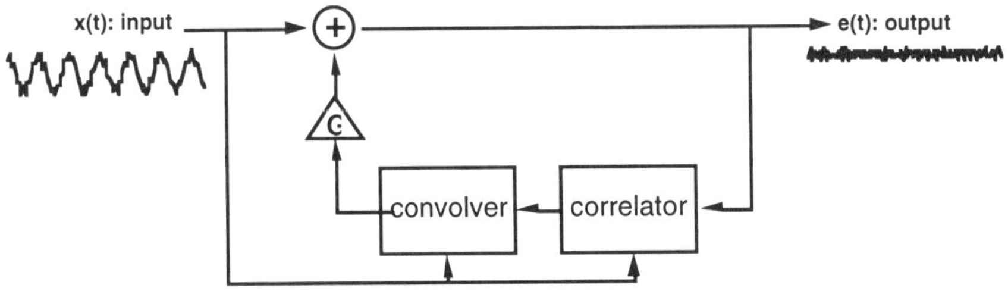

Figure 5
Adaptive Filtering System

When the filter has adapted to a particular set of signal and interference present at its input, the output will consist of a sum of the desired signal that is very nearly unaffected and a suppressed version of the interference. The degree of suppression is sometimes called the null-depth since the overall filter transfer function at this point is basically an all-pass network with deep notches at the intefering frequencies.

Among other factors, the null-depth performance of the adaptive system depends rather critically on the dynamic range characteristics of the devices that are used. Consider, in particular, the role of the photorefractive correlator in limiting the ultimate null-depth performance in the simplified case where the system input consists solely of a narrowband interference. When the system has adapted to the interference input, the correlator must reliably yield the correlation function of its two inputs, which are the system input signal and the system output signal. To yield an output in which the input interference power is suppressed by some factor, the correlator must be able to tolerate a signal power level disparity in its two inputs by the same amount. The correlator then imposes an upper limit on the overall system null-depth performance as a result of its own limited dynamic range. The correlator dynamic range must therefore be quantified both theoretically and experimentally.

5. Dynamic Range of Photorefractive Crystals

We define the dynamic range of a given photorefractive crystal in a holographic apparatus to be the range of power levels over which one writing beam can be varied in relation to the other while writing and maintaining a photorefractive grating in the crystal. Because gratings in photorefractive crystals can be read out nondestructively at the appropriate nonabsorbing wavelengths, the detectability level can be set quite low. This definition is consistent with the definition of the dynamic range of the photorefractive correlator since each input to the correlator modulates a corresponding writing beam incident on the photorefractive crystal as shown in Figure 2.

In reference to Figure 6, which describes a simplified apparatus used to experimentally measure the dynamic range of variuous photorefractors, the dynamic range is defined more precisely to be the range over which the signal intensity I_2 can be varied, keeping the other intensity I_1 constant, while maintaining a detectable grating, the strength of which is quantified by the diffraction efficiency

$$\eta = \frac{I_{diff}}{I_{read}}, \tag{4}$$

where I_{diff} and I_{read} are the diffracted and input reading intensities, respectively.

Figure 7 shows a schematic of the experimental arrangement used to measure the dynamic range of gratings in various photorefractive crystals (M, mirror; BS, beamsplitter; S, shutter; P, calcite polarizer; NDF, neutral density filter; HWP, half-wave plate; F, color filter; B, light baffle; PMT, photomultiplier tube). An argon-ion laser operating at the wavelength λ of 514.5 nm and at the fundamental Gaussian spatial mode is split into two paths by the beamsplitter BS that supplied two writing beams. The writing beams are ordinary polarized to suppress the formation of fanning gratings. The outside full angle between the writing beams was fixed at 37° for all the samples. The power of one writing beam (I_1) was fixed at 35 mW, while the power of the other beam (I_2) was varied over ~ 10 orders of magnitude (35 mW to ~ 7.5 pW) by a combination of calcite polarizers (P_1 and P_1'), a half-wave plate (HWP_1), and a set of neutral density filters (NDF_1). The photorefractive grating was read out by a He-Ne laser operating at λ=632.8 nm (fundamental Gaussian mode), and the diffracted signal was detected by a photomultiplier tube (PMT). The reading beam power was adjusted to maximize the diffracted signal while minimizing the destructive effects of the readout process; the power was typically ~ 200 μW. Also, since the reading beam was extraordinary polarized to maximize the diffraction efficiency, there was a potential complication of the reading beam interacting with its diffracted beam that may result in false diffraction efficiency measurements. This problem was avoided, however, by adjusting the reading beam intensity low enough that such an interaction was not significant. Additional experimental details can be found elsewhere [17].

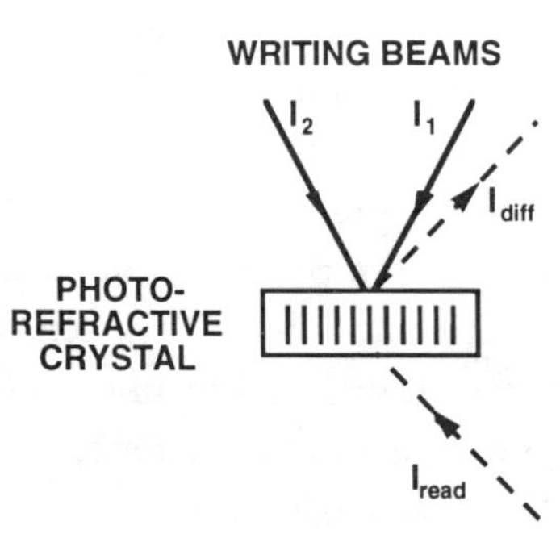

Figure 6 Simplified Holographic Set Up

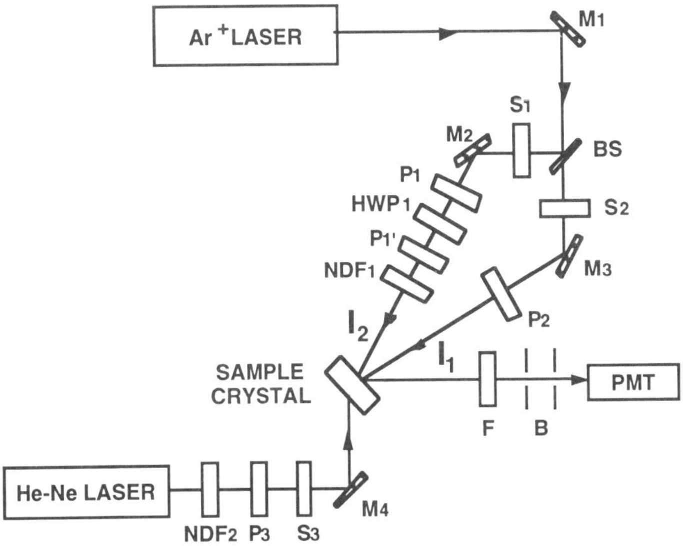

Figure 7

Dynamic Range Measurement Apparatus

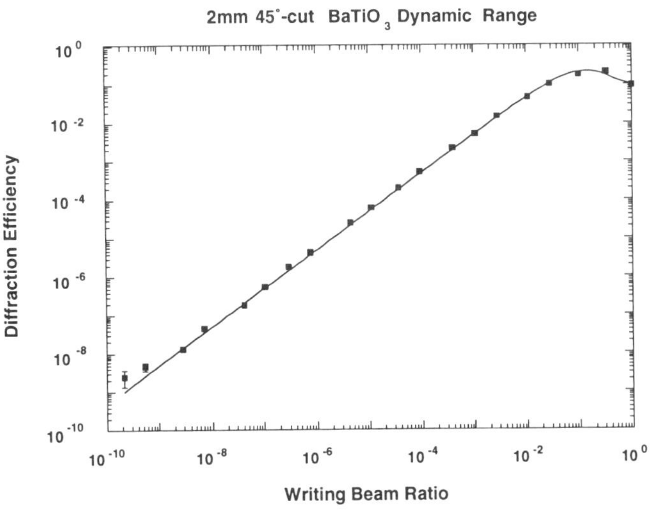

Figure 8

An example experimental result: log-log plot of grating diffraction efficiency vs. writing beam ratio for $BaTiO_3$. The dots are the experimental data while the solid lines are the fits to the Kogelnik's theory given in Eq. (1). As discussed in the text, the observed dynamic range (greater than 90 dB) was limited mainly by the detector noise.

An example experimental result is shown in the log-log plot of the diffraction efficiency versus the writing beam ratio for a 2 mm thick $BaTiO_3$ crystal in Figure

8. The observed experimental data are shown as dots and the solid lines are fits to the Kogelnik's theory [18] of grating diffraction, assuming that the photorefractive grating amplitude is proportional to the modulation depth of the incident fringe pattern; the theory is in good agreement with the observed data. Figure 8 shows the total dynamic range of this crystal to exceed 90dB. Other ferroelectric crystals such as SBN and BSKNN yielded similar results [17]. It is important to note also that the error in the weakest data point corresponding to a writing beam intensity ratio of almost 10^{-10} was dominated by the measurement noise from the laser/detector combination and does not necessarily represent the inherent photorefractive noise so that the dynamic range values that were obtained represent lower bounds [17].

6. Photorefractive Two Beam Coupling

Distinct from the direct approach of using photorefractive devices (i.e., the time integrating correlator) in place of conventional components in established architectures, an emerging class of photorefractive systems incorporates more of the functionality of the filter into optical processes ocurring within the crystal. In particular, we will show how the electronic feedback loop shown in Figure 5, which was used to implement adaptive filtering can be realized within the photorefractive crystal by a unique phenomenon known as photorefractive two-beam coupling. After briefly discussing the coherence selective property of two-beam coupling, we will describe a photorefractive/acoustooptic implementation of a continuously tunable notch filter. We will then proceed to a generalization of this system to address the more complex problem of adaptive notch filtering.

Two-beam coupling is a nonlinear optical process by which energy can be exchanged between two mutually coherent light waves intersecting in a photorefractive medium [9]. This occurs most efficiently in crystals where the dynamic photorefractive index grating is 90° out of phase with respect to the intensity interference grating that produces it. [9] The energy exchange is unidirectional, with the direction of energy flow determined by the crystal parameters such as the crystal orientation and the sign of the photoionized charge carriers. Customarily the beam that loses energy is labeled as the pump and the beam that becomes amplified as the probe. Because the photorefractive effect is a time-integrative mechanism [9] where the index grating is produced at a characteristic rate proportional to the total average incident intensity, the two beams of light must be mutually coherent. The degree of mutual coherence that is required varies with the overall optical intensity and the choice of the crystal. At nominal intensities of 1 W/cm^2, the time constant of photorefractors can vary widely from 10 μsec to 1sec depending on the particular crystal that is used, so that the difference in the optical frequencies of the two interacting beams must be within the corresponding range (100 KHz to 1 Hz) to experience two-beam coupling. This requirement is illustrated in Figure 9 where the first case shows

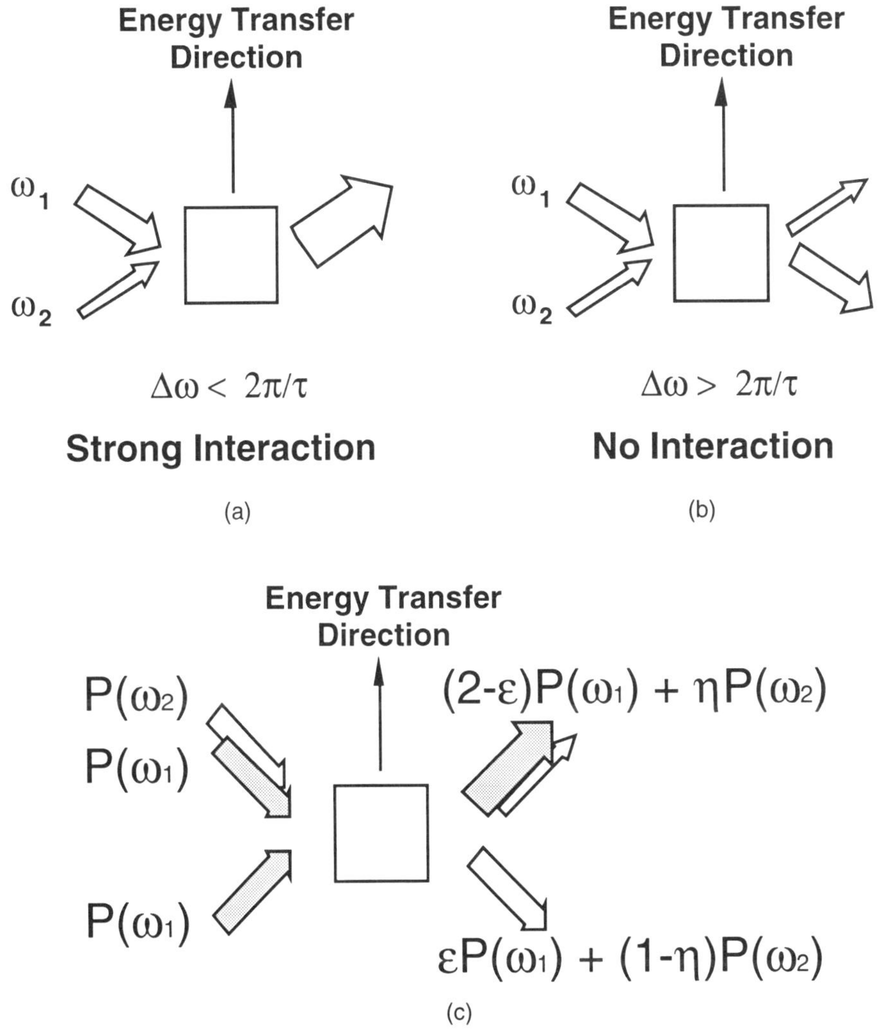

Figure 9
Coherence Selective Two-Beam Coupling
a) coherent case: ω_1-ω_2 $< 1/\tau$, b) incoherent case: ω_1-ω_2 $< 1/\tau$, c) mixed case: ω_1-ω_2 $> 1/\tau$, $\varepsilon << 1$.

energy transfer from the pump beam to the probe while the second shows no energy transfer, the distinction in the two cases being the frequency difference of the probe and pump beams. In the case of mixed coherence shown in Figure 9c, where the probe has one component at frequency ω_1 and the pump has components at both ω_1 and ω_2 where the difference ω_1-ω_2 exceeds the photorefractive response rate, most of the power at the common frequency ω_1 is diverted into the probe output, with the power at ω_2 dividing between the two output ports. The photorefractive crystal then performs a coherence selective function in which components that are common to both probe and pump are diverted into the probe output, while the pump output contains primarily the components that are not present in both beams.

7. Continuously Tunable RF Notch Filter Using Two-Beam Coupling [19]

We are now in a position to describe the system concept that performs the useful function of notch filtering where the notch location is continously tunable (Figure 10). The RF input signal is used to drive the acoustooptic modulator AOM1 which modulates the pump beam (oriented to lose energy in the two-beam coupling interaction for notched-out frequency components). A local reference signal is generated at the desired notch frequency and is used to drive AOM2 which modulates the probe beam (oriented to gain energy from the pump beam). If the input signal does not contain any frequency component near the notch frequency, then the probe and pump beams are mutually incoherent to within the response time of the photorefractive effect in the material and pass through the crystal without any interaction. If, however, the input RF signal contains an intereference term at a frequency close to the notch

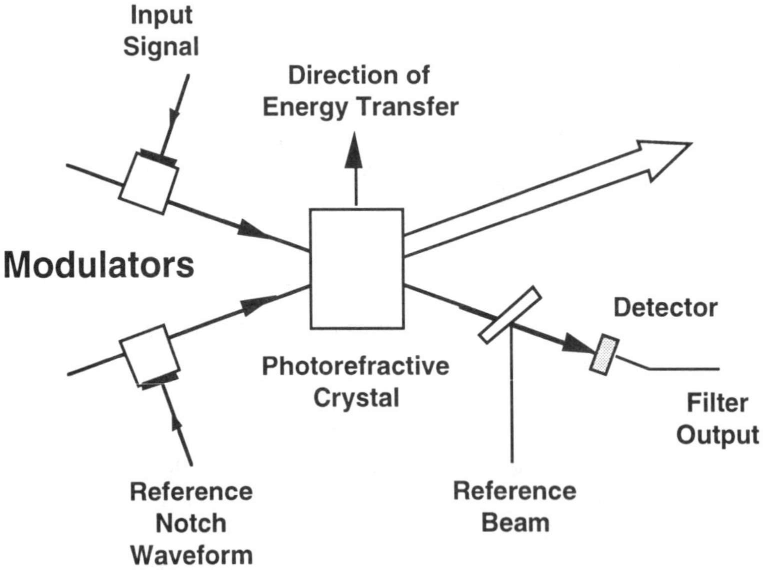

Figure 10
Tunable Notch Filter Concept

frequency (to within the photorefractive response rate) then the energy in that frequency component is transferred from the pump to the probe beam. This process yields an output pump beam (the system output) in which the interference component is strongly depleted.

Figure 11 shows a typical result of the notch filter in operation. This particular experiment [19] utilized a 2 mm thick $BaTiO_3$ crystal as the two-beam coupling medium. In the figure, $P_{sig} \propto I_S(L)$ and $P_{noise} \propto I_I(L)$. The change in the signal and noise components of the probe beam intensity is shown as a function of the presence of the pump in the crystal. As shown in the lower trace, the intensity of the noise component is greatly reduced by virtue of two-beam coupling transferring the energy of the noise component from the pump beam to the probe. In an ideal situation, the intensity of the signal component in the pump beam should not change whether or not the probe beam is present. However, as shown in the top trace of Figure 11, the signal level actually increases when the probe beam is present. This effect can be explained by a stimulated scattering phenomenon, called beam fanning, [5] arising from the signal component of the pump beam. In photorefractive crystals, the same mechanism that is responsible for two-beam coupling can cause an incident beam to spatially fan out, by the process of the incident beam amplifying scattered noise inside the crystal. This beam fanning effect is reduced when a second beam, mutually incoherent with the first, is present in the crystal by reducing the strength (modulation depth) of the fanning beam gratings. Thus, as shown in Figure 11, the intensity of the signal beam transmitted through the crystal is reduced due to beam fanning, but recovers significantly when beam fanning is suppressed with the presence of the probe beam. When the probe beam is turned on, the two components of the pump beam respond within ~ 100 msec, as shown in Figure 11. This response time is determined by the total intensity of the beams, including the pump and the probe, and the gain coefficient of two-beam coupling [20]. However, when the probe is turned off, there is a rather long decay of the signal and noise components of the pump beam. This is another manifestation of beam fanning where the transmitted beam is slowly losing power because the energy is being redirected. The beam fanning decay time in the pump beam when the probe is turned off is much longer than the turn-on time because, in part, the total intensity is lower with the absence of the probe.

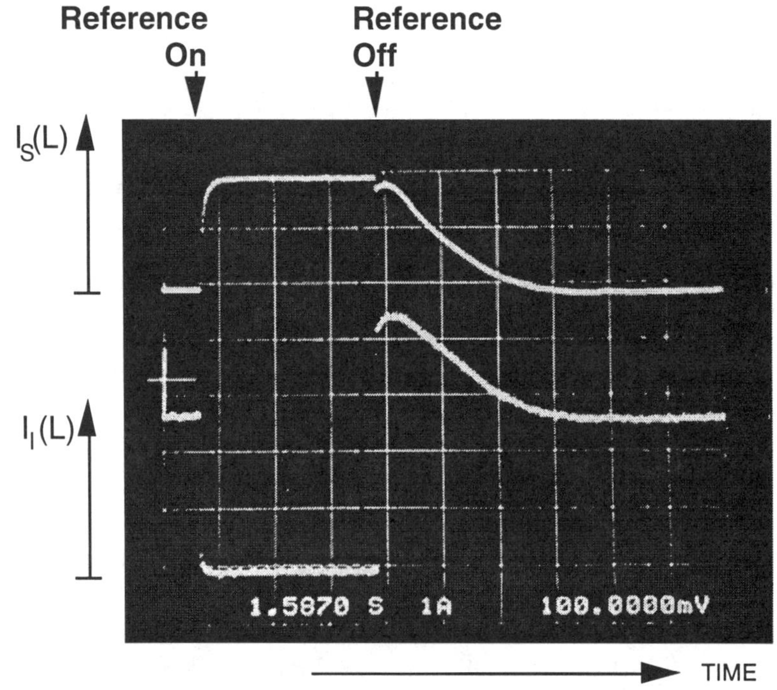

Figure 11
Tunable Notch Filter Experimental Result

8. Adaptive Notch Filter Using Photorefractive Two-Beam Coupling [21]

We can now generalize the notch filtering concept to address the more complex problem of adaptive notch filtering described in Figure 5. The signal processing problem is to suppress strong narrowband interference components in a received signal without knowing a priori the frequency location of the interferer while passing the broadband components. The new method for adaptive RF notch filtering is illustrated in Figure 12. Two mutually coherent laser beams (I_1 and I_2) probe two different locations of an acoustooptic deflector (AOD). The two locations are chosen such that there is an acoustic (temporal) delay T_a between them, and the AOD is driven by a rf signal to be processed. The two diffracted outputs of the AOD are arbitrarily assigned as the probe and pump beams (probe travels toward the direction of energy transfer in Figure 12) for the next stage of photorefractive two-beam coupling. When the two beams are combined in a photorefractive crystal, the components that are mutually coherent will interact to transfer energy from the pump to the probe, while the components that are mutually incoherent will not interact.

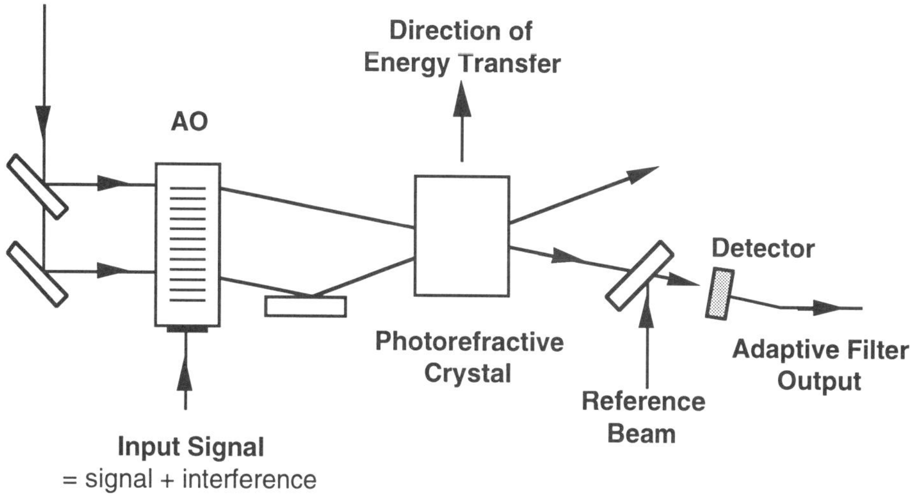

Figure 12
Adaptive Notch Filter Concept

To better clarify the experiment, consider the electrical input signal to the deflector to be $s_{in}(t) = n(t) + b(t)$ where n(t) represents the narrowband interference component and b(t) the broadband signal component. If we choose the acoustic delay T_a and the photorefractive response time τ such that $1/\Delta f_n > T_a$ and $1/\Delta f_n > \tau$ where Δf_n is the bandwidth of n(t), then the narrowband components will directly couple with one another. Likewise, the inequalities $1/\Delta f_b < T_a$ and $1/\Delta f_b < \tau$ where Δf_b is the bandwidth of b(t) will assure that the broadband components do not interact directly one with the other.

Finally, the output of the filter is obtained when the depleted pump beam (I'_{pump}) is heterodyned with a local reference beam. The reference beam is derived from the laser beams (either I_1 or I_2) incident on the AOD. Ideally, the heterodyned signal will recover $s_{in}(t)$, the electrical input signal to the AOD, minus the component that is mutually coherent in I_{pump} and I_{probe}. In practice, however, the output of the filter is given approximately by

$$s_{out}(t) \approx b(t) + \eta b(t-T_a) + \varepsilon_1 n(t) + \varepsilon_2 n(t-T_a) \tag{5}$$

where η is the diffraction efficiency of the photorefractive grating formed by the coherent component of the signal (i.e., the interference term) and the coefficients ε_1 and ε_2 characterize the interference cancellation performance of the system. Although the broadband components do not directly couple with one another, each diffracts off (fully Bragg-matched) the stable grating formed by the narrowband components with an efficiency η. An unwanted echo of the desired signal exists in the output signal (the term with η in Eq. 5), but it can be filtered out electronically since $\eta < 1$ and the delay T_a is fixed and known.

The details of the experiments that were performed can be found elsewhere [21]. We simply give some example test results obtained for the system that was implemented using a flint glass AOD and a single crystal SBN grown at Rockwell Science Center. Our first system test measured the cancellation performance of a sinusoidal interference with no broadband components. This is typically used as a rough measurement of the system performance before broadband components are introduced [14,16]. The interference at 82.4 MHz in frequency was applied to the acoustooptic deflector and the system output (coherently detected pump modulation) was monitored with a spectrum analyzer with the probe blocked (filter inactive) and unblocked (filter active). The inactive and active filter outputs shown respectively in Figures 13(a) and 13(b) demonstrate approximately 40 dB of cancellation. The time response of the system was tested by monitoring the output pump intensity as a function of time when the input (interference only) is suddenly switched on. This time constant will be some monotonically decreasing function of the total average intensity falling on the crystal (the two-beam coupling time constant, *not* the photorefractive response time) [14]. The time response shown in Figure 14 corresponds to approximately 5.5 msec at an average total intensity of 10 W/cm^2. If the interference frequency were suddenly changed (within the system bandwidth governed by the acoustooptic deflector bandwidth), the system automatically adapts to the new frequency at this rate.

SCP-0530-E

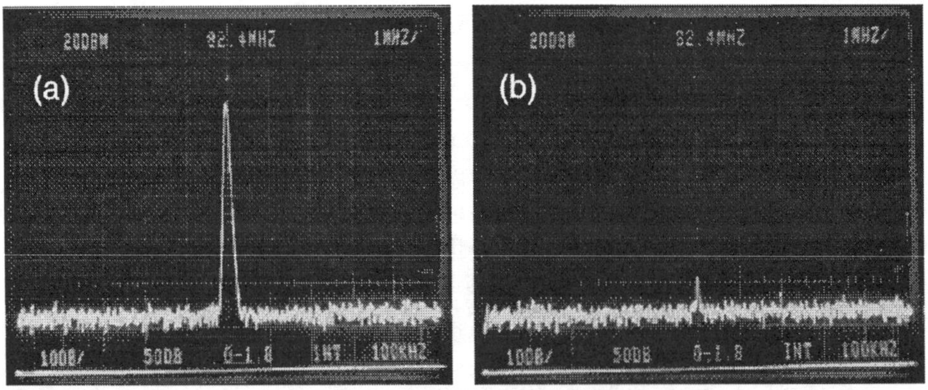

Figure 13

Experimental Results of Interference Suppression

a) output of inactive filter, b) output of active filter showing 40 dB suppression

The system was then tested with an input consisting of both narrowband (sinusoidal) interference and broadband signal by simply using double sideband amplitude modulation of a carrier with a random sequence signal (zero-mean, uniform distribution at 1 MHz sampling frequency). The unsuppressed carrier serves in this case as the interference. Figures 15(a) and 15(b) respectively show the inactive and active system output spectra for an input consisting of an interference at 82.4 MHz and the random sequence signal. In all cases, the suppression of the interference was approximately 40 dB.

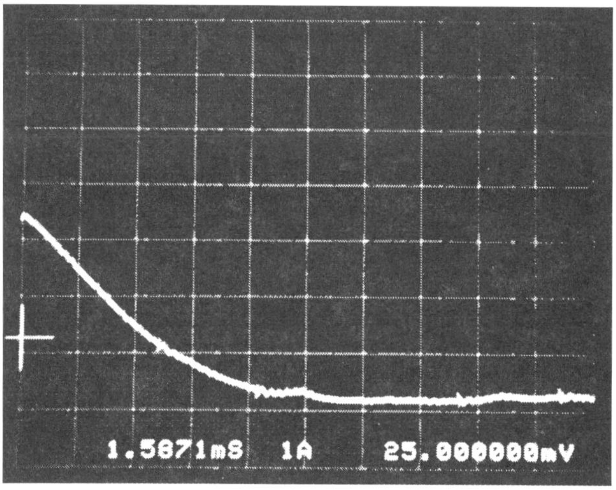

Figure 14
Filter Time Response

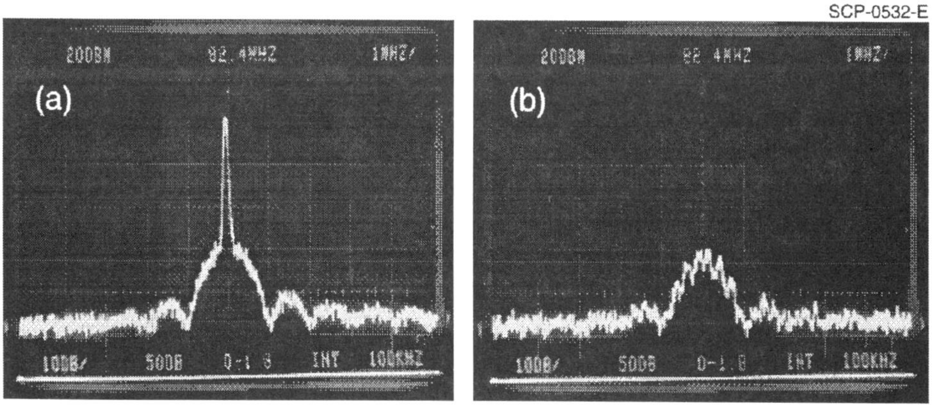

Figure 15
Experimental Results of Interference Suppression with PN Signal
a) output of inactive filter, b) output of active filter showing 40 dB suppression

Acknowledgement

The author gratefully acknowledges the important technical contributions of Tallis Y. Chang.

References

1. J. J. Amodei, D. L. Staebler, Appl Phys. Lett. **18**, p.540 (1971).
2. F. H. Mok, M. C. Tackitt, H. M. Stoll, Opt. Lett. **16**, p.605 (1991).
3. D. Psaltis, D. Brady, K. Wagner, Appl. Opt. **27**, p.1752 (1990).
4. D. Z. Anderson, D. M. Lininger, Appl. Opt. **26**, p.5031 (1987).
5. J. Hong, S. Campbell, P. Yeh, Appl. Opt. **29**, p.3019 (1990).
6. D. Psaltis, J. Yu, J. Hong, Appl. Opt. **24**, p.3860 (1985).
7. J. Hong, S. Hudson, J. Yu, D. Psaltis, Proc. SPIE **789**, p.136 (1987).
8. R. M. Montgomery, M. R. Lange, Appl. Opt. **30**, p.2844 (1991).
9. N. V. Kukhtarev, V. B. Markov, S. G. Odulov, M. S. Soskin, V. L. Vinetskii, Ferroelectrics **22**, p.949 (1979).
10. P. Yeh, Appl. Opt. **26**, 602 (1987).
11. M. H. Garrett, J. Y. Chang, H. P. Jenssen, C. Warde, Opt. Lett. **17**, p.103 (1992).
12. R. A. Vazquez, M. D. Ewbank, R. R. Neurogankar, Opt. Comm. **80**, p.253 (1991).
13. R. M. Montgomery, "Acoustooptical Signal Processing System," U. S. Patent No. 3,634,749 (1972).
14. B. Widrow, P. Mantey, L. Griffiths, B. Goode, Proc. IEEE **25**, p.2143 (1967).
15. W. B. Davenport, W. Root, *Random Signals and Noise*, McGraw-Hill, New York (1958).
16. D. Psaltis, J. Hong, Appl. Opt. **23**, p.3475 (1984).
17. T. Y. Chang, J. Hong, F. Vachss, R. McGraw, JOSA-B **9**, p.1744-1751 (1992).
18. H. Kogelnik, Bell Syst. Tech. J. **48**, p.2909 (1969).
19. J. Hong, T. Y. Chang, Opt. Lett. **18**, No. 2, p.164-166 (Jan. 1993).
20. A. Hermanns, C. Benkert, D. M. Lininger, D. Z. Anderson, IEEE J. Quant. Elec. **28**, p.750 (1992).
21. J. Hong, T. Y. Chang, "Adaptive rf notch filtering using photorefractive two-beam coupling," submitted to IEEE J. Quant. Elec. (Mar. 1993).